DESERT DEVELOPMENT

PART 2: SOCIO-ECONOMIC ASPECTS AND RENEWABLE ENERGY APLICATIONS

ADVANCES IN DESERT AND ARID LAND TECHNOLOGY AND DEVELOPMENT

Volume 1 APPLICATIONS OF SCIENCE AND TECHNOLOGY FOR DESERT DEVELOPMENT
Edited by Adli Bishay and William G. McGinnies

Volume 2 REALISTIC PLANNING FOR ARID LANDS
Natural Resource Limitations to Agricultural Development
By W. Gerald Matlock

Volume 3 DESERTIFICATION OF ARID LANDS
By H.E. Dregne

Volume 4 IMPROVEMENT OF DESERT RANGES IN SOVIET CENTRAL ASIA
Edited by Nina T. Nechaeva

Volume 5 DESERT DEVELOPMENT
Part 1: Desert Agriculture, Ecology and Biology
Part 2: Socio-Economic Aspects and Renewable Energy Applications
Edited by Adli Bishay and Harold Dregne

This book is part of a series. The publisher will accept continuation orders for this series, which may be cancelled at any time and which provide for the automatic billing and shipping of each title in the series upon publication. Please write for details.

DESERT DEVELOPMENT

PART 2: SOCIO-ECONOMIC ASPECTS AND RENEWABLE ENERGY APPLICATIONS

PROCEEDINGS OF THE SECOND INTERNATIONAL DESERT DEVELOPMENT CONFERENCE HELD ON 25-31 JANUARY 1987 IN CAIRO, EGYPT

Edited by

Adli Bishay
The American University in Cairo
Cairo, Egypt

and

Harold Dregne
Texas Tech University
Lubbock, Texas, USA

harwood academic publishers
chur • london • paris • new york • melbourne

Harwood Academic Publishers

Poststrasse 22
7000 Chur
Switzerland

Post Office Box 197
London WCZE 9PX
United Kingdom

58, rue Lhomond
75005 Paris
France

Post Office Box 786
Cooper Station
New York, New York 10276
United States of America

Private Bag 8
Camberwell, Victoria 3124
Australia

Library of Congress Cataloging-in-Publication Data

International Desert Development conference (2nd : 1987 : Cairo Egypt)
Desert development : proceedings of the Second International
Desert Development Conference held on 25-31 January 1987 in Cairo,
Egypt / edited by Adli Bishay and Harold Dregne.
p. cm. -- (Advances in desert and arid land technology and
development, ISSN 0142-5889 : v. 5)
Includes bibliographical references.
contents: pt. 1. Desert agriculture, ecology, and biology --pt.
2. Socio-economic aspects and renewable energy application.
ISBN 3-7186-0526-0 (set). -- ISBN 3-7186-0396-9 (pt. 1) --
ISBN 3-7186-0522-8 (pt. 2)
1. Desert resources development--Congresses. I. Bishay, Adli.
II. Dregne, H.E. (Harold E.) III. Title. IV. Series.
GB611.I5173 **1987**
333.73 615--dc20 90-4391

 Printed in the United States of America.

CONTENTS

EDITORIAL ANNOUNCEMENT

Since the initiation of this series in 1979, Professor William G. McGinnies and I have worked together on the identification of important contributions in the field, as well as shared other responsibilities as editors of the series. Recently, however, for health reasons, Professor McGinnies relinquished a number of activities, one of which was his editorial contribution to this series. We are indeed indebted to Bill for his valuable ideas and contacts, which made it possible to achieve a truly international circulation for this series. We wish him long and happy years of retirement.

After a thorough search for an equally distinguished and internationally known scientist, it is now my privilege to announce that Dr. Harold Dregne, Horn Distinguished Professor Emeritus of Soil Science and former director of ICASALS, has accepted to share the editorial responsibilities of the series. Among many of his achievements, Professor Dregne is senior consultant to UNEP, USAID, FAO and other national and international agencies. I have known Harold since 1977 when I invited him to be a plenary speaker at the First International Desert Development Conference held in Cairo in 1978. For the past ten years, we have worked together on a number of assignments, including the chairmanship of the International Desert Development Commission initiated in 1987.

Adli Bishay

PREFACE

This volume of the series covers the proceedings of the Second International Desert Development Conference held in Cairo on 25-31 January 1987. Although all papers were refereed after presentation at the conference, the number of those accepted for publication was large enough to require the division of the proceedings into two separate parts.

Included in Part 2 are 3 plenary papers by M. Kassas, Harold Dregne and Merle Jensen; 3 contributed papers in the area of systems research; 32 papers in desert agriculture; and 7 papers in the areas of ecology and biology.

As general editors of the series, as well as editors of volume 5, we would like to thank Dr. Youssef Wali, Egyptian Minister of Agriculture and Land Reclamation, for chairing and supporting the conference. We would also like to acknowledge the assistance of the US National Science Foundation (NSF), the US Agency for International Development (AID), the International Development Research Center of Canada (IDRC), the Near East Foundation (NEF), the United Nations Environmental Program (UNEP), the Egyptian Ministry of Electricity and Energy, the Egyptian Supreme Council of Universities, the American University in Cairo and Texas Tech University.

Thanks are due to the scientific committee of the conference: Drs. Adel El Beltagy, Yousef Hamdi, Hosny El-Lakany, Anhar Hegazi and A.M. Aboul Naga, who have contributed long hours of hard work to coordinate the evaluation of different papers by international referees. Thanks are also due to H.E. Mr. Badr Hammam of the Egyptian Ministry of Foreign Affairs, Dr. Adel El-Beltagy, Mr. Mahmoud Dessouki, Mrs. Atiat El-Menshawi and other members of the International Center of Agriculture, as well as Dr. Hosny El-Lakany and other members of the Desert Development Center of the American University in Cairo for their invaluable contributions to the organization of the conference.

The proceedings of this conference included papers from many countries, institutions and backgrounds. It is indeed a pleasure to acknowledge the supreme efforts of Dr. Betty Wass of the American University in Cairo, who has been responsible for reviewing or revising most of the papers after their acceptance for publication.

PREFACE

The preparation of the final manuscripts of the two volumes was the result of dedication and long hours of work by one young woman, Nancy Wadie of the AUC Desert Development Center, to whom we are greatly indebted. She was assisted by Mrs. Vicky Elabd, who was responsible for proofreading.

Adli Bishay
Harold Dregne

ERRATA

The second paragraph of the "Preface" should read:

Included in Part 2 are 3 plenary papers by Gerald Thomas, David Nygaard and Dennis Hayes; 6 contributed papers dealing with integrated approaches; 16 papers covering energy and architecture; and 14 papers on socio-economics and desert communities.

ARID LANDS DEVELOPMENT: STRIKING A BALANCE BETWEEN ECONOMIC AND ECOLOGICAL ISSUES

GERALD W. THOMAS
President Emeritus, New Mexico State University

Abstract The development of arid and semi-arid lands has been driven historically by the need to provide goods and services for an increasing population. The primary constraints in the past have been economic and yet, the long term success of these activities will depend on adequate consideration of the more subtle environmental issues, such as air population, water pollution, climatic change, and other perturbations in these sensitive ecosystems. This paper focuses on the serious nature of these environmental issues in arid situations, with some recommendations for obtaining a balance between these issues and the more pervasive economic driving forces. Attention is also given to the importance in resource management to provide both "an incentive to conserve" as well as "an incentive to produce".

INTRODUCTION

The development of arid and semi-arid lands has been driven historically by the need to provide goods and services for an increasing population. The primary constraints in the past have been "economic" and yet, the long-term success of these activities will depend on an adequate consideration of the more subtle environmental or "ecological" issues. While it is possible to accommodate both economic and ecological concerns, it is obvious that desert development projects to date have not adequately considered the ecological dimension in either the planning or the development phases. This neglect is particularly true for agricultural development programs.

The ecological consequences of development -- particularly in desert environments -- are frequently ignored due to lack of knowledge about the function of large ecosystems. Also, in many cases, environmental concerns are purposely ignored because they appear to be too costly and the increased cost might jeopardize the political support for the development project or program. Experience has shown that ignoring the ecological dimension in the planning or development phases leads to more costly corrective measures after the project begins operation. Sooner or later, the politician and the economist must add an appropriate ecological analysis to the feasibility studies for all food production programs.

Perhaps at this point, I should clarify my concepts of "Ecology". Ecology is a legitimate field of science defined as; "The relationships among all organisms and the environment." I underline "all organisms" because mankind must be considered the most important part of the formula due to the human impact on the environment as well as the ability of humans to modify the environment to fit specific needs.

Ecology is not a concept of environmental "protection." Man has already influenced every part of the world's environment, including the most remote polar areas. The challenge now is to understand these impacts and to adjust development activities based upon research. The key to our relationships with the environment is "management" not "protection" per se. My own field of range science could be called "applied ecology".

I believe in the axiom, "Never does nature say one thing and wisdom another." But, to let nature take its course in a protectionist atmosphere will neither benefit mankind nor maintain the values we are trying to protect. Even the so-called "wilderness areas" must be "managed" in order to maintain the natural beauty and the biological populations of the area. Also, we have learned the hard way that deserts are not an unlimited sink for man's pollutants, but are indeed fragile ecosystems which react to man's involvement.

While it is possible to obtain a balance between economic pressures and environmental concerns, we have not yet attained that objective. If we do not make proper adjustments soon to accommodate the ecological dimension in arid land development programs, we will eventually discover that these same ecological issues have become serious economic issues -- perhaps too late to prevent irrevocable damage to the sustainability of the development project. Perhaps I can illustrate some of these interrelationships by an examination of world-wide food production systems.

THE ECOLOGY OF FOOD PRODUCTION AND UTILIZATION

Food production has risen dramatically in the last five years -- both world-wide and on a per capita basis. Africa remains the major exception, where recent drought, combined with an extremely high population growth rate and other factors has kept food production below minimum needs. If we exclude most of Africa, the potential for feeding both the present world's population, and the projected increase beyond the turn of the century, looks very promising. The United Nations Food and Agriculture Organization (FAO) stated that at the end of 1986 there would be an unprecedented carryover of 448 million tons of cereal grains, representing 26 percent of annual production.[1] This does not mean that there will not be millions of people continuing to suffer from malnutrition due to unequal distribution and poverty. Nevertheless, the technology is on hand to feed the world's people.

As early as 1960,[2] I started emphasizing that the important question was not, "Can the World Feed Itself?", since this was obviously possible from a technological viewpoint, but rather, the real question should be, "Can the World Afford to Feed Itself?". In other words, will the world's environment withstand the pressure of the technological changes necessary as we attempt to feed more and more people? What is the environmental cost of eating? I have spent most of my professional career examining this question. I hope my observations will help to bring a more balanced perspective to the process of development.

POPULATION PRESSURE

There are two factors in the population picture that impact on resource requirements and environmental pollution:

(1) Growing numbers of people - the United Nations and Population Reference Bureau[4] projections still indicate about 7.8 billion by the year 2020 -- with an optimistic possibility of eventual leveling about the mid-21st century at nearly 12 billion, and;

(2) Higher levels of affluence combined with the growing aspirations of the masses. We know that wealthy people place more pressure on the resource base than poor people. You and I require more units of land, water and energy than the poor people in Indonesia or Bangladesh. We also create increased levels of pollution.

THE NUMBERS PROBLEM

I will cite only five examples to illustrate the impact of population numbers, per se.* Kenya is now the fastest growing country in the world with an average number of living children per female of 8.2. The population of Kenya is projected to double in 15 years. The present population momentum will carry Nigeria to 1/2 billion people sometime beyond the turn of the century. Mexico and Latin America's population growth rates are so rapid that it is difficult to provide employment for the masses or to sustain a desirable "quality

---

* Data from various publications of the Population Reference Bureau, Washington, D.C.

of life." China's population, despite the emphasis on the one-child family, will gain an additional 288 million by 2020. Finally, a team of university officials returning from India in 1985 reported that India, with the present reduced emphasis on family planning, will surpass China as the most populous country in the world in about 50 years.

In 1987 we will have to plan for a net gain in population numbers of about 83 million, with the highest growth rates in the less developed countries. At average levels of grain consumption (320 kilograms/person) we will need to add about 13-15 million metric tons of cereal grains each year to meet this annual projected population increase.[4]

LEVELS OF AFFLUENCE

But, while the numbers alone are important, it is the second factor in the population picture, i.e. levels of affluence and/or greater aspirations of the people, that is of most concern as we look at worldwide environmental pressure.

Today, there are about 40 countries of the world with per capita annual incomes of less than $300. Many of the countries that I have visited in Africa have per capita incomes of less than $200 per year. In 1983, Mainland China was reported to have a per capita GNP (Gross National Product) of $290[5] -- not wealthy by any standards. But, China, like many developing countries is undergoing change.

What will happen to the world's environment as another billion people raise their standard of living? What if China moves the per capita GNP from $290 to the world average of about $3,000; or eventually to the level of the American middle classes which is perhaps 40 times that of China? A billion poor people are one problem -- a billion wealthy people another. What I am trying to say is that the impact of population pressure is not only a numbers problem but perhaps more importantly, a level of affluence problem. The relationship between per capita income and food requirements can be easily demonstrated. The relationship between per capita income and environmental pollution can also be demonstrated.

A SYSTEMS APPROACH

In this analysis of the food system, I am including the three main segments:

(1) The Supply Sector which includes providing the seed, chemicals, machinery and many other resources necessary for production.

(2) The Production or on-farm sector.

(3) Processing and Distribution of farm and livestock products.

These three sectors are involved in getting food to the consumer. However, an adequate environmental analysis must include some post-consumer consequences such as waste disposal, sewage treatment and partial recycling.

One logical approach to relating people's food needs to the environment is to consider two sides of the equation:

(1) The productive capacity of the environment with special emphasis on land, water, climate and energy.

(2) The absorptive capacity of the environment, that is, the impact of the technologies associated with the food system on air and water pollution, climatic change, soil erosion, and shifts in biological populations.

While it is possible to look at both of these environmental constraints, it is important to recognize that there are critical interrelationships between the two. All too often, decisions are made at one point in the system without understanding the impact on other points. A decision to reduce pollution may increase the per capita resource requirements. There are good examples of the interactions between the productive capacity of the environment and the absorptive capacity in the energy field. For example, early attempts to reduce pollution with the catalytic converter resulted in poorer mileage per gallon and therefore raised the energy resource requirements for the same distance covered.

THE LAND RESOURCE

The first question is, How much land is really needed to supply our food needs? Regardless of whether we live in the city or in the country, your food needs and mine are associated with a certain land base. How much land, what kind of land, and where is it located?

I first became interested in this question when I went to work for the U.S. Soil Conservation Service right after World War II. In our search for techniques to promote soil conservation, we tried to convince American farmers and ranchers that they should reduce erosion in order to preserve our valuable land base. At that time, we estimated that about 1.2 hectares of cultivated land was necessary to support each person with a satisfactory level of living. If you looked at this requirement in relation to the world's arable land, it was obvious that there was not enough good land to sustain the world's people at a desirable standard of living. That was 1946.

This approach to selling soil conservation helped. But, what has happened to that 1.2 hectare "requirement" in the past 40 years? The major impact of new technologies in the agricultural sector came after World War II. Hybridization, a revolution in agricultural chemicals (particularly pesticides and fertilizers), irrigation, a major move toward mechanization, significant improvements in food processing and distribution, and many other new technologies swept through the food sector. We no longer "require" 1.2 hectares of arable land per person.

I have changed my estimates of land requirements per capita many times in the past 4 decades. However, the basic relationship between man and the land remains and continued population growth makes it imperative that we conserve this remaining valuable land base. None of the new technologies on the horizon will separate mankind from dependence on land for food production even though the land requirements per capita will vary with the quality of land, the level of productivity, the kinds of crops produced and our concepts of an "adequate" standard of living. Also, with increased international trade in food crops, some people's food ties may be to distant lands, making it difficult to establish a relationship.

In any analysis of land resources, it is necessary to separate the land into two broad categories: (1) arable or cultivated land, and (2) uncultivated range and pasture suitable for grazing by livestock and wildlife. Of course, in some desert environments with little or no rainfall, some land may not produce any food crops unless imported water is available. Nevertheless, from an environmental standpoint, this "unproductive" open land may have value in terms of biological populations or as a "sink" for pollutants.

In relating per capita food needs to land requirements, it is easier to work with the arable land base. More research has been conducted on cultivated lands because of the higher production potential. Also, most references to the status of world food supplies are based on cereal grains produced on cultivated lands. Grain is the best single indicator of food status, but any analysis using cereal grains alone can be misleading, since balance in dietary requirements is often more important than total energy. Using grain as the primary indicator of world food status ignores the important role of livestock and many critical horticultural crops.

If we calculate the productivity of any particular desert development project based upon the number of people the project will support, we must realize that these figures will change over time. One hectare may support the food needs of 8-10 people at a per capita income level of $500 per year but only 2-4 people as the standard approaches the world average of $2,760.[5]

A brief examination of the "absorptive capacity" of the land quickly reveals major problems of soil erosion, vegetation change and chemical pollution.

Soil erosion has been and continues to be one of the most serious problems in all countries. Probably the most comprehensive treatment of this subject in the U.S. was Neil Sampson's book "Farmland or Wasteland."[6] For a world viewpoint, several books and articles have been published by the World Watch Institute on this topic including the "State of the World" series. Personally, I was somewhat surprised to find in my recent visit to China that soil erosion was a major problem in China even though each acre of land is subjected to very careful hand labor. The new orientation of China toward more "economic incentives" for farmers will increase the problem of land deterioration. Lester Brown[7] stated it well when he said,

> "The very practices that lead to excessive erosion often yield short-term production gains or even surpluses, creating an illusion of progress."

In 1980, I had the privilege of working with the Rockefeller Foundation on an assignment which took me to ten countries south of the Sahara, from Senegal to Ethiopia. My report, and a comprehensive review of the literature on food production and desert encroachment was published under the title "Profile of a Fragile Environment".[8] It is obvious from this and other studies that the problem of man-caused or man-accelerated desertification, particularly in sub-Saharan Africa is seriously reducing our potential to feed future generations.

Change in biological population on our land base is often an overlooked concern. Overgrazing by animals, burning, and cultivation, change the vegetation complex and affect other biological populations. The emphasis on "endangered species" may cause amusement among certain sectors of our society, but these trends are indicators of a more serious ecosystem disturbance. Conserving biological diversity is important indirectly to mankind as well as directly to food production systems.

> "The genetic diversity within the handful of crop and livestock species that feed humanity -- so-called germplasm resources -- holds much of the potential for improving agricultural performance."[9]

Regardless of our progress in genetic engineering and improved biotechnology, we cannot replace the built-in balances in biological populations that are a part of natural ecosystems. How much longer can mankind continue to divert more and more photosynthetic material to his or her own use and thus substract from biomass available to other biological populations? Where, how, and when should humans "bleed" the system for their food and fiber needs? Closer to the "Producer" organisms (vegetation), or closer to the "Decomposer" organisms?

ENERGY FLOW AND UTILIZATION

For the near term, that is, for the next two to three decades, the availability and cost of energy will probably remain the most critical factor limiting world food production and distribution. Even in the less-developed countries where agriculture is less "energy intensive", the supply of fertilizers and petrochemicals, the food delivery systems, and the dependence on mechanization (though limited) are critical to agriculture development.

The design of "appropriate technologies" for the agricultural sector, based upon energy constraints, remains one of the most popular topics for discussion in international agricultural circles. The recent "oil glut" has obscured our vision of the long-term energy problem. Energy is particularly crucial to the future of irrigated agriculture. Energy costs or petroleum curtailment may lead to a shift in some countries from irrigation back to the more risky rain-fed agriculture at a time when we need the additional crop yields associated with supplemental irrigation.

A comprehensive approach to an ecosystem analysis of energy in the Food System goes far beyond a simple discussion of petroleum or electricity. It is necessary to consider food, wood, and all forms of organic material also as energy. Therefore, in this analysis, two major energy flow patterns must be studied:

(1) The capture of solar energy by the vegetation and the movement of this food energy or biomass through the ecosystem until a part of it eventually reaches the consumer as a food product, and

(2) Energy subsidies or "cultural energy" required to produce, process, and deliver the food to the ultimate consumer. This latter source of energy includes not only petroleum, hydroelectric power and coal, but it includes horse power, oxen power, man power and certain other energy subsidies.

In an examination of these two relationships, i.e. solar energy (biomass) flow from capture to dissipation in relation to the energy subsidies required to deliver biomass to the consumer, it is obvious that we are operating the system on non-sustainable resources. I made this statement at the first Desert Development Conference here in Cairo in 1978 and I now repeat it in 1987.[10]

What has changed in the energy analysis in the last nine years? We know more about the energy costs for the supply sector; we have increased effective solar energy capture by food plants through genetic engineering, and we have developed more energy efficient techniques for food processing and distribution. While petroleum was expensive, we researched some alternatives such as solar, geothermal and bioconversion. However, as oil and coal supplies again became more available, most countries discontinued their search for alternatives to these fossil fuels. <u>As a result, we remain as dependent, or perhaps more dependent, on depletable resources in the food sector as we were two decades ago</u>.

We are changing the energy flow patterns in two ways:

(1) More refined and processed foods, greater consumption at distant points from the source of production, and a more variable (but perhaps better balanced) diet is expensive from the standpoint of photosynthetic biomass utilization.

(2) The trend toward mechanization in the agricultural sector is continuing world-wide in spite of the emphasis on "appropriate technology" and the underemployed man-power available in most countries. This trend creates a greater negative balance in the ratio of "renewable" to "depletable" energy sources.

There is, however, one area relating to energy where we have made significant progress: that is, in our understanding of the impact of fossil-fuel burning on climate and the air environment. Awareness so far has <u>not</u> led to many changes in our approach to energy use, but perhaps this will come with time. At least the scientific community has pointed out the dangers associated with increased carbon dioxide levels, problems with acid rain, depletion of the ozone and the new concern about atmospheric trace gases.

A small change in climate, which will occur unless we change many of our economic development activities, could have significant impact on regional agricultural development programs, even to the point of affecting the Nile basin.

WATER RESOURCES FOR FOOD PRODUCTION

Water may be more critical in the long term than either land or energy. We can, and must find solutions to the energy problem. We can, and will, determine ways to operate with a smaller relative land base; but the amount of water in our system is fixed. There is no substitute for water. Water is a renewable resource. Man uses it as it moves through the hydrologic cycle, usually pollutes it to a certain extent, and feeds it back into the system. While we can reduce the dependence upon water by increasing the efficiency of water use, there is a very limited supply which must be husbanded with great care as the world population increases. There can be no desert development projects without fresh water supplies.

Irrigated agriculture is becoming much more important with time, as the world's population increases the demand for food and fiber. The recent growth in irrigated land has not only been in the traditional irrigated areas, but in the moderate to high rainfall zones as a risk-reducing factor. For example in the United States, Nebraska, traditionally a rainfed-crop producing State, has become the fifth most irrigated state in the nation. Also, irrigation is becoming more important in the less-developed areas of the world -- perhaps the major hope for many of these poor countries.

On a global basis, roughly 40 percent of all increases in developing country food production in the last two decades has come from expanded irrigation.[11] Since 1950, the irrigated acreage world-wide has increased from 94 million to 261 million hectares.[12] Approximately a third of the total world food harvest now comes from the 17 percent of the cropland that is irrigated. The irrigated acreage in the United States comprises about 15 percent of the total harvested crop production. The trends toward more irrigated cropland will likely continue in spite of the increased energy costs, the problems of underground depletion, and serious problems of salinity and other forms of population.

As our standard of living rises so does our per capita use of water for domestic, industrial, and agricultural purposes. An individual needs only about 2 liters daily for drinking, but water use rises rapidly with the level of income. My study in sub-Saharan Africa indicated that the average water consumption from wells was 1-15 liters per person per day for home use, about 20 liters for cattle and about 3 liters for sheep and goats. In the United States, our average home use is about 681 liters (180 gallons) but the wealthy residents of Beverly Hills, California use more water per capita than the residents of Compton, California, with lower average earnings. To sustain the present life style in the U.S. for food, industrial and other purposes, the average American now uses over 9000 liters per day.[13]

Our largest per capita water requirement is for food. For example we spend about a metric ton of water to produce a pound of bread and on some of our Southwestern rangelands over 100 tons of water are associated with the production of a pound of beef.[14] Much of the water "associated with" the production of range meat is dissipated bv undesirable weeds and brush or evaporates from the unprotected soil surface. such statistics provide a convincing argument for better water management in the food sector. It is time to assemble better data on such water use and to design systems for food production which value water, keeping in mind the concern of the desert nomad.

Agriculture cannot compete for water against municipalities, business and industry. These uses can afford to pay more for water and will continue to purchase water rights away from the food sector. We see this transfer every year in New Mexico and in other agricultural areas, and this moves land out of food production just as surely and as effectively as direct transfer to housing or highway construction. Dawson states that water is a marketable quantity and "water flows uphill to money."[15] Dawson further stated that "short-run economic returns will continually guide the resource allocation decisions....therefore, agriculture will be the residual economic user of water."

The major concerns about the water resource fall into the following categories:

(1) Periodic drought and highly variable rainfall patterns create risk to rain-fed crop production and range livestock operations. Changes in the overall climate associated with air pollution from agriculture and industry will increase our concern about water deficiencies and increase the risk to the food sector in arid regions.

(2) The continued depletion of most underground aquifers will seriously impact on irrigated agriculture in less than 50 years.

(3) Salinity and pollution from agricultural chemicals is a continuing problem. World-wide, we are experiencing some reversion of irrigated land back to dryland. In some cases, these lands will have less production potential than in the original state. Fortunately, many water pollution problems can be corrected with a technological but economically costly input. We have not as yet developed techniques for desalting sea or brackish water at the economic levels required for food production.

SUMMARY

The peoples of the planet Earth are faced with an apparent paradox -- that is, more units of renewable, and non-renewable, natural resources are required per person as we develop and as we seek to attain a "satisfactory" standard of living. Dr. Lester Brown, with the Worldwatch Institute expressed his concern as follows.[16]

> "If the question asked is 'Can we produce enough food to sustain, at adequate and acceptable dietary levels, the projected population increase between now and the end of the century?', then the answer is yes. But if we ask, 'What are the social and ecological consequences of the rapid advances in farm technology which enable us to sustain past and future increases in population?', we get a different answer. Unfortunately, although we can produce more than enough food, we are only beginning to realize that the social costs are painfully high and the ecological consequences only beginning to appear. If we succeed in producing more than enough food but destroy our environment in the process, what will be gained?"

Even with all of our sophisticated research, there is still an appalling lack of understanding about natural and man-influenced ecosystems. We know far too little about energy flow in the agricultural sector, about nutrient cycling, about weather and the hydrologic cycle. We do not understand fully the interactions between rangeland and cultivated land. Too often we attempt to correct a problem at one point in the system and create havoc at another.

In the development of arid and semi-arid lands for food production and for other human uses, we continue to see an appalling lack of attention to ecological concerns. Somehow, we must try to attain a better balance between economic objectives and environmental constraints. This balance will be attained when we learn how to place proper monetary values on the basic resources as well as a clean environment and when we can demonstrate that environmental degradation will affect the long-term economic returns from development programs. The challenge is to bring together the expertise of the two great "Ecos" -- economics and ecology in the planning, development and operation of food production programs.

REFERENCES

1. FAO at Work. Oct./Nov. Food and Aariculture Oraanization of the United Nations.

2. Gerald W. Thomas. October 1973. Can the world afford to feed itself? the environmental cost of eatinq. Vital Speeches of the Dav. Vol. XXXX, No. 1.

3. Population Todav. Dec. 1987. Vol. 14, No. 12. Population Reference Bureau, Washinaton, D.C.

4. Gene A. Marthia and Larry Deaton. 1985. Whv people eat what they eat around the world. In U.S. Aariculture in a Global Economy. Yearbook of Aariculture.

5. Mary Kent Mederios, and Carl Haub. 1985 world population data sheet. Population Reference Bureau, Washinaton, D.C.

6. R. Neil Sampson. Farmland or Wasteland. Rodale Press, Emmaus, Pa.: 422 p.

7. Lester R. Brown, et.al. State of the world-1986. Worldwatch Institute, Washinaton, D.C. Published by W.W. Norton and Co., New York and London.

8. Gerald W. Thomas. January 1986. Profile of a fragile environment: the Sahalian/Sudanian zones of Africa. International Publications Series 0001-81, New Mexico State University.

9. Edward C. Wolf. 1985. Conserving Biological Diversity. In State of the World-1985, Worldwatch Institute, W.W. Norton and Co., New York and London.

10. Gerald W. Thomas. 1979. Energy flow for agricultural ecosystems in desert environments. In Advances in Desert and Arid Land Technology and Development. A. Bishay & W. McGinnies (eds.). Vol. 1 Applications of Science and Technology for Desert Development. Harwood Academic Publishers.

11. Robert S. McNamara. March 4, 1980. Food as a determinant of world development. Congressional Roundtable, Washington, D.C.

12. Sandra Postel. December 1984. Water: rethinking management in an age of scarcity. Worldwatch Paper 62.

13. Council of Environmental Quality and the Department of State. 1980. The Global 2000 Report to the President, U.S. Government Printing Office.

14. Gerald W. Thomas. 1977. Environmental sensitivity and production potential on semi-arid lands. In Frontiers of the Semi-Arid World, Texas Tech University.

15. George R. Dawson. June 1981. Water for agriculture in a competitive environment. Proceedings of the Twenty-Sixth Annual New Mexico Water Conference, New Mexico State University, WRRI Report 134.

16. Lester R. Brown. 1985. A false sense of security. In State of the World-1985, Worldwatch Institute, W.W. Norton and Co., New York and London.

DESERT DEVELOPMENT, IS IT ECONOMICALLY VIABLE?

DAVID F. NYGAARD*
Winrock International Institute for
Agricultural Development

Abstract In order for desert development to be economically viable, it must meet a number of conditions. First, it must use its resources efficiently and find goods and services that it has a comparative advantage in producing. Second, it must be fully integrated into the economic system in which it exists. Finally, it must get out of the laboratory and off the experiment station and test its technologies under actual conditions in the real world. The paper argues that desert development can be successful and discusses ways to improve its prospects for success. These include using resources that are abundant, e.g., sun energy, while saving those that are scarce, e.g., water; drawing adequate financial resources into desert communities; building human capital as well as infrastructure in desert areas; but, in doing so, avoiding reliance on government subsidies.

INTRODUCTION

Desert development -- is it economically viable? Yes, but! And there are three buts, three conditions each of which must be met if desert development is to be successful. Desert development must:

1. Demonstrate technical and economic efficiency in the use of its resources, and demonstrate a comparative advantage in the goods it produces and the services it

---

* Currently, Regional Director, The Ford Foundation, Cairo, Egypt.

provides. Desert communities must pay their way.

2. Be fully integrated into the economy in which it exists. Desert development does not take place in isolation and it must fit into a broader economic system.

3. Test its technologies, techniques, and systems in the real world as quickly as possible. It must get out of the laboratory, off the experiment station and into the field.

Stated somewhat more starkly, for desert development to be economically viable we must:

-- make it different,

-- make it fit, and

-- make it work.

Before looking at these three issues in greater detail, I would like to step back and ask a broader and more general question that is facing us.

Why are we holding this conference? While I do not wish to impose my views on the conference organizers, I would like to suggest for the purpose of this presentation a working hypothesis on the rationale for the conference, at least why I was interested in participating in this gathering. We are holding this conference because in the past desert development has failed. However, there have been recent technological innovations that may work in desert environments and many scientists believe there is great potential in developing these environments. Due to the population pressure in many countries in the world, there is a need to tap the deserts in order to provide more people with a better life -- more productive, healthier and happier.

If this statement is true, and I doubt that many of you will disagree with me about this, it seems then to be my task to determine and describe the economic conditions under which desert development will succeed. And, the best way to achieve that is to find ways in which desert development differs from development in other ecological areas, and then to seek ways to take advantage of those differences.

There are a number of questions of interest:

On differences:

-- How are desert communities different?
-- What resources are different?
-- What unique or particularly appropriate techniques are available?
-- What products can desert communities produce?
-- What are the differences in the way desert people live?
-- What are the quality of life differences?

On comparative advantages:

-- What are the comparative advantages of desert communities?
-- What are the important linkages with the rest of the economy?
-- In particular how do people, financial capital or money, and goods and services fit together in making desert development communities work?

And, finally on integration:

-- What help do we need and what help can we expect from the outside -- primarily the government and the international donor community?

-- And, with that help, how can we make it work?

I have set myself an impossible task of trying to address those questions, and I warn you that I will not be able to deliver on all of them. But, these are the tasks facing desert developers. They must address all of them simultaneously and answer them satisfactorily. In order to make my task slightly smaller and thus substantially easier, I limit my comments to agricultural issues and to examples from Egypt.

In concluding this introduction, I share with you two observations I have made in preparing this paper. The first is that the literature on the economics of desert development is both skimpy and weak. There is indeed a large literature supply on technical topics, e.g., drip irrigation, solar energy, soil salinity, etc. However, there are few economic studies and, in particular, little economic analysis of successful desert development projects in the developing world.

The second observation is that the solutions and techniques offered in this literature are hi-tech. The successful examples are new communities depending on sophisticated technology with capital intensive and labor-using technologies. Therefore, it seems to me to be most appropriate to be holding this conference here in Egypt where desert development is not concerned only with hi-tech solutions and techniques, but emphasizes traditional aspects as well. When I first visited the Desert Development Center (of American University in Cairo) in 1982, that is what struck me -- the blend of the old and the new. Development is going on as if people mattered and I liked that. The task is to see if we can find some economic principles and concepts that will make this kind of development work.

STEP I: IS IT DIFFERENT?

First, let's look at the process of developing new and appropriate technology. Hayami and Ruttan in their seminal book on agricultural development[3] demonstrated in a convincing manner that both Japan and the U.S. had achieved impressive rates of agricultural development, but each had taken a very different path to that development. Japan with a shortage of land and a surplus of labor developed technologies that saved land and used labor. The U.S. on the other hand, had a shortage of labor and an abundance of land. As a result, U.S. technologies tended to be big machines -- tractors, combine harvesters, etc. -- which were labor-saving and land (and capital) using. In examining other countries, Hayami and Ruttan found a clear relationship between relative factor endowments (resources: human, financial, and physical) and the technologies that were developed in each country.

Since price is a reasonable measure of scarcity, one concludes that the relative value of these inputs (or factors) strongly influences economic development. And, the lesson for the development community is that the introduction of new technologies will be successful only if they meet this condition.

What does this imply for desert development in Egypt? It should encourage researchers and developers to search for differences between, for example, delta agriculture and desert agriculture, in particular identify and measure differences in the availability of important factors in agricultural production and their costs. If there is a difference it will affect the kinds of technologies that could be tried, the choice of crops and the livestock mix, the pattern of land use, the essential thrust of infrastructure development, etc.

While I have little data to support me, I suspect that there are substantial differences between these characteristics in the delta and in desert areas. The value of land is one. Land in the desert is certainly more abundant and much cheaper than land in the Delta; it is also less productive. Labor on the other hand is relatively scarce in the desert. At least it has been difficult in the past to encourage people in Egypt to settle in the land reclamation areas. Natural energy sources, namely the sun and wind, are in relative abundance. And, of course, water is more scarce and will be more expensive in the desert.

STEP II: DOES IT FIT?

Here then lies one clue for developing new technology and introducing new crops. Desert communities should choose technologies and crops that will most likely be produced more cheaply in the desert than in the delta. This allows them to create a trading relationship that will benefit both sides. They will produce products which they can produce more cheaply, in which they have a comparative advantage, and trade them for produce from the delta which delta farmers can produce more cheaply. While this may seem obvious, it is still surprising how often development projects are designed to import inappropriate equipment -- usually from the country of the donor.

In addition, in the recent debate on food security, the confusion over the principle of comparative advantage becomes important because in the literature on desert development, desert communities tend to be designed to be self sufficient, independent of other sections of the economy. However, I would argue that desert communities will need to find their technological and economic niche within the total economic system.

The development of desert communities requires the inflow of two things from the rest of the economy: people and money. These are treated in turn.

People. What would it take for you to be willing to move with your family from Cairo to South Tahrir? Or, perhaps, more importantly, what would it take to encourage a farmer from the delta to move to the Salhia project area? Let me suggest that it will take a promise of a better life. There is little question that the main component of that is a higher wage, but there are other components, too:

-- lower cost housing or better housing,

-- cheaper land or more land and space,
-- better schools,
-- better working conditions,
--.improvements in other quality-of-life factors.

In practice, it will take some combination of the above. One of the major challenges for desert developers is to create those conditions.

One could argue that population growth in the delta will eventually force people to move elsewhere. That may happen, but I would not call that development. I prefer to think that we can create conditions that will "pull" people in or draw them to desert communities because they are offering prospects for a better life.

Capital. In a similar way, we need to draw financial resources into the development process. These may come from a private sector, e.g., personal savings, from the government, or from the international donor community.

In Professor Kassas' paper, he spoke of private investment decisions and desert development. Kassas said that at current rates of 13.5% on savings accounts, one can earn more money in a bank than in investing in desert development. Yet, he hopes that some of us would still invest in desert development. If that is the case, desert development is not economically viable in Egypt, and I do not think it will work. However, before I accept that as true I would like to ask two questions:

1. What does the bank do with its money that it can afford to pay savers 13.5% interest?

2. Is the calculation of a lower rate of return in desert development accurate? Perhaps the social rate of return is higher than the private rate, and I suspect that planners and policy makers need to take a careful look at this issue.

There are many reasons why the two rates could be different. One reason leads nicely into the next section on government intervention and policies for desert development. If prices of food are controlled, which means that the prices paid to farmers are artificially low, the private rate of return on investments in food production will be lower than the social rate of return. The same thing would be true if the local currency were overvalued in international markets, if agricultural exports were taxed or if agricultural production were taxed. I stress this example because in most countries in the developing world there is an urban bias in national economic policies that inhibits the development of the rural economy. This is true in Egypt. In most developed countries the reverse is true, i.e., there is an agricultural-rural bias in national policies.

If researchers would use an accurate calculation of the net social rate of return, desert development can be compared with the banks' interest rates. If the bank rate is still higher, desert development with current technologies is not viable. Then another strategy must be found; and the challenge falls on researchers to find cost-reducing technologies. Otherwise, desert development will depend on a constant source of government subsidy: this is something that few economists recommend. It is an essential task for promotion of desert development to create and demonstrate new technologies that will lower the cost of agricultural production in desert areas so that desert development can compete for available private funds.

Economists in general have a strong preference to use public funds to finance things that fall into the realm of "public goods". It is often difficult, for the private sector to finance these because they cannot capture the returns on their investment. Public goods include roads, research, education, communication, electricity and health care.

In addition, governments of developing countries have tended to perform poorly in doing things the private sector does well. Marketing agricultural products is an example. While there are some exceptions, marketing parastatals in sub-Saharan Africa have been notoriously inefficient in handling agricultural products. Governments should resist using their scarce financial resources for these kinds of projects.

STEP III: WILL IT WORK?

What are appropriate investments in desert development? The first is research. The progress that has been made in research in Egypt in desert development since my last visit in 1983 is impressive. Nevertheless, more research needs to be done before, for example, agricultural production in desert areas can compete with the delta. Research as an investment in the future needs government support, and, governments, while sympathetic, face tight budgets. Research budgets are always among the first to be cut and this is unfortunate. The work of Evenson (1977, 1979) looked at returns to investments in agricultural research and found impressively high rate returns, 40 to 60 percent, which are almost always higher than most alternative forms of investments in the agricultural sector.

Donors are willing to fund research, and research components are a part of most development projects. I think that a concerted effort on the part of desert development activists to increase the availability of research funds would be successful.

The second investment opportunity is in human capital development, i.e., investing in people or training people. I work at an organization that has a history of 34 years of training rural social scientists in Asia. It was formerly the Agricultural Development Council, now recently merged with two other development organizations to become Winrock International Institute for Agricultural Development. The returns to investments in people, while hard to measure, also appear to be impressive. In the area of desert development, I think there are two types of training that are needed. The first is scientists. Fellowship opportunities to do bachelor's, master's or Ph. D. study in the agricultural sciences with a focus on desert agriculture is an important way to increase the number of researchers in this field.

Secondly, more people need to be trained at the technical level. Because desert development is experimenting with sophisticated technology, a higher caliber of general education is needed by laborers and more specialized training is needed for technicians. Investments in education in rural areas will help stem the rural-urban migration that is so characteristic of the third world. Desert development needs "in-migration", not "out-migration". Education is an important ingredient to developing Egypt's deserts.

A third area in which desert development needs public funds is in developing the necessary infrastructure and building rural institutions. Roads are the most obvious example and are critical to integrating desert communities with the rest of the economy. One area in which desert areas may have a comparative advantage is in production of high value crops, e.g., vegetables, for export or sales in urban areas. Without efficient transportation this apparent comparative advantage cannot be exploited. Similar reasoning applies to other desert infrastructure.

The final area of government intervention is important, but also the most controversial, and complicated -- it is government policies. D. Gale Johnson, renowned economist at the University of Chicago, addressed this issue recently in a conference on Development of Rainfed Agriculture Under Arid and Semi-Arid Conditions. He sets several guidelines for developing appropriate policies for agricultural development in arid areas; they are relevant for desert development, too.

First, Professor Johnson argues for a balanced policy approach ". . which emphasizes incentives to produce through appropriate prices for outputs and inputs, the development . . . of new and improved production techniques and the encouragement of factor markets for labor and capital that will, over time, result in equality of returns to comparable resources in agriculture and non-agriculture".[4] The argument stems in part from the urban bias of government policies that I mentioned earlier in which policies work to the detriment of people in rural areas (desert communities included) and farmers in particular.

His second policy guideline is to take advantage of opportunities for international trade, and domestic macroeconomic agricultural policies should stimulate trade prospects. However, too often policies, such as overvalued local currency, act to the disadvantage of agricultural production and exports. This also emphasizes the difference between food self-sufficiency and food security although the two are frequently and mistakenly seen as identical. Food security assures a country an adequate supply of food, but does not prohibit international trade or specialized agricultural production. The same applies to desert communities. I admire the approach of the Desert Development Center in its attempt to use all of its resources including recycling of waste products, etc. That, however, does not mean that its goal should be to become self-sufficient. Trade based on comparative advantage of desert systems is a key to successful development.

Government policies that fit Johnson's guidelines and address issues of concern are agricultural research, human capital development and rural infrastructure which were mentioned above, but also include agricultural credit, tax policy, marketing, input subsidies (water pricing being particularly important), and resource conservation. These are not discussed further, but there are obvious links between this list and the requirements for desert development.

These are not discussed further, but there are obvious links between this list and the requirements for desert development.

Most of the matters discussed here require data bases on desert development in order for better development policies to be implemented. Unfortunately, such data bases do not exist, and will not exist until we take experimentation to farmers' fields. In doing so, there is much to learn from Farming Systems Research (FSR). FSR has demonstrated clearly the value of ex ante studies in answering many of the questions I have posed. Furthermore, they have demonstrated the value of on-farm trials in providing quick feedback to researchers, socio-economic data to social scientists, and moving appropriate technologies to the testing and diffusion process.

FSR is an interdisciplinary team approach that takes a holistic view of the farm enterprise. Farmers are members of the team. FSR researchers have made significant gains in developing simple and quick survey methods and a sound approach for testing new technologies on farmers' fields.

It seems to me that given the sophisticated techniques that are called for in desert development, a systems approach is needed. Whether these methods are directly applicable to desert development or not is not such an important question. What is important to conduct research experiments in real desert situations in order to be assured that priority is given to the right problems. Economists can play a useful role in this process. Conflict in labor requirements, shorter growing seasons, bottlenecks in input markets or output marketing constraints are examples of problems that were identified in FSR more quickly than would have been the case with traditional research models. The same analysis is appropriate for desert development.

CONCLUSION

I have attempted to challenge desert developers to search for technologies that use abundant factors and save on those that are scarce; to find new technologies and strategies that fit the environment; and to develop production systems that will allow desert communities to find their niche in the larger economy. Desert development will fail if it requires large capital outlays and demands continual government subsidies due to inappropriate choices.

Desert communities must draw in new financial resources and additional labor, and they must be able to pay for these inputs. Desert communities have special ecological and economic characteristics, and there should be some products in which they specialize. They should export crops in which they have a comparative advantage and import those in which they do not.

More research and more investment in human capital is urgently required particularly in rural areas. Finally, the governments and donors must give priority to building the necessary infrastructure and rural institutions' government policies are part of this process. The final determinant depends on production systems that are economically sound.

Is desert development viable --

Yes, but . . . !

REFERENCES

1. Robert E. Evenson. 1977. Comparative evidence on returns to investment in national and international agricultural research institutions. In Thomas A. Arndt, Tana A. Dulrymple, and Vernon W. Ruttan, (eds.), Resource Allocation and Productivity in National Agricultural Research. University of Minnesota Press: 237-64.

2. Robert E. Evenson, Paul E. Waggoner and Vernon W. Ruttan. September 1979. Economic benefits from research: an example from agriculture. Science 205: 1101-07.

3. Yujiro Havami and Vernon W. Ruttan. 1971. Agricultural Development: An International Perspective. Johns Hopkins Press, Baltimore, Maryland.

4. D. Gale Johnson. 1980. Policy issues in rainfed agriculture. In Ted J. Davis, (ed.). Development of Rainfed Agriculture Under Arid and Semiarid Conditions. The World Bank.

5. G. Edward Schuh and Helio Tollini. 1979. Costs and benefits of agricultural research: the state of the arts. World Bank Staff Working Papers: 360.

SOLAR ENERGY AND DESERT DEVELOPMENT

DENIS HAYES
RenewAmerica, Palo Alto, California, U.S.

Abstract During the last ten years, renewable energy sources have experienced striking advances. Several of these new technologies may prove to be of special long-term importance to arid regions. For example, if the costs of solar cells continue to decline, photovoltaic will eventually provid an economical source of electricity that can be used to refine silicon and fashion it into new photovoltaic cells. If the lifetime of solar cells continuous to lengthen, a well-maintained silicon photovoltaic cell may eventually be able to produce more than a hundred times as much electricity in its lifetime as was required to manufacture it. A single field of photovoltaic panels could thus be used to breed a hundred more such fields.

Other technologies may also hold special promise for arid regions of high insolation. Using high temperature solar devices to power endothermic reactions, it may prove possible to produce cheap exportable fuels with high energy density.

Renewable energy technologies might thus allow deserts to play a vital role in the world's long-term energy future, exporting desert sunshine as embodied energy.

INTRODUCTION

For most of human history, it was possible to address the future with a high degree of confidence. If Leonardo di Vinci had flown through a tesseract -- a time warp -- and returned a couple of hundred years later, he would not have been dazzled by a host of fundamental changes.

Today, however, we cannot speak with certainty about even a 25-year interval. The year 2015 will doubtless witness developments that are currently unimagined by even our most creative intellects. We don't know where we are going, or what the consequences will be of getting there. Indeed, on most subjects the range of opinion is probably greater among recognized experts than among the lay public.

In no field is this more true than for energy.

Energy forecasts are judgments made today about tomorrow, using data gathered yesterday. To protect against foreseeable snags, energy projections use sensitivity analysis. To protect against the unexpected, they incorporate extra fudge factors. But all these protections do not yield much real safety.

Because their forecasts yield so little certainty, energy decision-makers have been forced to operate by the seats of their pants. They have come to view future demand for energy as an unpredictable, exogenous variable that lies wholly outside their control, or even their ken. Consequently, they devote virtually all of their attention to tinkering with the details of their existing sources of supply. In their preoccupation with such day-to-day "tactics", they lose sight of strategy. Their role is like navigating an automobile through Cairo traffic, very carefully avoiding an accident, but in the process forgetting their destination. This talk on the other hand, will deal almost entirely with destination.

One overwhelming fact must govern our thinking about energy. Eighty percent -- 80 percent! -- of all the petroleum that ever will be consumed on our planet will be used during the lifetimes of people who are currently alive. The world runs on oil, and the oil is running out.

We must begin building -- not just studying and planning, but building -- a post-petroleum world. Our energy program must be governed by a vision that is anchored in reality, but we cannot wait for all the engineering details to be complete before we begin. We must begin to judge our energy-related investments, not just in terms of whether they appear to solve some urgent crisis, but also in terms of whether they lead us toward, or away from, the realization of that vision of a post-petroleum world.

Indisputably, the most important steps, are in the realm of increased energy efficiency. The Desert Development Center here in Egypt has demonstrated how to build affordable buildings that remain comfortable in the scorching summer sun without requiring any electricity for cooling. Now those buildings must be commercially replicated in similar climates around the world.

Similarly, we know, technically and economically, how to reduce the energy required by lightbulbs by 80 percent; by refrigerators by more than 80 percent; of various industrial processes by 50 to 90 percent. We have built dozens of 100 mile-per-gallon automobiles. Such energy-saving equipment must become commonplace.

Conservation will stretch out our oil supplies, allowing a more gradual and well-planned transition. However, no matter how efficient we become, a swift transition will be needed. The global transition from oil must be complete within 40 years. This may be the most awesome technical challenge our species has ever faced.

Moreover, the particular choices we make will have far-reaching consequences. The coal age made possible the industrial revolution, with sweeping effects in politics, and economic organization. The age of oil, with its jet planes and automobiles, shrank the planet and fundamentally reshaped our cities. Our next energy choices may affect the very habitability of the planet.

We tend to think of energy sources as neutral and interchangeable, but they are not. Some fuels promote a global greenhouse effect by adding to atmospheric CO_2; others do not. Some energy sources are inextricably linked to fissile materials that can be used to make atomic bombs; others can cause no problem more serious than a leaky roof. Some sources are inherently centralized; others are modular and allow decentralized application. Before asking what energy sources we wish to pursue, we must first ask what kind of world they will need to power.

THE INFORMATION REVOLUTION

The world is in the early stages of an information revolution. We are advancing rapidly toward a post-industrial era in which the greatest growth will be in information-related services.

Talk of an "information revolution" is often greeted with bemused skepticism. Data processing and communications strike the casual observer as remote from the central concerns of life. One cannot eat information. Neither can one wear it, drive it, or live in it. Intuition suggests that the principal business of the world is more likely to involve food, clothes, shelter, and mobility than video games and electronic spread sheets.

Yet, some of the new information technologies could be truly revolutionary -- in the same sense that the stream engine, the cotton gin, and the automobile were truly revolutionary. These new technologies have the potential, for good or bad, to fundamentally reshape human society.

Responsible talk of a "post-industrial" world does not imply that industry will disappear. One could speak in the same way of the early 20th century as "post-agricultural." Of course, agriculture did not wither away during the industrial revolution. Instead it prospered; we grow more food today than ever before. But the great new areas of growth and employment have been in the manufacturing sector.

In a similar fashion, the "post-industrial" future will not be devoid of concrete, steel and textiles. But the service sectors in general, and the information areas in particular, are likely to be the principal engines of new economic growth. The energy consequences, as well as the social consequences, of this shift may be even greater than those that accompanied the industrial revolution.

The current revolution hopes to do for our brains what the industrial revolution did for our brawn. It will, through computers, vastly increase our ability to store and analyze data. It will, through communications advances, magnify our senses. Its impact will reverberate throughout all aspects of our world, including the amount and quality of energy we require.

Recent developments in communications allow far-reaching changes in the way that societies are organized. For example, a combination of personal computers, high-quality printers, component audio and video equipment, cable systems, and satellite transmission already make it possible to perform a vast array of activities without leaving one's home. Fiber optics promise to increase the world's wired communications capability a thousand-fold in the years ahead.

Even as such advances in communications technologies accelerate the transmission of information, breakthroughs in storage technologies are dramatically increasing our abilities to retain and manipulate that data. Existing devices can store all the information in an encyclopedia on a 14-inch disk. The next generation of technology is expected to store upwards of 200,000 books on one side of a disk the size of a phonograph record.

A modest computer can display any of the text very quickly on a color screen, and it can print hard copy if the user wishes it. The same small disk can incorporate motion pictures, graphics, and even music as appropriate. Storage that is currently available in only the largest, most expensive data base computers will tomorrow be available for a few thousand dollars in devices the size of hat boxes.

All of this can be linked up into communications networks. Most of tomorrow's systems will be "videotext" networks -- allowing two-way communications and multiple direction conference hook-ups. One American data base currently offers full-text versions of more than 50 major magazines, ranging from Scientific American to Modern Bride. Another has an electronic shopping feature that allows the consumer to compare prices among hundreds of discount stores carrying about 50,000 items.

These technologies already exist, and their applications are growing rapidly. What might this mean for society?

New communications technologies can facilitate working at home in many service industries. This could eliminate much wasted commuting time, reduce automobile pollution, conserve fuel, and cut-back the capital investments needed in office space and parking lots. If it costs a firm less to have a worker work at home, the resulting savings can be invested in new ways to produce new goods and services, creating new jobs.

A system could be designed to automatically "shop around" all the stores in the network for the best price for a particular product. The person could specify what color, size, and brand he wanted, find the cheapest item, and order it through the network. Theoretically, this could yield the perfectly informed choice of Adam Smith's idealized marketplace.

Another sweeping change could lie in the entertainment field. Widespread use of cable transmission, large-screen thin film television monitors, and superb stereo acoustical peripherals may well bring about changes in recreational patterns as well. High quality video disk recordings, coupled with large screen displays, already allow the owners of such systems to view movies or concerts with higher resolution pictures and clearer stereo sound in their living rooms than can typically be obtained in theaters and concert halls. An American rock group, the Grateful Dead, recently charged an identical amount to cable viewers of a concert as it charged for tickets to attend the concert.

Even as huge audiences have built up for televised athletic events, similar audiences could perhaps be found for symphonies, plays, first-run movies, etc. With widespread use of video recording technologies, people could automatically record events as they were broadcast and enjoy them whenever convenient.

The information revolution will being dramatic changes in the amount and kind of energy use needed for economic growth. Consider, for example, the computer. ENIAC, the world's first electronic computer, was switched on at the University of Pennsylvania in 1946. It was as large as a small house. Its 18,000 vacuum tubes, 7,000 resisters, and 10,000 capacitors consumed enough power to run a locomotive. Today, a chip with more computational power than ENIAC costs $5. It is smaller than a fingernail. It is connected by wires 1/30 the thickness of a human hair. And it runs for days on the energy contained in one small pen light battery.

As such chips are integrated into an expanding range of uses, yielding "smart" automobiles, "smart" furnaces, "smart" appliances, and "smart" industry machinery, the world will experience great changes. One consequence will be that many tasks will require much less energy than at present. The information revolution will allow us to save huge amounts of energy without sacrificing comfort, convenience, or productivity.

DESERT DEVELOPMENT

What does the coming transition bode for the world's deserts? Let me become even more speculative here.

A large part of the world may jump from the agricultural stage of development to the post-industrial stage. It may skip the industrial revolution -- the age of steel -- and move directly to the information era -- the age of silicon: silicon semiconductor chips for integrated circuits; silicon for fiber optics; silicon for photovoltaic devices for decentralized energy production. This age of silicon could be an era of re-ascendence for the world's deserts.

The world's history is rooted in deserts. The earliest civilizations, three of the world's major religions, and several of history's most successful armies, were born in deserts. But in recent decades, deserts have drawn more international attention for their failures than their successes. Declining agricultural production in the face of mounting populations have led to widespread malnutrition, and starvation.

Arid and semi-arid regions comprise about 40 percent of the earth's land, inhabited by about 14 percent of the world's population. These desert areas are growing larger. The eminent Egyptian ecologist, Mohammed Kassas, estimates that about 6.7 percent of the earth's land -- a collective area larger than Brazil, has been turned into deserts by human activities: deforestation, over-grazing and unwise farming practices. Perhaps 90 million people today live on lands already rendered almost useless for agriculture because of such factors as soil erosion, dune formation, and salt encrustation.

Many are flocking to overcrowded cities to compete with other landless people for scarce jobs. Cairo, for example, now has more inhabitants than the entire country of Australia. It is manifestly desirable to create a desert development strategy that will yield sustainable development offering a decentralized alternative to this urban implosion.

One important component of a comprehensive desert development strategy could involve harnessing sunlight to provide commercial energy. This could allow the Middle East and North Africa to remain large exporters of energy, even after their oil runs out.

These regions hold some comparative advantages in a solar-powered era. Sunlight delivers about three times more sunlight per acre each year in North Africa than in northern Europe. Seasonal variation in Egypt is about 40 percent; in Stockholm the variation from December to June is about 3,000 percent.

Deserts tend to have huge expanses of uninhabited, unutilized land. Even if large tracts were set aside for preservation in their natural state, much land would remain for solar collection.

Some of the collected energy could be exported directly as energy. For example, photovoltaic-generated electricity could be sent long distances with virtually no losses along superconducting transmission lines. Chemical fuels produced photochemically, or through thermal processes, or harvested as complex hydrocarbons from desert plants, could be exported in tankers that currently haul oil. Photovoltaic or photoelectrochemical devices could dissociate water to produce hydrogen that could be transported in gas pipelines.

Another complementary strategy could involve performing energy-intense refining and manufacturing tasks in the desert. The resulting products may be thought of as having "embedded" energy. For example, a huge amount of energy is needed to refine bauxite into aluminum. Once the aluminum has been refined, only 5 percent as much energy is needed each time it is recycled. When aluminum is purchased, much of price represents the energy embedded in it. If aluminum refining were done in the desert, the exported aluminium would already contain this embedded energy.

Similarly, the silicon in photovoltaic cells must be refined in an energy-intense process. A photovoltaic cell may produce more than 20 times as much energy over its lifetime as was needed to manufacture it. So a photovoltaic breeder in a silicon-rich desert can export 19 photovoltaic cells (with their embedded energy) for every cell it must make to replace itself.

Is all this possible? Certainly. Is it likely? Different people may have different answers.

One thing is certain. Such a transformation will not occur spontaneously. It will require a major commitment by the world's arid nations -- particularly those with discretionary wealth from their depleting oil resources.

Perhaps most important, it will need a base of concrete experience upon which to build. It must begin with exactly the sort of solid steps represented by the solar work being reported at this conference.

THE DDC: FROM CONCEPTION TO REALIZATION

ADLI BISHAY, DIRECTOR
The American University in Cairo,
Desert Development Center

Abstract The paper summarizes the work of the Desert Development Center (DDC) of the American University in Cairo (AUC). Founded in 1979, the DDC conducts research and demonstrations based on a systems approach integrating biological, technological, and community aspects of desert development. The DDC integrated approach advocates that a desert should be treated as a desert (no outside manure or silt added), and that desert development should be based on a balance of appropriate indigenous methods with modern technology. Any trials for improving productivity should be environmentally compatible with desert conditions and economically replicable under Egyptian social and technical constraints.

The DDC has two sites at Sadat City and South Tahrir where activities are conducted in desert agriculture, renewable energy, desert architecture, and socio-economic aspects of desert development. Recently, work has begun with farmers who have settled on reclaimed lands in neighboring areas.

INTRODUCTION

Egypt suffers from continual population growth and dwindling resources. With Egypt's population* concentrated in the Nile Valley which constitutes only 4% of the total area of the country, development of her deserts to make them productive and habitable is an urgent need that is widely recognized. The Egyptian Government, in order to

---
* 50 million in 1986

solve the problems of food and housing for the continuously increasing population has decided to expand horizontally (in the desert) as well as continuing the present successful vertical improvement of its cultivable land. According to the Government of Egypt, a minimum of 500,000 feddans will be reclaimed in the next five years.* Unofficial estimates for long range reclamation plans exceed five million feddans, which would constitute a doubling of the land presently under cultivation in Egypt.

The first International Desert Development Conference, held in Cairo in 1978, adopted a recommendation to initiate an applied research, demonstration and training program in desert development at the American University in Cairo (AUC), and called upon the Government of Egypt and different international agencies to support such a program. In 1979, AUC launched a program to conduct research and undertake demonstration and training in the field of desert development (Figures 1 and 2). The Desert Development Center (DDC) adopted a systems approach integrating (a) biological aspects including appropriate drought-resistant varieties, cropping systems, livestock, and desert soil development; (b) technological aspects including agricultural technology, irrigation, water management, agro-industries, renewable energy technologies, and passive solar architecture; and (c) community aspects including infrastructure, socio-economics, management, desert housing, and aesthetics.

Fundamental to the DDC integrated approach is the concept of treating the desert as a desert (no silt or manure from other areas is added). In addition, trials for improving soil productivity and initiating new communities should be environmentally compatible with desert conditions and economically replicable under Egyptian social and technical constraints.

---

* As stated in Pres. Mubarak's speech on Labor Day, May 1986.

FIGURE 1

THE DDC INTEGRATED APPROACH TO DESERT DEVELOPMENT

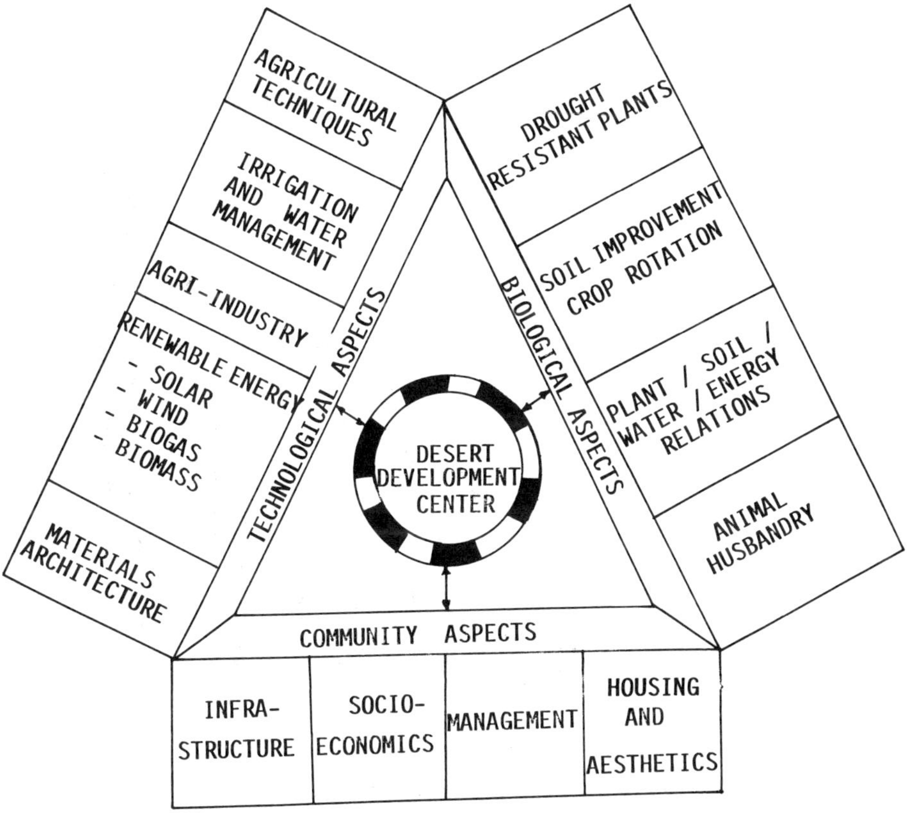

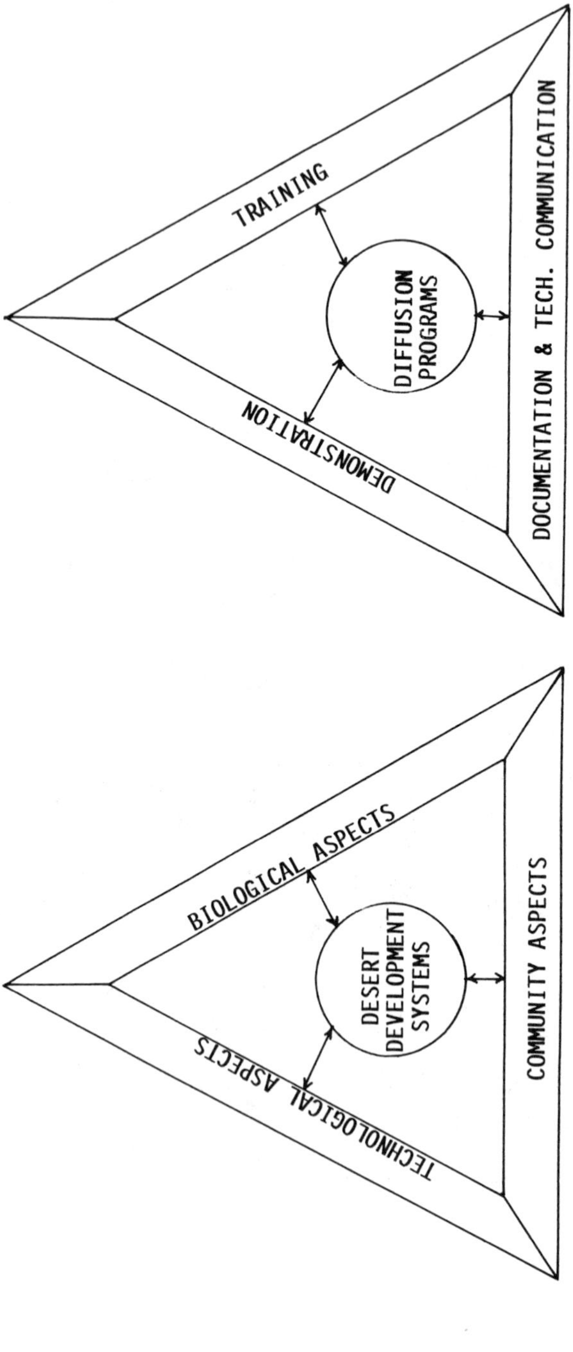

FIGURE 2: THE AUC DESERT DEVELOPMENT CENTER APPLIED RESEARCH AND DIFFUSION PHASES

The AUC program takes advantage of two factors. First, it profits from those creative and efficient technologies -- so suited to the region's harsh environment -- which have evolved over the millennia in Egypt and other parts of the Arab World. And secondly, it draws upon the equally creative and sophisticated modern technology recently developed and applied to agriculture, arid land development, and harnessing of renewable energy. It is this integrated approach and the balance of indigenous methods with modern technology which makes AUC's approach unique.

With virgin desert land as well as material and moral support generously donated by the Government of Egypt, and with financial and technical support from a number of international, Egyptian and American agencies*, the idea which was conceived in 1978 is, we believe, approaching maturity at this time.

DDC faculty and their associates are currently active in the areas of desert agriculture, renewable energy, and desert architecture as well as social and economic aspects of desert development. While most of the work was initially conducted on the two DDC sites at South Tahrir and Sadat City, representing rural and urban desert development respectively, our work has recently been extended to neighboring farmers who have lived on reclaimed desert land at South Tahrir for periods between five and twenty years. Demonstrations on farmers' lands conducted by an interdisciplinary group of researchers in cooperation with the farmers themselves are adding a new dimension to our work.

Our experience to date strongly supports the original concept of an integrated approach. Within this framework, we are continuously improving our

---

* IDRC, AID/FRCU, NEF, Ford, UNDP, World Bank, FAO, ARCO Solar, Samuel Rubin Foundation, Sheikh Saleh Kamel, Goelet Enterprises, NIMOS, and Misr Iran Development Bank.

techniques to suit the type of resources available, the needs and aspirations of the farmers, and the different problems and constraints facing them. Technical and economic evaluation of our studies since 1979 have lead to the initiation of specific desert farming systems, including a soilless system.

GOALS AND OBJECTIVES

The DDC's ultimate goal is to demonstrate alternative patterns which can be replicated by others for the establishment of viable desert communities in Egypt. Such communities should offer a quality of life attractive enough to encourage migration of families from the overcrowded cities and villages to desert areas. The current work with farmers who have settled in desert lands reclaimed over the last five years is a step toward this goal. In working toward the ultimate goal, on-going activities of the DDC aim to:

1. Search for technologies appropriate for desert agriculture where water and nutrients are scarce and energy is expensive, and identify desert farming systems which are technically feasible, economically viable and socially acceptable.

2. Harness renewable energy resources abundant in the desert and develop technologies appropriate for desert environments. This includes passive solar architecture for desert housing, in addition to solar, wind, and biogas technologies.

3. Identify and demonstrate desert community development strategies which are feasible given local resources that are socially acceptable, and which protect the natural environment and enhance the quality of human life.

4. Disseminate the results of the above activities through: demonstration sites (on DDC land and on the farmers' lands); technical and popular publications and audio-visual presentations; and consulting; training; and on-site experience for farmers, students, technicians, scientists, businessmen, and policy makers. In order to strengthen this activity, plans are being formulated to create a desert development data base encompassing agricultural, technological and community aspects of desert development, using information resulting from the above on-site and on-farm activities, as well as those obtained from other relevant sources.

DESERT AGRICULTURE AND DESERT FARMING SYSTEMS

The DDC approach to the problems of practicing agriculture in the desert, where water and nutrients are scarce and energy is expensive, stipulates that "a desert should be treated as a desert". Essentially, no silt or manure* should be transported to a desert site from other areas. Technologies, practices, crop varieties and rotations appropriate for desert agriculture are expected to be quite different from those adopted in the Nile Delta and Valley.

Due to the prevailing environmental conditions in the desert, windbreaks and shelterbelts are necessary and their establishment was one of the first activities undertaken when initiating the DDC desert farms. Some of the trees and shrubs which can be grown in desert areas may be used for fodder or fuel, and some varieties are salt-tolerant.

The development of desert soils at DDC sites includes identification of soil resources,

* Unless it is properly treated against any infestations; in which case this practice may not be economically viable.

assessment of soil development under various agronomic practices, identification of major factors limiting soil productivity, studies to improve soil management and enhance soil development, and development of techniques and methodology for training and extension.

The scarcity of water in desert areas necessitates proper water management as well as energy and water conservation. The latter can be achieved through application of special irrigation techniques. Drip, biwall, bubbler, and low pressure sprinkler systems (e.g. side roll) are being used at the DDC sites. Under no circumstances should flood irrigation (the predominant irrigation system used in the Nile Delta and Valley) be allowed in desert agriculture.

Field and fodder crops, plant varieties, agronomic practices and rotations appropriate for desert environments are being tested and demonstrated. Some fruits and vegetables and selected rootstock/variety combinations and cultural practices are also being introduced at the two DDC sites. A selection of arid zone plants and salt-tolerant plants as well as plant protection measures against weeds, nematodes, and other potential plant diseases are being tested under desert conditions.

While the growing of fodder crops in the limited area of the Nile Delta and Valley may be subject to reservations based on the critical shortage of food crops in Egypt, it is our belief that a fodder/livestock/biogas sequences is most appropriate for desert development. Most fodder crops, like animal manure, improve desert soil productivity. Barki sheep and two types of hybrid cattle are important components of one of the DDC farming systems which is being demonstrated and evaluated for its economic viability.

The concept of <u>Desert Development Systems</u> adopted by AUC gives equal stress to biological, technological and community aspects of desert development. When agricultural aspects are emphasized, the terminology "Desert Farming Systems" will be used.

While conventional farming systems research aims to solve farmers' problems, our Desert Farming Systems approach begins with early stages of land reclamation and endeavors to help farmers to avoid problems as well to solve them if they do occur. It should be stressed, that the AUC approach covers both experimental station research and on-farm research; one leads to the other at the initial stages and the two activities are parallel at a later stage. This two-pronged approach is a valuable means for demonstration of different technologies and systems, as well as training of farmers and students.

Our experience to date has shown that at least six alternative <u>farming systems</u> are appropriate for the Egyptian desert environment. These are:

1. A Crops, Multipurpose Trees, Sheep System with appropriate rotations, soil, water, and fertilizer management.

2. A Horticulture, Fodder, Sheep System with appropriate fodder, fruits, vegetables, and possible intercropping, water and fertilizer management.

3. A Fodder, Livestock, Biogas System with perennial and/or annual crops, meat and dairy processing.

4. An Agroforestry System, i.e. appropriate integration or intercropping of trees and agriculture crops and/or animals.

5. An Agro-Industrial Crops System, i.e. appropriate oil, fiber or aromatic plants, extraction techniques, and processing of by-products for animal feed or biogas.

6. A Controlled Environment Agriculture System based on appropriate horticultural crops, protective structures and nutrient solutions

(with or without the use of an inert medium)

Agricultural technologies adopted for the implementation of any of the above systems have to be appropriate to the desert environment where water, energy and nutrients are major constraints. If proper soil management is not viable, soilless agricultural technologies may be adopted (e.g. Nutrient Film Technique, [NFT]). Under extreme conditions, controlled environment agriculture may be necessary to achieve high plant density, maximum crop yield, relative indifference to seasonality, efficient use of water and fertilizers and minimal use of land area. Desert agricultural technologies and practices should, in most cases, be different from those adopted in the Delta and Nile Valley, and should be based on appropriate environmental and resource management.

Four of the above-mentioned farming systems have been demonstrated on the DDC sites for periods of two to six years, while components of the other two (agro-forestry and agro-industrial crops) have been successfully tested and the systems will soon be initiated. In all cases, economic monitoring and evaluation, social acceptance, and successful demonstration on the farmers' land are prerequisites for adoption by small farmers or agro-industrialists. Training and extension activities are needed to achieve this goal.

In the last few years a number of factors, mainly economic, have affected the farming systems in the Nile Valley and Delta leading to a decrease in food production (wheat, corn, lentils, etc.) and an increase in fodder (especially berseem clover) for animal production. National and international agencies have warned against this tendency. It is our belief that this problem can be solved in two parallel ways, namely: a) by changing the pricing policy for food products to encourage farmers to grow more food crops, and b) by adopting desert farming systems based on fodder and/or horticultural crops. Studies at the DDC strongly favor the latter approach.

Recently, the technology of "Protected Structures" has spread on the Nile banks and in many of the areas characterized by "good" soil. This technology, and its advanced version of "Controlled Environment Agriculture" is more appropriate in the desert where good soil is a major constraint. Furthermore, before expansion on a large scale, proper management of these systems should be achieved, under our Egyptian conditions. Specifically, summer problems associated with high temperatures in the protected structures should be solved and/or the economics of planting for only the winter season should be established. In our view, the adoption of hydroponics or soilless agriculture (with or without the use of an inert medium) under controlled environment structures is one potentially attractive system for desert areas - especially those adjacent to new cities - to produce some of their vegetable and fruit requirements.

AN EXAMPLE OF A DESERT FARMING SYSTEM: "FODDER/LIVESTOCK/BIOGAS"

This "MAXI Linkages Project" was initiated in cooperation with Alexandria University and sponsored by the U.S. Agency for International Development through the Egyptian Supreme Council of Universities. The Fodder/Livestock/Biogas farming system has been chosen for the following reasons:

a) A large number of fodder crops are leguminous, thus fixing atmospheric nitrogen. This is very valuable in desert soils which are usually deficient in nitrogen and most of the macro- and micro-nutrients. Furthermore, grasses used as fodder crops (e.g., sorghum, sorghum-sudan grass, and millet) can also improve sandy soil characteristics through the formation of sod which decreases leaching of nutrients and soil erosion.

b) One of the major reasons for insufficient meat and dairy products in Egypt is the insufficiency of fodder and feed for animals around the year.

c) Priority has been given to growing crops for food and fiber (cotton) in the Delta and Nile Valley. This makes it necessary to promote growing fodder crops in newly developed desert lands.

d) While fodder crops may be partly used for green manuring, animal manure is an important source for fertilizers and organic matter needed for improving desert soil productivity

e) Animal manure can also be used for the production of biogas; a valuable source of energy for cooking and lighting. On the other hand, the residue formed during the process of biogas production is an excellent source of fertilizers.

f) Most of the present work in Egypt deals only with one aspect or another of this system, and none combines the three activities (fodder, livestock and biogas).

The current governmental policy for desert development adopts two parallel approaches aimed at (a) absorbing a proportion of the increasing population through providing new employment opportunities, settlement incentives, and satisfactory living conditions in new desert communities, and (b) maximizing the efficiency of intensified capital investments: The first approach encourages the initiation of small desert farming units, while the second approach requires the establishment of large scale agro-business firms.

In the MAXI Linkages project, models for both of the approaches advocated by the government are being investigated through building and monitoring of three model farming units (20 acres each) and a demonstration agro-business unit (100 acres) at the DDC South Tahrir Site:

Three Model Farming Units at South Tahrir

Based on the original theme of the project, fodder has been given priority as a crop because of its value in improving virgin soil. However, some farmers may be interested in some cash or economic crops. Based on our experience in South Tahrir, the cropping patterns for the three farms (each 20 feddans) were designed as follows:

Farm No. 1: 100% fodder crops (15 feddans alfalfa, 5 feddans perennial grasses). The fodder is basically planted to feed 14 cows on the farm.

Farm No. 2: 70% fodder crops, 15% vegetables and 15% medicinal plants. These are distributed as follows:

10 feddans alfalfa,
4 feddans perennial grasses,
3 feddans horticulture,
3 feddans medicinal plants.

The fodder area in this farm is designed to feed 64 head of sheep.

Farm No. 3: 50% fodder crops, 30% field crops, and 20% fruit trees. These are distributed as follows:

6 feddans field crops (e.g. peanuts),
6 feddans annual fodder (3 leguminous and 3 cereal),
4 feddans alfalfa,
4 feddans fruit trees.

This cropping pattern is designed to provide enough feed for 12 sheep and 2 cows in addition to other horticultural and field crops.

Implementation of the three farming units was initiated in the spring of 1985. Activities on these farms had to be slightly modified during the early stages of the project. For example, the areas allocated to perennial grasses in Farm No. 1 and 2 are planted temporarily with annual fodder grasses until enough seeds of perennial grass are available.

The Agribusiness Unit: Alfalfa Production

Alfalfa, grown in an area of 100 feddans, represents the main diet component for the milking animal units of agribusiness in South Tahrir. Continuous monitoring of alfalfa growth, yield, and quality is essential to study its productivity, with respect to animal feeding requirements. It is important also to study the major factors affecting such productivity under South Tahrir conditions. A number of cultural practices were used to establish the alfalfa stand on the five plots of twenty feddans each. General production trends show that narrower distances between planting rows and seeding in one direction produce higher alfalfa yields. Alfalfa yields one cut per month in spring and summer. The present yield per cut represents a reasonable production level (2-3 tons/cut/feddan).

To test the quality of the alfalfa produced in this unit, chemical analyses for total dry matter, protein, fiber, ash, silica and major nutrient contents were carried out. Data for the major constituents of alfalfa foliage, especially that of crude protein content, exhibit standard levels. On the other hand, NPK analysis showed the following trends:

a) Total nitrogen is above the adequate level; this reflects the activity of root nodules and fixation of atmospheric nitrogen.

b) The phosphorus level varies from stand, to stand, with lower levels associated with poor stands. A close relationship appears to exist between the level of phosphorus in the plant and the state of its growth.

c) Potassium is generally within the adequate range. However, when the K content fells below the lower limit, plant growth is unsatisfactory.

These observations suggest that appropriate fertilizer recommendations should be applied to alfalfa in order to achieve good growth and higher productivity. This includes the use of micro-nutrients in addition to phosphorus and potassium.

Table 1 presents the alfalfa yield in the agribusiness area (plots 2, 3, 4, 5 and 6) for year two of the project. It is noted that alfalfa production in the first six months was low. As mentioned above, new fertilizer practices, including some micro-nutrients (iron, manganese, zinc) were applied beginning in March 1986, which resulted in marked increases in alfalfa yield.

Evaluation of the different cropping patterns in the three model farming units as well as the productivity of the agribusiness unit is underway.

It should be noted that the MAXI project also demonstrates the application of the DDC integrated approach where different components of agriculture, energy, architecture, economics, and social aspects interact in each model farm. Renewable energy, passive solar architecture and socio-economic aspects of this project will be discussed in the relevant sections of this paper.

RENEWABLE ENERGY TECHNOLOGIES

In concert with the DDC Systems Approach, renewable energy technologies, agriculture, and industry should be linked to achieve one of the DDC's major goals, namely, demonstrating the most viable means of creating new energy-independent desert communities. The approach is shown schematically in Figure 3.

TABLE 1: Alfalfa production (tons/feddan in the Agribusiness area for one year (September 1st, 1985-August 31st, 1986)

Cut	Plot No.	Yield per cut (Tons per Feddan) in different plots)				
		2	3	4	5	6
1		1.11	0.45	1.00	1.30	0.96
2		1.08	1.05	1.07	1.31	0.96
3		1.11	1.12*	1.08	1.20**	0.75
4		1.41	1.88*	1.28	2.10**	1.86
5		1.79	2.52*	1.64	0.94**	1.90
6		1.81	2.52**	1.30	2.07**	2.08
7		2.58	2.66**	2.57	3.11**	2.25
8		3.28	2.40**	2.65	2.92**	2.42
9		3.07	2.41	2.57	2.38**	2.42
10		3.17	2.08**	2.09	2.57**	
TOTAL		20.41	19.09	17.25	19.9	15.6

* Equivalent fresh fodder.

** Average of the whole plot.

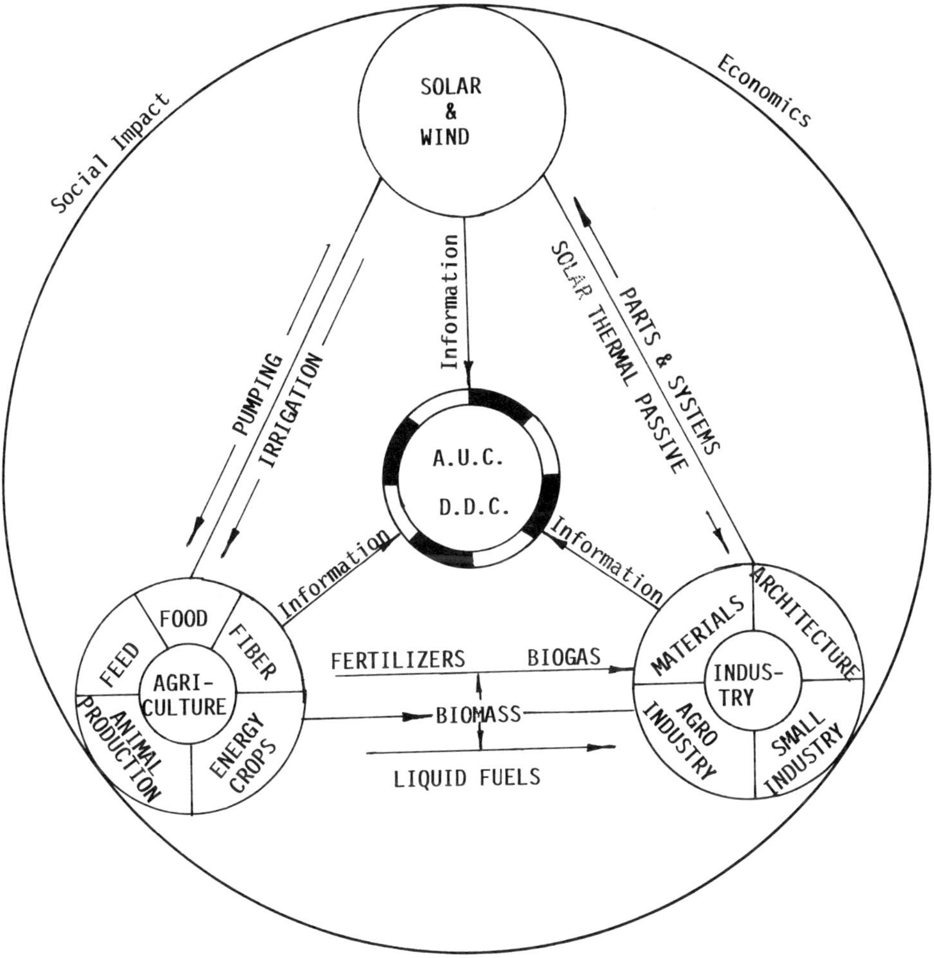

FIGURE 3: Application of renewable energy technologies for desert development

Figure 3 shows that solar and wind technologies are used to provide energy required for pumping and irrigation. The resulting biomass and energy crops in turn are a source of energy for the pilot industries such as building materials and small agro-industries. Passive and active solar systems are also used in the architecture of the pilot housing and industrial facilities. Products of the pilot industries might include parts for the energy technology equipment. Together, these interrelated activities, based on renewable energy technologies, seek to demonstrate how energy-independent desert communities may be based on a carefully integrated infrastructure.

In designing renewable energy systems, the following criteria have to be taken into consideration: (a) technical feasibility, (b) replaceability, (c) economic viability, (d) social acceptability, and (e) suitability for desert development.

The main DDC site for renewable energy activities is at Sadat City (a new desert city being built halfway between Cairo and Alexandria). An area of 25 acres in this City was assigned to the DDC to house its headquarters building and to conduct research and demonstration in solar and wind applications for desert development. The other DDC site at South Tahrir (500 acres of desert land, 40 km east of Sadat City) houses the biogas unit, since a major part of DDC agricultural and animal husbandry research and demonstration is conducted on this site.

Energy management receives considerable attention at the DDC as it is a key element in attaining economic viability for renewable energy sources. Because of the importance of monitoring meteorological data at the site, this activity is discussed as part of energy management. At the Sadat City site of the DDC, a microcomputer-based data logger has been operational since October 1984. The system, Hewlett-Packard 3054 DL, has been programmed to collect and analyze data without operator intervention for an uninterrupted period of one week. Data collected during this time includes measurement of parameters related to solar energy, storage in batteries, water pumping, water usage, wind energy and the indoor and outdoor environment.

SOLAR ENERGY

The incident solar radiation (insolation) is monitored on site in two ways: a) global solar insolation, i.e. the total solar energy received in a day at the Sadat City Site; measured by an EPPLEY Pyranometer b) The actual solar energy incident on the array plane (inclined towards the south); monitored by a simple PV cell installed in the array plane.

Three systems employing photovoltaic conversion of solar energy are currently operational at the Sadat City site of the DDC; and a fourth system is being installed:

i- A 10 Kwp system* with an inverter (DC to AC) to power the DDC Headquarters Building and a submersible AC deep well pump.

ii- A 3 Kwp system* which supplies power to a DC deep well pump (progressive cavity) as well as a DC irrigation (booster) pump.

iii- A 700 wp system which powers the recirculating pump (also a DC driven pump) used in the stand-alone Nutrient Film Technique (NFT) experiment.

iv- A 7 Kwp system is part of the Solar-Wind hybrid which will supply the necessary power for three model farms, 5 acres each.

While the first two systems have been installed for some years, the third system became operational

---

* Donation from ARCO Solar through the World Bank and Egyptian Ministry of Electricity and Energy.

only recently. The latter unit was designed and implemented by the DDC. The fourth system (7.0 Kwp) arrived recently and will be discussed under the section dealing with the Solar-Wind System. All four systems have electrical storage capability in the form of batteries, each with an appropriate storage capacity. All currently operating systems are monitored by a data logger, based on current and voltage measurements.

Everyone concerned with renewable energy knows that the user must adapt to the resource. This means that some form of load management is usually necessary in the use of renewable energy systems. While to non-specialists this may imply shutting off the load when the resource is scarce, it is equally important in good load management to make use of the resource when it is plentiful. Load management schemes for the Sadat City Headquarters Building posed very special problems. The variety of activities in the building necessitated a non-automatic load management scheme although an automatic system was technically feasible. The building contains living quarters for a number of people as well as a several laboratories, a library, and an office.

A simple load management scheme was devised whereby the decision to operate a piece of equipment is left to the operator. However, appropriate information as to the availability of the resource is provided to allow the decision to be made on the basis of the actual situation. The medium through which the information is conveyed is by means of a "traffic light indicator" where red, yellow and green lights have conventional significance. The traffic light indicator which is composed of small desktop units is placed in various places throughout the building where users of electrical equipment can, at a glance, make the appropriate decision. The LED's (light emitting diodes) are, of course, switched on and off by the computer according to prescribed set points.

WIND ENERGY

Since wind energy is site specific and may differ for different sites at the same general area, it was necessary to install a wind anemometer at the DDC Sadat City site. The anemometer used is a sophisticated type capable of measuring the three components of wind velocity, U, V and W.

The wind-speed frequency curves for daytime, nighttime, and a daily average (for both day and night) were plotted for each month (averaged over four weeks) from April 1985 to September 1986. Averaged over six months, these wind-speed frequency curves were plotted, and the results for the period April - September 1985 showed that the most frequent wind speeds at the day and the night are equal to 5 and 3 m/s with percentages of time of 16.8 and 22.6, respectively. Comparison of the average wind-speed frequency curves with diurnal curves for each month reveals that the most frequent wind speed (e.g. 3.5 and 3.0 m/s for April and August) is considerably lower than the maximum wind speed (6.4 and 7.2 m/s for April and August, respectively).

Selection of Wind Turbine for Sadat City Site

To date, the wind measurements at the Sadat City site provide a representative and adequate basis for selection of a suitable wind turbine. The site exhibits wind speeds varying from 0 to a maximum of 12 m/s. (See Figure 4 which shows the wind duration curve for six months). It was decided to base the choice on a rated wind speed of 7 m/s. Referring again to Figure 4, wind speeds higher than this value prevail for almost 17% of the time (with inclusion of winter data this goes up as high as 20% of the time).

A survey of various commercially available wind turbines showed that values of cut-in speeds of the order of 2-3 m/s are generally used. This value has been adopted for the system already installed at the Sadat City site. Figure 4 shows that with

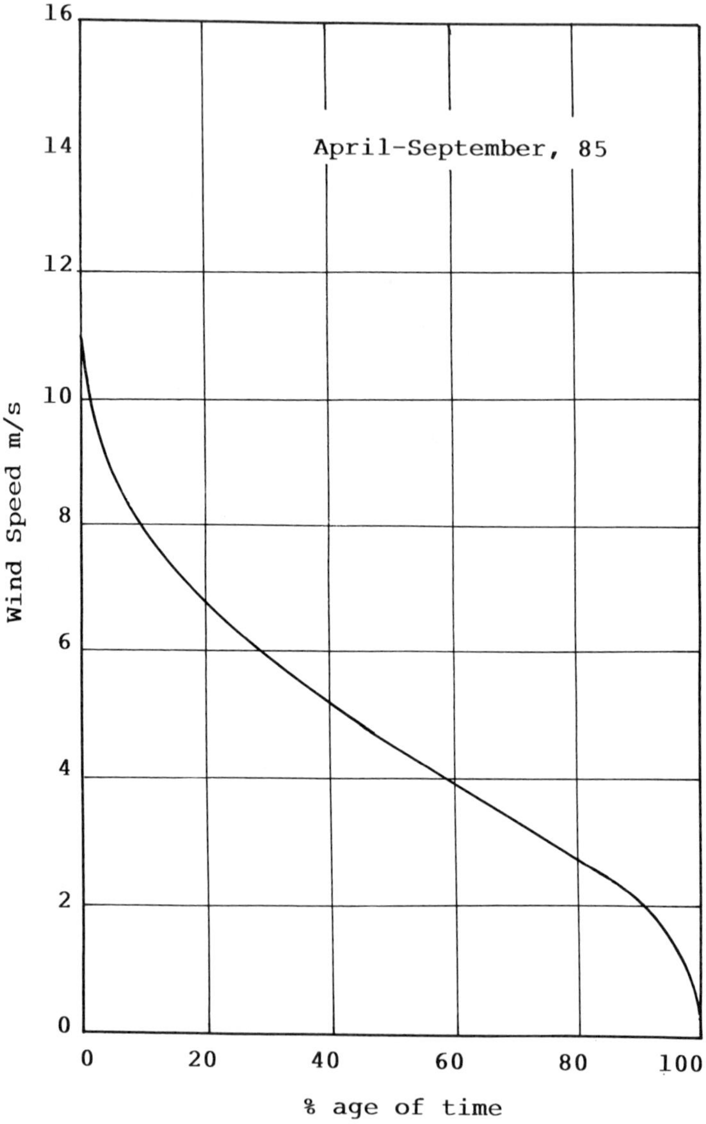

FIGURE 4: Wind Duration Curve for Six Months

this value of cut-in wind speed, the wind turbine will not run for about 20% of the time.

The selected system (Bergey BWC EXCEL-R/48 - 48VDC) has just been installed and is expected to provide about 2.7 kw at a rated wind speed of 7 m/s and a cut-in wind speed of 2.5 m/s. With one Grundfos AC pump, the water pumping system will supply 1.8 lit/s at a head of about 45 m with the motor voltage at 220 volts. The motor's power is 2.0 HP (1.49 kw). At speeds slightly higher than 7.0 m/s, two pumps can be operated.

SOLAR-WIND SYSTEM

The site data clearly indicate that a considerable amount of wind and solar energy can be utilized in the Sadat City test-site. To utilize this amount of available energy, a hybrid solar-wind system was designed together with a storage system to supply the required electric power for pumping water for a head of 45 meters for 24 hours a day. The solar photovoltaic subsystem supplements the wind turbine discussed in the above section in order to form the envisioned hybrid system.

The wind-solar hybrid system has a nominal power of a 10 kw wind machine at 12.1 m/s. The solar photovoltaic system has 7.0 kw peak power (133 modules, Model M55, ARCO Solar). The hybrid system will supply electric energy to the storage system (batteries) and/or directly to the pumping system to ensure 24 working hours per day.

BIOGAS TECHNOLOGY

Biogas technology which is the anaerobic fermentation of animal droppings, crop residues, and human habitation wastes can provide the desert agriculture system with clean fuel, "biogas", and a very valuable organic manure, "biogas manure".

The system of fodder/animal/biogas, which is one of the main farming systems adopted by the Center, is now being implemented at the DDC South Tahrir farm. The resources initially available on site are the dung of 22 milking cows and the droppings of 200 sheep. The animals are not kept under shed all the time. The "collectable" dung represents about 50% of the total amount generated by the animals. A comparison between the expected amount of biogas production based on (a) the typical Indian digester, (b) the heated Indian type digesters, and (c) the screened dung plus heating plus the use of fixed film procedures, clearly indicates the superiority of (c). The incorporation of these procedures leads to a biogas production at the rate of approximately 0.8 m^3/kg volatile solids added and a substantial reduction of digester volume to approximately 15 m^3 for the expected available substrata.

Efficient biogas generation requires a constant temperature of 35° C inside the digester. Since the ground temperature fluctuates during different seasons of the year and is certainly much less than 35° C, especially at ground depth greater than 1.5 meters, the rate of heat transfer from the digester to the ground was calculated for different operating conditions. The results indicate that for soil conditions around the digester, the rate of heat transfer can vary from 100 to 200 watts depending on the assumed ground temperature. Conduction heat transfer from the digester, however, is not the only possible heat transfer mechanism. Heat is transferred by convection from the top part of the digester which is also used for collection of the biogas. The convection heat transfer rate was also calculated.

Calculations were also carried out to evaluate the effect of gas holder insulation on the rate of heat transfer. A 10 cm thick styrofoam layer of insulation reduced the heat transfer rate by 92%. For practical considerations, however, the actual thickness of the insulation layer to be used on the gas holder has been limited to 7 cm of styrofoam.

In line with the philosophy of using renewable energy sources as much as possible, passive solar energy is used to pre-heat and maintain the digestible material in the reception pit at the required temperature level of 35-40° C. In order to accurately monitor temperature changes of the digesting material in the different parts of the biogas unit, 14 heat sensors (diodes) were located in the inlet compartments, in the digester, and in the manure storage pit. The aim of this monitoring is to quantify energy inputs through solar heating and minimizing energy losses during the course of fermentation.

The construction of the biogas system has recently been completed. Evaluation of the efficiency of biogas and biogas manure production is underway.

PASSIVE SOLAR ARCHITECTURE (DESERT HOUSING)

The objective of this activity is to test, demonstrate, and evaluate alternative designs for desert housing. The general concept is the choice of appropriate bioclimatic designs, which suit the climatic, environmental, cultural, social, and economic conditions for desert communities. Passive solar techniques were adopted in order to satisfy basic housing requirements. It is complemented, whenever needed, with either solar active, wind energy or biogas techniques.

The DDC Headquarters Building at Sadat City

Funds in the amount of L.E. 100,000 were granted by the Egyptian Government for the construction of the first DDC building at the Sadat City Site. This building uses traditional Islamic and Coptic low-cost construction techniques (domes, vaults, courtyards), locally available materials (adobe bricks), passive solar design, and provides training and demonstration for future Egyptian desert settlers.

The design of a building, as well as the materials used in its construction, are major determinants of the environmental conditions inside the building. The DDC Headquarters Building at Sadat City has been designed and constructed to incorporate many passive solar architectural features. Based on qualitative observations, the environment inside the building is judged to be quite pleasant. On the other hand, for the purpose of quantitative evaluation of the building performance, several parameters are continuously monitored. These are the temperatures of the inner and outer surfaces of the walls, the temperature inside the building, the ambient temperature outside the building, the global insolation and the wind velocity and direction.

Average temperature variations inside and outside the room are recorded weekly as to temperature variations at different sensor positions. The temperature inside the building remains fairly constant despite the extensive temperature changes outside the building. These data are being collected on a continuous basis. Similar data will be recorded and analysed in relation to design, orientation and material for the detached and semi-detached houses in the three model farms at South Tahrir.

Alternative Plans for Detached Desert Housing at South Tahrir

Three houses are to be constructed at the DDC South Tahrir site as a component of three model farms. One of the farms is to have a single house (detached), while the two other adjacent farms are to have a duplex (semi-detached housing).

Three architectural plans were drafted for the detached house. These plans were sent to the College of Architecture, University of Arizona for energy analysis using a CALOR computer program. The results favored alternative "two". In preparing the final design of the detached house for Farm No. 1, alternative II was slightly modified and the construction showed a number of improvements including double walls for the East and West facades, domes and vaults for the roof, trombe wall for the south side, and malkaf for the north side (see Figure 5).

The thermal characteristics of walls for a passive solar building have been studied, taking into consideration the U value, the solar heat factor and the time lag for different wall configurations including air space. A finite-difference transient heat-conduction program for studying the thermal performance of composite building envelopes was developed. The program recommended a wall consisting of 1/2 clay brick (12.5 cm) plus a non-reflective air space (12.5 cm) plus a clay brick (25 cm) for the east and west walls of the house on Model Farm (1). Additional passive cooling and heating designs were incorporated in this model building; viz Trombe wall on the south side for winter heating, and the Malkaf system for summer cooling. Power lighting and electrical domestic needs are provided by a set of photovoltaic panels used as roofing for the cattle shed adjacent to the building. A small wind mechanical pump will be installed to provide drinking water from a well (static depth at about 15 meters). It is expected that a small biogas system will be added to provide necessary energy for cooking, while a solar water heater will be used for heating water.

DESERT COMMUNITIES AND SOCIO-ECONOMIC STUDIES

As mentioned earlier, a major long-range goal of the AUC Desert Development Center is to develop and demonstrate economically viable systems that can be used by government and private enterprise in creating desert communities in Egypt. These communities must offer an attractive quality of life to encourage families to migrate from overcrowded cities and villages to desert areas. The DDC Desert Community unit seeks to identify

FIGURE 5

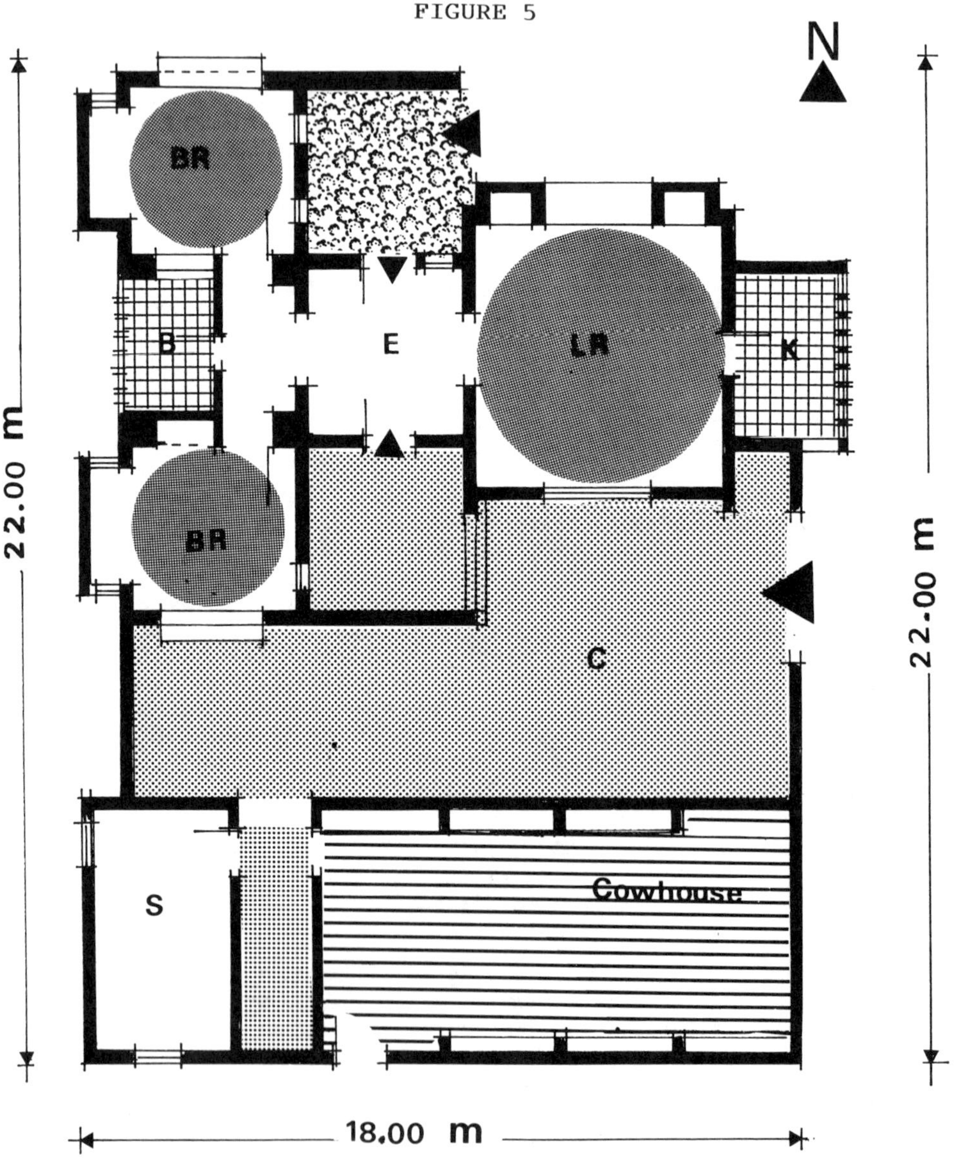

desert community development strategies that are feasible with local resources, are socially acceptable, protect the natural environment, and enhance human life. The DDC disseminates its findings through demonstration sites, consulting, training and actual on-farm experience where the farmers play a major role through (a) allowing the agronomists to demonstrate their results (and even sometimes to test potentially useful varieties and techniques) on the farmer's land (b) sharing the implementation of these activities on the farmer's land with the agronomists, technicians, or engineers, and finally (c) giving the farmers major responsibility of implementing the results on their own land. In working towards this long range goal, it should be emphasized that the "Desert Community Unit" is specifically concerned with "people" as the prime objects of and participants in desert development ... people for whom the new technologies, facilities, systems and policies must be suited if they are to succeed.

It is important to consider what types of individuals can most satisfactorily work the land or perform other functions, and on what terms. This involves studies of secondary school and university graduates in agriculture, smallholders and landless peasants, and setting up criteria for selecting each for DDC on-farm demonstrations. Such studies were initiated in 1985.

Keeping in mind the long range goal of community development, members of the "Desert Community Unit" have been analyzing, for the last three years, social, and economic experiences of settlers on reclaimed lands in South Tahrir. The results of this analysis are used by the unit to help develop socio-economic research frameworks for biological and technological activities within the AUC Desert Development Center.

ON-FARM DEMONSTRATION AND EXTENSION

The settlers in the desert are the final judges in evaluating the appropriateness of desert agricultural practices. If practices are not compatible with social norms and cultural values, they will not contribute to solving the problems which are at the base of Egypt's land reclamation policy. Providing food and a livelihood for its burgeoning population is the basic issue. The DDC is assessing social acceptability and economic viability of its work while conducting activities with the settlers on lands surrounding the DDC.

In 1984, the DDC conducted a socio-economic survey* of 100 graduate farmers living in the neighborhood of the DDC sites. An objective was to identify major problems that these farmers face. The farmers specified lack of water, shortage of agricultural labor, and plant diseases as their three main problems in crop production.

In March 1985, the DDC initiated an On-Farm Demonstration Program with the aim of helping farmers solve recognized problems. The demonstration program concentrated on plant diseases because the DDC had the knowledge and capability needed to alleviate this problem on a few farms. Specifically, the DDC hoped to train farmers to use appropriate practices to combat a high infestation of nematodes and weed infestations and to begin to diffuse successful practices of plant protection on a larger scale.

In the first year of the program, 43.5 feddans of groundnuts belonging to six farmers were sprayed with herbicides and nematocides. On the treated land, the yield ranged between 16 and 20 ardabs per feddan. The average yield on the control areas located on the same farms was less than half that of the treated areas. Similar results were obtained

---
* Funded by Ford Foundation.

in fields of peas that were also treated for weed and nematode control.

In the second year of on-farm demonstration, the DDC notes that the farmers who participated in the original program are keen to continue, and many other farmers wish to be included. The small-scale on-farm demonstration program continues for nematode control on groundnuts, weed control in peas, citrus, and other crops and a comprehensive program for citrus and for field crops* has been initiated. The DDC has also prepared a bulletin on plant protection that has been distributed to about 500 farmers.

Within the popular Egyptian theme of reclaiming the desert, the DDC strongly supports desert development that advocates much more than land reclamation. Our studies on lands reclaimed since the early sixties (South Tahrir) confirm that there are major constraints affecting soil improvement and crop yield. These include poor infrastructure, (especially electricity), unavailability of necessary inputs (fertilizers, herbicides, nematocides, etc.), and major problems due to plant diseases, product yields, and marketing. In addition, DDC scientists observe widespread use of inappropriate techniques for irrigation, soil, water, and fertilizer management and for cropping practices. It is our belief that before embarking upon reclaiming millions of additional acres, which would require hundreds of millions of dollars, "priority should be given to solving the major constraints on the old new lands" and to identifying the appropriate desert farming systems and technologies which would be technically feasible, economically viable, and socially acceptable under our Egyptian environmental conditions.

The need for on-farm demonstration and extension services for desert farmers suggests

---

* Different aspects of each of these programs are funded by IDRC and by the Ford Foundation.

continuous enrichment of the present on-farm demonstration program and development of the outreach function of the DDC as a whole making use of the experiences gained from the farming systems which were described briefly.

To those at the DDC desert sites, and more importantly to those at other sites where the DDC program may help in the design of communities, it is of crucial importance to implement programs which aim at economic, social, and political advancement, encouraging cooperation among settlers to help solve mutual problems and execute local development projects.

Finally, the true measure of success of the AUC integrated approach to desert development is to reach the optimum combination of technical, agricultural and socio-economic practices which would attract Egyptians who are currently living in over-populated areas to participate in the formation of new desert communities where they would enjoy a better quality of life.

THE CONTRIBUTION OF THE JACOB BLAUSTEIN INSTITUTE FOR DESERT RESEARCH TO THE DEVELOPMENT OF THE DESERT NATIONALLY AND INTERNATIONALLY

LOUIS BERKOFSKY AND SHABTAY DOVER
The Jacob Blaustein Institute for Desert Research, Ben-Gurion University of the Negev, Sede Boqer, Israel

Abstract The contributions of the Jacob Blaustein Institute for Desert Research to the development of the desert, by means of scientific research, national and international cooperation, and planning for regional development are described.

INTRODUCTION

The Jacob Blaustein Institute for Desert Research is an interdisciplinary research institute whose aim is to study the scientific aspects of desert evolution, and those aspects of the desert relating to future settlement, i.e., to create techniques and methodology for enhancing the development of the region.

Figure 1 lists the Units of the Institute. The relationship of the Sede Boqer Campus to the Ben-Gurion University via science and education cane be seen in Figure 2.

CONTRIBUTIONS TO STUDIES OF SCIENTIFIC ASPECTS RELATING TO FUTURE DEVELOPMENT

This discussion is divided broadly into air, water, soils, plants, animals, people, architecture and energy, and ecology.

FIGURE 1. Research Units of the Jacob Blaustein Institute for Desert Research

1- Desert Meteorology

2. Desert Hydrology

3- Salinity and Water Engineering

4- Plant Adaptation Research

5- Desert Run-off Farms

6- Algal Biotechology Research

7- Controlled Environment Desert Agriculture

8- The Isan Center for Comparative Medicine in Arid-Zones

9- Ecophysiology and Introduction of Desert Plants

10- The Mitrani Center for Desert Ecology

11- Applied Environmental Microbiology

12- Arid-Zones Resources Economics

13- Applied Solar Calculations

14- Desert Architecture

15- Research Implementation in Arid-Zones Development

16- Social Studies Center

FIGURE 2. Sede Boqer - A Campus for Science and Education

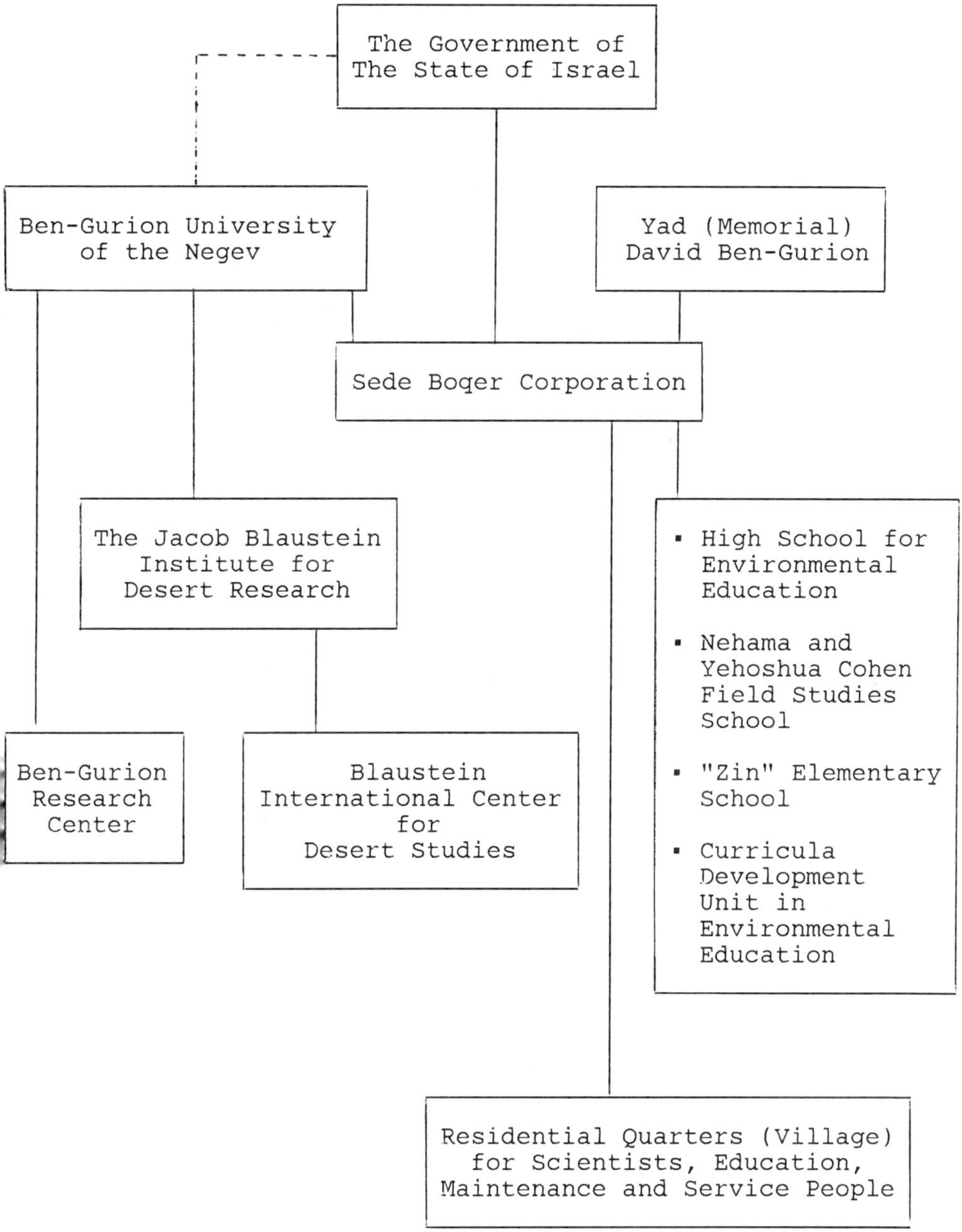

DESERT METEOROLOGY

The properties of the desert atmosphere are documented by means of observations near and in the ground, on towers, and from balloons and satellites above the ground. Aspects studied include the energy balance of the atmosphere near the ground (the disposition of the solar and infrared radiation), the transport and disposition of dust (and its relation to wind erosion), the circulation of the desert atmosphere by means of mathematical models, the properties of desert clouds and rainfall with a view to rainfall enhancement and prediction, and socio-economic problems arising from drought and other climatic fluctuations.

HYDROLOGY

Also studied is the location, depth, quality, and utilization of the Nubian sandstone aquifer, an aquifer which lies beneath the Negev, the Sinai and beyond, and contains large quantities of warm, brackish water. Hydrological surveys are carried out. There has been some cooperation with El-Matariya Institute for Desert Research in Cairo (established by the World Bank in the 1950's). The efficiency of Nubian aquifer waters is studied by means of sophisticated econometric methods.

DESERT WATER HARVESTING

The methods of the ancient Nabateans are used to collect the run-off from the small amount of rainfall (90 mm/yr.) to effectively increase the retention of rainfall, which is trapped in small lakes for agricultural use. Trees grow well with no additional irrigation.

APPLIED MICROBIOLOGY

Waste water is recycled for use in agriculture and industry - always applying drip irrigation techniques.

Near the Dead Sea, bacteria were clogging underground drainpipes with ochre in an intensively irrigated area. A solution to this problem was found by creating anaerobic conditions with the aid of siphons installed at outlets. These raised the groundwater level. Also, within weeks the clogged drainpipes were cleaned by the raising of the groundwater level around the pipe. As long as the groundwater is not too close to the surface, an elevated outlet can be applied, and can be used to design underground drainage systems free of ochre all year round. This is especially suited to arid zones where water containing magnesium and iron can be found and where other methods for raising groundwater levels are inapplicable.

SOILS

The subsurface saline water is used to develop technology for growing plants in sandy soils. Drip irrigation can supply nutrients according to plants' needs, and can overcome poor and salty soil. The salinity problem in the soil is solved by washing the soil and by changing the nutritional regime. The use of pigeon guano, which is rich in nitrogen, can overcome the effects of salination. Guano is widely used by Egyptian farmers.

PLANTS

Plants from other deserts are introduced and their ecophysiology and that of local plants is studied, i.e., the effects of stress due to salinity and drought. Algae, which require high solar radiation and warm brackish water are grown. A sophisticated greenhouse has been developed. It is a closed system into which can be pumped carbon dioxide, thereby accelerating the growth rate of plants. Food crops are intercropped with trees to develop desert agroforestry techniques. Plants are developed which require only a small amount (100 g) of soil for transport to other arid regions.

ANIMALS

Both terrestrial and marine animals are studied. The physiology and economic feasibility of new species of animals have been investigated for this region - Dorper Sheep from South Africa (which thrive well here). The camel is being studied as a source of protein, by treating female camels to induce more frequent conception. The health of these and other animals, such as Arabian horses is carefully monitored. The growth of fresh water fish, such as eels, tilapia and catfish, in brackish water is the subject of experiments. These all represent economically feasible varieties.

PEOPLE

Man and his community in the desert environment are studied as to the social problems; the dwelling conditions; medical services; education, transport, communication, municipal, commercial, and cultural services.

DESERT ARCHITECTURE AND ENERGY

Locally available materials are used (to make adobe bricks, for example) in the design of buildings using solar energy for heating of water, air in buildings, and industrial processes. Under the auspices of the Ministry of Energy, which is building a National Test Center, the efficiency of solar collectors will be compared. Wind energy is also being studied with the aim of developing a National Energy Center at Sede Boqer.

DESERT ECOLOGY

A combination of many of the topics described above is embodied in the desert ecosystem, in which is studied the relation between organisms and their environment. Predator - prey relationships, foraging problems under varying conditions, energy balance of organisms and their relation to meteorological conditions are studied. One might say that the desert ecosystem is a complete interdisciplinary system, which is how all of the Blaustein Institute's activities are viewed.

THE BLAUSTEIN INSTITUTE AS A NATIONAL CENTER

Other research institutes and universities throughout Israel work with the Blaustein Institute in joint research projects. Some involve several groups. For example, there is one such group consisting of six researchers from the Technion, Tel Aviv University, Hebrew University, and Ben-Gurion University, established to study the effect of meteorological conditions in natural dusty atmospheres on transmission of radiation in the Negev.

To bring such groups together in Sede Boqer requires laboratory and living facilities. An effort is being made to create the physical conditions to facilitate the research programs through the Planning and Grants Committee of the Council on Higher Education.

THE BLAUSTEIN INSTITUTE AS AN INTERNATIONAL CENTER

Under the Blaustein Institute for Desert Research is the Blaustein International Center for Desert Studies. This Center invites post-doctoral people, scientists for varying periods, and qualified students. International conferences are also hosted.

The Institute has signed many exchange agreements with universities and institutes abroad. In the U.S. alone, it has agreements with UCLA, Cal-State University - Fresno, New Mexico State University, University of Arizona, International Center for Arid and Semi-Arid Lands Studies of Texas Tech University, State University of New York at Stony Brook, and the University of Maryland. It has also cooperated with the Navajo Indians in Arizona and now has an exchange program with the Indian nation. In Europe, it has agreements with Technische Hochschule Darmstadt, is working on an agreement with Catholic University, Louvain, Belgium, and with Spain; and has programs with many national research councils.

There also exists the US-Israel Binational Science Foundation (BNSF), Binational Agricultural Research and Development Foundation (BARD), Binational Industrial Research Foundation (BIRD), and the Binational Science Foundation with the Federal Republic of Germany.

The Institute is a very active participant in the U.S. AID Cooperative Development Research Program, which funds joint research between Israel and less-developed countries.

The Institute is a member of IFIAS (International Federation of Institutes for Advanced Study). IFIAS recently approved a project proposed by an Institute member an official project - The Greening of the Desert. By means of this project, other IFIAS member institutes will be encouraged to work on training and transfer of technology developed at the Blaustein Institute for arid and semi-arid zones with water, soil and agriculture problems.

CONTRIBUTION OF THE INSTITUTE TO REGIONAL DEVELOPMENT OF THE ARID ZONES

A major contributing factor to regional development in the Sede Boqer area is the fact that the researchers in the Institute live in the desert, many of them at Sede Boqer itself. Thus the Institute contributes to the demographic statistics of the Negev. Indeed, Sede Boqer and Elat are the only two settlements in the Negev with a positive migration balance.

The Institute supplies manpower for education and other community activities, both at Sede Boqer and in surrounding areas such as Yeroham and Beer-Sheva.

At Sede Boqer, residential areas are shared by university and non-university people, thus avoiding segregated communities.

The Israeli government is placing a great deal of emphasis on the development of the Negev. Indeed, Sede Boqer has been designated as a focus of this activity. In this regard, there are plants to designate the Jacob Blaustein Institute for Desert Research as a Research and Development Center, based on techniques developed here. Stronger contact will be established with other Negev research institutes, such as the Marine Biology Institute in Elat.

NATURAL RESOURCE MANAGEMENT AND AGRICULTURAL DEVELOPMENT IN THE ARID AREAS OF EAST AFRICA

MOHAMED T. EL-ASHRY AND BONNIE J. RAM
World Resources Institute, A Center for Policy Research, Washington, D.C., U.S.A.

Abstract The soil, forest, and water resources of arid East Africa are being depleted at an alarming rate resulting in reduced agricultural productivity and desertification. The degradation can be attributed to many factors. These include: rapid population growth, recurring droughts, ill-advised government policies, misguided actions by international assistance organizations, and years of overexploitation of a fragile ecosystem.

Deforestation, overgrazing, and the expansion of rainfed agriculture have severely degraded many watersheds and accelerated soil erosion. In some places, up to 450 tons per hectare of soil erode annually, and in the Ethiopian Highlands, a billion tons of topsoil erode each year. In all, about 7 million square miles are threatened by desertification in sub-Saharan Africa.

A critical component of East Africa's resource deterioration is the destruction of its forests. Almost 6 million acres of dry tropical forests were destroyed each year between 1980 and 1985. In Ethiopia, the forested area has shrunk from 16% of the land area in the 1950s to 4% in the 1970s. Once damaged, the ecosystem supported by the forest collapses; the soil loses its nutrients, soil erosion increases, and reservoirs and irrigation systems are filled with sediment. For example, the capacity of Khashm El Girba dam in Sudan was reduced due to silting from 1,300 million m^3 to 778 million m^3 by 1976 and is expected to reach 500 million m^3 in 1997.

The broad land-use problems that contribute to resource degradation cannot be solved by addressing one sector alone. They require an integrated approach to agriculture, forestry, energy, and rural development, combined with economic incentives and a strong education program. This approach, in turn, requires adequate information on the resource base and on the land's capacity to support different activities.

INTRODUCTION

Recently, discussions of arid lands issues turn readily to the current problems of land degradation and the food crisis in sub-Saharan Africa. The continent, two-thirds of which is arid, semi-arid or subhumid, embodies more than any other, the challenges of arid lands development. Such challenges stem mainly from the physical and biological constraints imposed by nature such as inadequate water, thin soils, sparse vegetation and frequent droughts. However, droughts are only partially to blame for Africa's famines. The latest drought has been exacerbated by overexploitation and mismanagement of a fragile resource base. Growing human and animal populations and inappropriate development policies have led to the degradation of forest, soil and water resources.

Africa is unique in the less developed world. In the past decade, it has experienced a 20% decline in per capita food production as compared to an increase of about 15% in both Latin America and Asia.[1] Almost all of sub-Saharan Africa, but particularly a huge arc of countries extending across the Sahel from Mauritania to Ethiopia and then south to Mozambique, is caught in a dangerous cycle of rapid population growth, inability to afford technological advances in farming, expansion of agriculture onto marginal and fragile lands, prolonged droughts, and too often ill-advised agricultural development policies and weak institutions.[2]

Annual population growth in Africa is the highest in the world today -- about 3.1%, up from 2.3% in the 1960s, and 2.8% in the 1970s. In contrast, food production increased 1.5% per year in the 1970s, and only 1% in the 1980s.[3] The problems of sub-Saharan Africa can best be described in terms of the "carrying capacity" of the region and the use and overuse of its fragile resource base. Over two-thirds of the land area of the continent is arid or semi-arid and rainfall is slight and quite variable. The soils are generally thin, have low fertility, and are easily eroded and compacted. Groundwater reserves are limited, and droughts lower water tables temporarily. Even more serious is the removal of trees and other vegetative cover which has suppressed the water table permanently in some places.[4]

However, given these limitations, indigenous cultures long ago developed production systems which are well adapted to the limits imposed by these fragile and arid conditions. Known as "shifting cultivation" or "bush fallow," these agricultural methods allow fields to rest and recover nutrients and moisture for lengthy periods after two or three seasons of crop production. In addition, some traditional systems have farming practices that complement livestock production, namely feeding livestock on crop residues and tree fodder, while using the animals' natural fertilizers for fields and gardens. Unfortunately, these traditional production systems have been increasingly disrupted by rapid population growth; changing social, political and economic policies; and inappropriate western-style development schemes.[2]

RESOURCE ISSUES

One of the most significant factors which limits the replacement of shifting cultivation with a more productive and intensive system of land use is a fragile soil base. Indiscriminate transfer of technology from temperate zones to tropical soils has seriously accelerated soil erosion and the conversion of forest into grass-savanna and sterile steppes. In short, many development projects in Africa have done little to restore the traditional production systems' equilibrium with the environment. Reliance on some cash crops, for example, actually increases soil erosion and decreases soil productivity. Cash crops, such as cotton, tobacco and peanuts, consume large amounts of nutrients and reduce soil fertility.[5] The thin topsoil which is left bare between growing seasons is particularly susceptible to erosion and compaction. (Table 1 illustrates the impact of soil erosion on crop production in Kenya.) In addition, with large areas of good soils used for cash crops, farmers are forced to cultivate increasingly marginal and vulnerable lands. At the same time, pastoralists are pushed off their traditional grazing land and end up overgrazing the limited and more fragile rangeland available to them. In the Kondoa Province in Tanzania, for example, the soils are too fragile to sustain the intensive agricultural and grazing activities. Some 150,000 hectares are so badly damaged by gully erosion that they cannot be rehabilitated.[6]

The impact of overgrazing on marginal lands is exacerbated by increased numbers of livestock. Moreover, land competition with mechanized farming areas has placed a greater burden on the shrinking resource base. Continuous cultivation of staple crops, such as sorghum can also mine precious soil nutrients. Monocropping and generally poor management in Sudan's mechanized farming sector has led to an infestation of weeds, a drop in crop yields and ultimately to the abandonment of farm plots.[7] Lack of an effective land use policy has brought 'suitcase' farmers to these areas of clay soils - some with short term leases and others unauthorized. These farmers in Sudan have no incentive to practice conservation measures or sustain these agricultural lands.[8]

Closely tied to Africa's agricultural problems is the continent's acute energy shortage. Many African countries depend on fuelwood and other forms of biomass for more than 90% of their energy and construction needs. Forests are being cut down at an alarming rate with devastating effects on the soil and the entire ecosystem. Studies by the National Energy Authority in Sudan suggest that by the year 2000, and at current rates of consumption, 77 percent of all tree cover in the northern regions (Northern, Central, Eastern, Khartoum, Kordofan and Darfur) will have been destroyed. In the Khartoum, Central and Eastern provinces, all tree cover could be lost this year (1987).[9] In Ethiopia, over 40% of the country was covered with forests. Two decades ago, 16% of the land was still afforested; now the area occupied by the forests is only 3.1%.[10] De-vegetation of the land for fuelwood and charcoal production is degrading millions of acres of Africa's watersheds, causing severe soil erosion, and disrupting the delicate water balance. In Kenya, the charcoal sources for Nairobi have moved to the slopes of Mt. Kenya over 200 km away. The fuelwood sources for Mogadishu, Somalia, are fully 500 km to the south.[11] Every year about 3.6 million hectares of forests are being lost in Africa; but these continent-wide rates underestimate the severe regional pressures on forest resources.[12]

A fragile ecosystem exploited beyond its carrying capacity eventually breaks down. Whether in the U.S. dust bowl or in some parts of Africa, the results can be tragic. Deforestation, overgrazing, expanding rainfed agriculture, and other practices have combined to degrade severely many watersheds and to accelerate soil erosion. Sudan, Ethiopia and Tanzania have some of the highest clearing rates of agricultural lands per family.[13] Figures for soil erosion as high as 450 tons per hectare per year are not unusual, and in Ethiopia's Highlands one billion tons of valuable soil are being stripped by erosion every year.[14] In all, about 6.9 million square kilometers are threatened by desertification in sub-Saharan Africa and 16.5% of Africa's existing rainfed cropland will be lost by the end of the century if conservation efforts are not taken.[15]

Because of all the problems facing rainfed agriculture, large-scale irrigation is often promoted as the answer to sub-Saharan Africa's food problems. A study by FAO shows a vast potential for irrigation in comparison to the limited amount that has actually been developed. Tanzania, where only 6% of a potential 2,300 hectares have been developed, is a case in point.[16] To some observers, it seems strange that an area of extremely unreliable precipitation has so little irrigation. However, there are a number of reasons for slow progress in expanding irrigated acreage. These include: the high cost of developing surface irrigation which many African countries cannot afford (potentially as high as US $25,000 per hectare as in Bura project in Kenya); the uncertainties of the groundwater resource; the lack of deep-rooted tradition and expertise in irrigation in most countries; the vast distances between rural villages; and poor communications.

In addition, increasing sedimentation, linked to indiscriminate agricultural clearing and degradation of the watersheds, has negatively affected reservoirs and irrigation systems. According to a recent World Bank report, the capacity of the Khashm El Girba dam in Sudan was reduced due to silting from 1300 million m^3 to 718 million m^3 by 1976 and is expected to reach 500 million m^3 in 1997. For the Roseires Dam, the rate of siltation recorded was more than two times greater than design estimates. In addition, this reduced storage capacity in the Roseires has resulted in decreased hydroelectric potential necessitating a costly heightening of the dam and dredging of accumulated silt.[17] Siltation problems in the Roseires reservoir stem in part from upstream activities in the Blue Nile watershed in Ethiopia. Policy formulation and effective strategies became more difficult in these cases where transnational river systems are involved.

FUTURE MANAGEMENT NEEDS

Despite its massive problems, Africa's situation is not hopeless.[18] Malawi and Zimbabwe have significantly increased maize production. In Ghana, "no till" methods have reduced runoff by 70% to 90% of that from conventional plowing methods.[19] In Kenya, agroforestry practices have increased grain production by 30-50% and, in the Machakos district, water-harvesting techniques have increased staple crop yields from two to seven times. Terracing of erodible slopes, organized by women's savings clubs in Kenya, has demonstrated that it can sustain production and conserve land resources. Throughout the region, successful examples of small-scale, community-based tree planting, terracing, and sand dune stabilization projects can be found. However, all of these successful methods and practices await widespread replication.

Helping sub-Saharan Africa increase its food production and prevent future disasters will be a long and difficult process. Clearly, a single solution, a "perfect" economic development scheme, or even a blueprint for African development is out of the question. Africa is simply too diverse to respond to monolithic solutions.[20] Future strategies must include the development of new agricultural systems and technologies designed for Africa's soils and climate, and appropriate to its culture, educational and financial conditions. This requires more agricultural research on improved seed varieties and on soil and water conservation techniques to be conducted in Africa itself. In that way, feedback from local farmers would be immediate and field tests could be conducted under realistic conditions. A few research centers already exist in Africa. With adequate funds, staff, and technical assistance, they could contribute significantly to agricultural productivity and resource management.

An intensive system of agricultural production can be developed either by improving the existing system of shifting cultivation or by transferring proven successful elements of production systems from other parts of the world. It is now obvious that attempts to adapt temperate zone practices to tropical soils have failed in many places. Research in West Africa has shown that the use of no tillage techniques, with enough crop residues left on the soil surface, offers an attractive compromise between shifting cultivation and permanent land use. Minimum tillage techniques, with the use of commercial or biological fertilizers, would help maintain a high production capacity and permit more intensive use of the land.[19] However, fragile soil in the arid tropics, which is one of the limiting factors in developing intensive agriculture, must be properly managed. This can be accomplished by minimizing surface runoff losses through the use of soil management practices or by directing the runoff water off the land through engineering methods. For example, mulching, the covering of soil around growing crops with crop residues or compost, reduces erosion and conserves the soil. On a 10 percent slope planted to maize, mulch cut runoff from 42 percent of the rainfall to only 6 percent. Soil losses were reduced even more dramatically by 97%, from 27 tons per hectare to less than one ton.[18]

The effectiveness of more conservation-conscious farming techniques is usually influenced by the quality and capability of extension services. Such services are generally lacking in many African countries. In some countries, such as Tanzania, rural extension services exist but have for some time been affected by agricultural policies favoring large, mechanized state farms at the expense of village-level food producers. In the future, it will be impossible to devise and apply solutions on an effective country-wide basis unless a sufficient number of knowledgeable and competent specialists are available.

Finally, these broad land-use problems cannot be resolved by investing in one sector alone. Instead, an integrated approach to agriculture, forestry, and rural development is needed combined with economic incentives and a strong education program. Thus far, most governments and international assistance organizations have failed to recognize the interrelationships between resource sectors, often enhancing one resource at the expense of another. Furthermore, resource management has not been a priority among most development planners. The words "integrated resource management" must now take on dual significance, referring to both integrated management of the land, water, forest, and energy sectors; and the integration of resource management into economic development plans. To do otherwise is to perpetuate resource degradation and human suffering.

TABLE 1. Percent reduction in production potential due to soil erosion in Kenya

	Low Input	Intermediate Input	High Input
	% Production Loss	% Production Loss	% Production Loss
Maize	74	61	52
Beans	68	56	49
Cassava	64	52	41
Bunded Rice	8	6	8
White Potato	73	55	51

Source: Shah and Fischer (1981) as reproduced in 21.

REFERENCES

1. UNEP. 1985. Report of the executive director. African Environ- mental Conference, Cairo, (December 16-18).

2. World Resources Institute and International Institute for Environment and Development. 1986. World Resources 1986, Basic Books, Washington, D.C., pp. 55-59.

3. World Bank. 1984. Towards Sustained Development in Sub-Saharan Africa, Washington, D.C.

4. L. Berry and M. Khogali. 1984. The Physical Resource Background to African Development, Background Paper, Committee on African Development Strategies, Overseas Development Council, Washington, D.C.

5. John Madeley. 1986. "Tobacco: a ruinous crop," The Ecologist. 16: 2/3.

6. Lloyd Timberlake. 1985. Africa in Crisis: The Causes, the Cures of Environmental Bankrupcy. Earthscan, London.

7. Hassan A. Musnad. 1983. Deforestation resulting from clearing land for agriculture, grazing pastures, and charcoal production. Preassessment of Natural Resource Issues in Sudan. IES, University of Khartoum and Clark University, Khartoum: 81-100.

8. L. Berry. November 1985. Land, people and resources in Sudan. Prepared for the World Resources Institute, Washington, D.C.: pp. 54-58.

9. World Bank. November 1985. Sudan Forestry Sector Review. Washington, D.C. Annex III.

10. Robert Lamb and Milas Seifulaziz. 1983. Soil erosion, real cause of Ethiopian famine. Environmental Conservation: pp. 157-159.

11. Daniel Finn. 1983. Land use and abuse in the East African Region, Ambio (12) 6: pp. 296-301.

12. UN/FAO. 1983. Tropical forest resources. Forestry Paper 30, Rome.

13. Peter Freeman. November 1986. Natural resources in sub-Saharan Africa: review of problems and management needs. U.S. Agency for International Development, Washington, D.C. Mimeographed.

14. UN/FAO. 1981. Rehabilitation of forest grazing and agricultural lands. World Food Program Interim Evaluation by WFP/FAO/ILO Mission of Project Ethiopia 2488. Rome, WFP.

15. Jack Mabbutt. 1984. A new global assessment of the status and trends of desertification. Environmental Conservation. (11) 2: pp. 103-113.

16. UN/FAO. 1985. Land and water division. Estimates of Areas Developed: Irrigation Potentials, Rome.

17. World Bank. November 1985. Sudan Forestry Sector Review. Washington, D.C. Annex VIII, p. 2.

18. Paul Harrison. Forthcoming 1987. African success stories. World Resources Report 1987, WRI and IIED, Basic Books, Part III.

19. R. Lal. November/December 1984. High yields, land-use and land abuse. International Agricultural Development.

20. M.T. El-Ashry. 1986. Resource management and Development in Africa. Journal '86, World Resources Institute, Washington, D.C., pp. 5-12.

21. Richard Ford. November 1986. Land, people and resources in Kenya. Prepared for the World Resources Institute, Washington, D.C., p. R-15.

GROUND WATER, SOIL AND RENEWABLE ENERGY RESOURCES: EVALUATION OF EAST OWEINAT AREA

H. KAMEL AND A. EL-OSTAZ
General Petroleum Company
Nasr City, Cairo, Egypt

Abstract The East Oweinat area is located in the southwest of Egypt between latitudes 22°00^1-24°15 N, and longitudes 27°00^1-30°00' E. It covers a total area of about 39,000 km^2, i.e. about 9,400,000 feddans. The results of the groundwater resource evaluation was that the aquifer-saturated thickness ranges from 100 mts to 700 mts in the intensive study area.

The average optimum rate of pumping per well was found to be about 300 m^3/h with an anticipated draw ranging from 6 mts to 27 mts. The T.D.S. of water ranges from 220-700 ppm. The model developed to evaluate groundwater resources concluded that a long-term extraction of 4.7 X 106 m^3/day which can irrigate an area of about 189,000 feddans is available.

The results of the study of soil resources illustrate various land tracts of arable land, which have been classified according to F.A.O. standards (1976). This study shows the existence of about 3.7 million feddans of promising sandy soils - an extensive area. About 1.5 million feddans which have a deep soil section (90 cm) are ranked as first priority.

Fortunately, a meteorological station was established in 1979 for measuring the sun isolation, wind speed and other meteorological data. The region climatological features, with exceptionally high mean wind speeds combined with solar irradiation, make the prospects particularly promising for using wind and solar energy as a primary energy resource.

General Petroleum Company (GPC) in cooperation with Solarex Corporation established a photovoltaic pumping system with peak power of 21 KW to irrigate an area of 10 feddans on the experimental farm at East Oweinat. Also G.P.C. operates a 3 kW photovoltaic pumping system originally supplied by Lucase to the Military Technical College.

The Egyptian-Italian Renewable Energy Settlement (EIRES) will be comprised of an integrated system of photovoltaic (PV), wind and biogas energy generation with a total annual load of 500,000 KWh to irrigate an area of 200 feddans.

INTRODUCTION

The area called East Oweinat is located in the southwest of Egypt, latitude 22°-24°15N and longitude 27°-30°E, and covers about 39,000 km^2 or 9,400,000 feddans. It is bounded by Darb El-Arbain road to the east, Gebel El-Oweinat to the west, El-Dakhla Oasis to the north, and the Sudanese-Egyptian border to the south (Figure 1). An aquifer underlies the East Oweinat area and extends south to latitude 20°N in Sudan. This aquifer covers an area of 220,000 km^2.

EVALUATION OF GROUNDWATER RESOURCES

An intensive hydrological study has been made of the East Oweinat aquifer. Gravity aeromagnetism, geoelectric, and shallow-refraction seismic geophysical surveys were conducted. The study also included drilling of test production wells and associated observation wells at eight locations, and a pump test as well as the determination of petrophysical properties of the aquifer (Figure 2). As a result, the saturated thickness of the aquifer was found to range from 100 to 700 meters in the area studied. Northwards, the thickness increases and reaches 1000 meters at the southern periphery of the Dakhla Oasis. The average optimum rate of

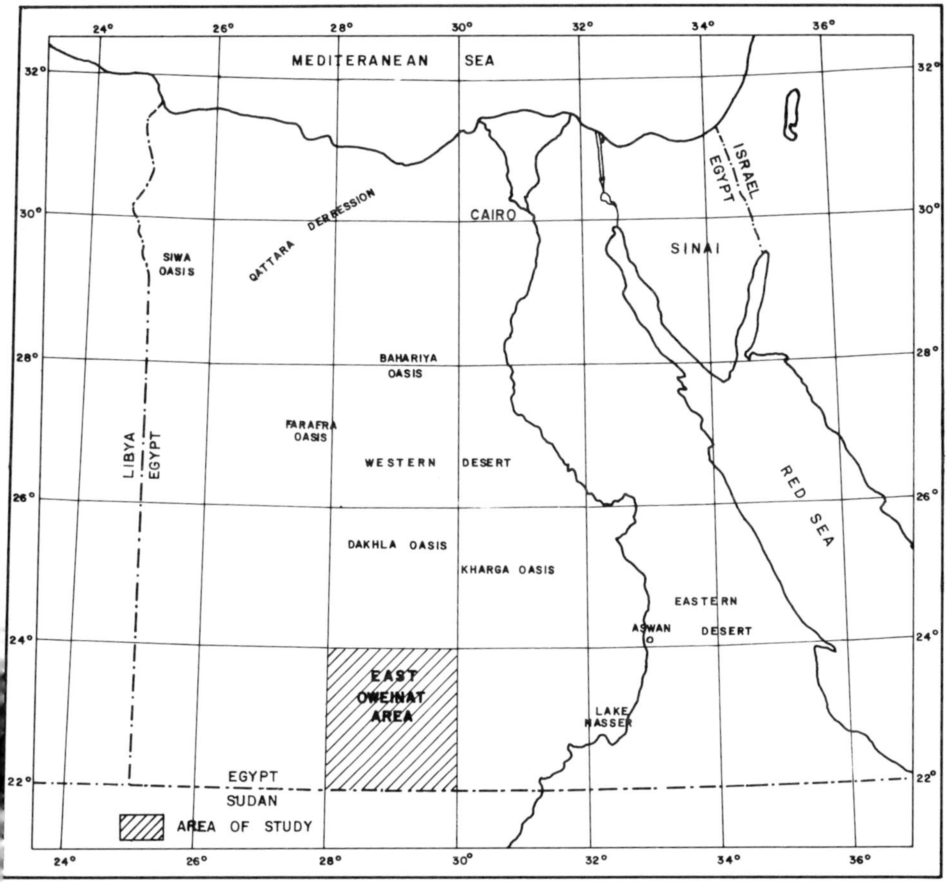

THE GENERAL PETROLEUM Co.
EAST OWEINAT AREA
LOCATION MAP
FIGURE 1

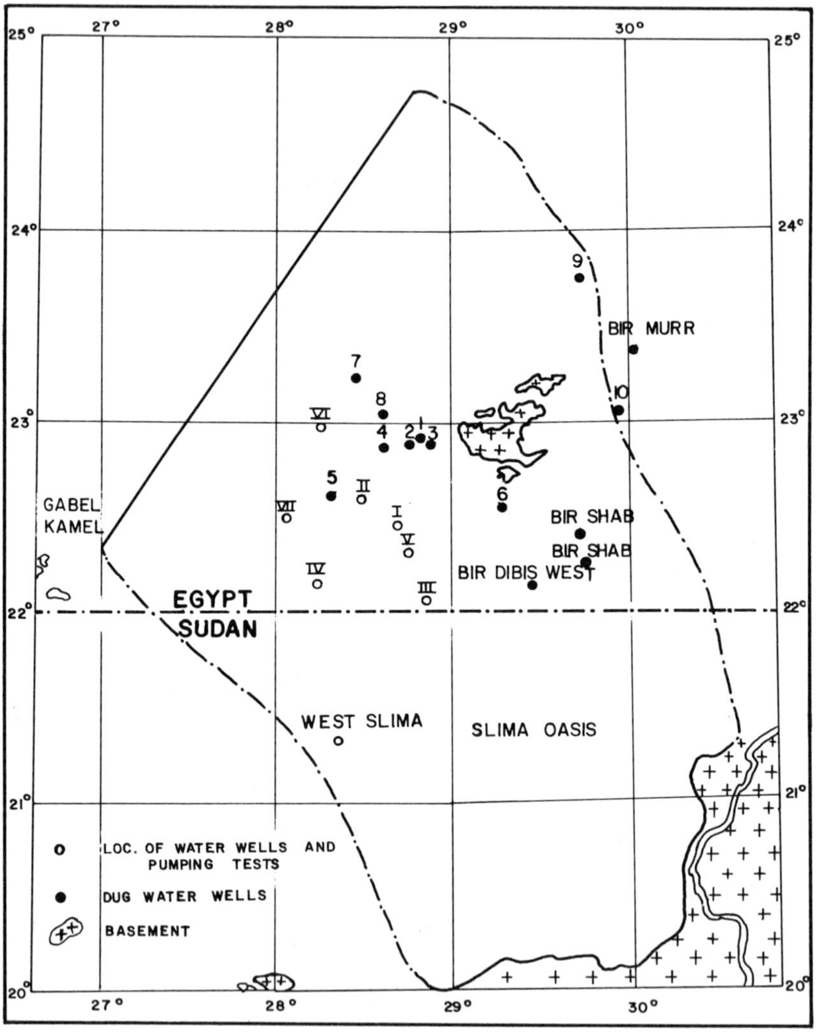

THE GENERAL PETROLEUM Co.
EAST OWEINAT AREA
MODEL AREA BOUNDARIES

FIGURE 2

pumping per well was found to be about 300 m^3/hr with an anticipated drawdown ranging from 6 to 27 meters. The water is suitable for agriculture and manufacturing purposes since the value of the Total Dissolved Salts (T.D.S.) ranges between 200 and 700 ppm.

For a precise evaluation of the groundwater availability in the East Oweinat area, a groundwater model was developed in cooperation with the Development Research and Technological Planning Center (DRPTC) of Cairo University. The model was calibrated for the present prevailing hydrodynamic conditions. The model was operated with three extraction plans to provide forecasts of the aquifer response for a time period of 100 years. The following constraints were imposed:

i. Safe groundwater system exploitation must be assured throughout the forecast period. Serious aquifer depletion of any part of the system must be avoided.

ii. The decline in potentiometric level should not fall below 50% of the initial saturated thickness of the aquifer.

iii. The extraction pattern should be based on the availability of cultivable land that has been classified as first-grade.

iv. The gross water requirement per feddan would be $7500m^3$/ annum for an irrigation period of 300 days/year.

The extraction of 4.74 by 10^6m^3/day was agreed upon as it was found to adequately satisfy the given constraints of the basic model. The anticipated drawdown in the land development areas will range from 35mts to 100mts with water being pumped from 58mts to 144mts below ground surface at the end of the 100 year forecast period (Figures 3 and 4). If the proposed extraction plan (4.74 X 10^6m^3/day) is adhered to and the gross water requirement per feddan ($7500m^3$/annum for an irrigation period of 300 days/annum) is satisfied,

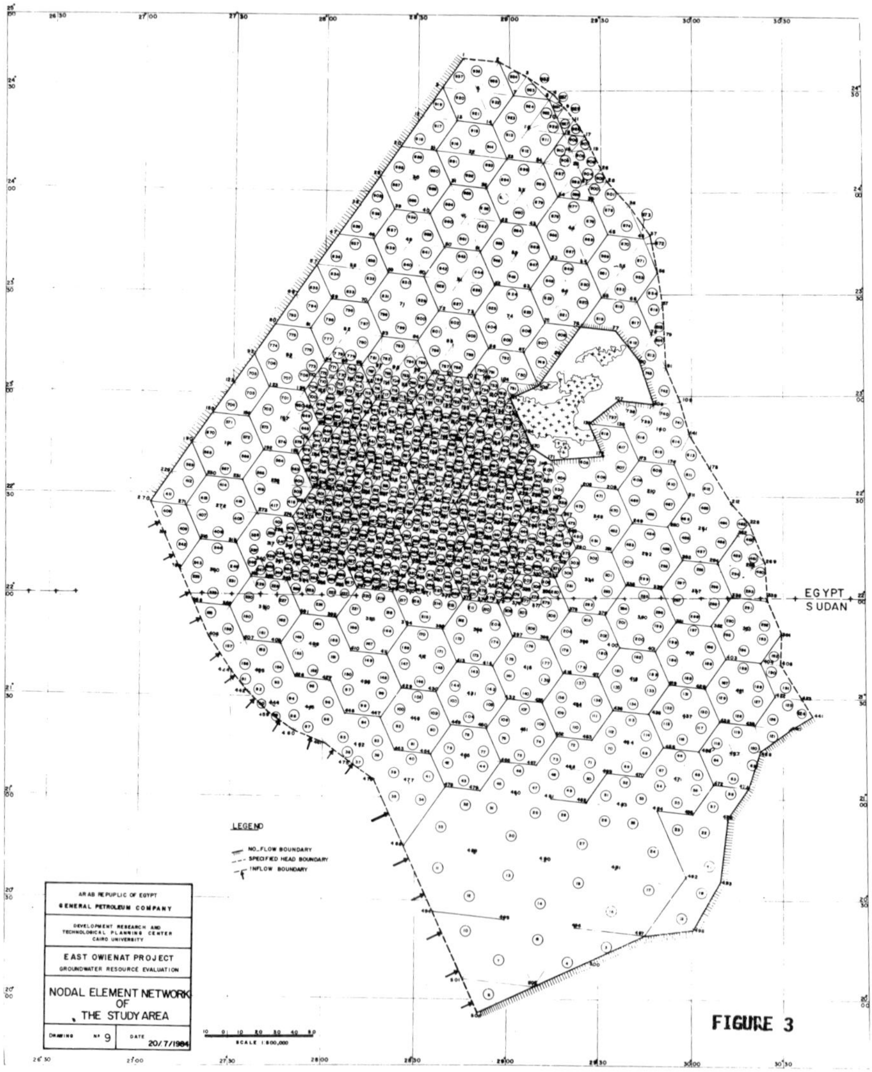
EGYPT
SUDAN
LEGEND
NO_FLOW BOUNDARY
SPECIFIED HEAD BOUNDARY
INFLOW BOUNDARY
ARAB REPUBLIC OF EGYPT
GENERAL PETROLEUM COMPANY
DEVELOPMENT RESEARCH AND TECHNOLOGICAL PLANNING CENTER
CAIRO UNIVERSITY
EAST OWIENAT PROJECT
GROUNDWATER RESOURCE EVALUATION
NODAL ELEMENT NETWORK OF THE STUDY AREA
DRAWING Nº 9
DATE 20/7/1984

FIGURE 3

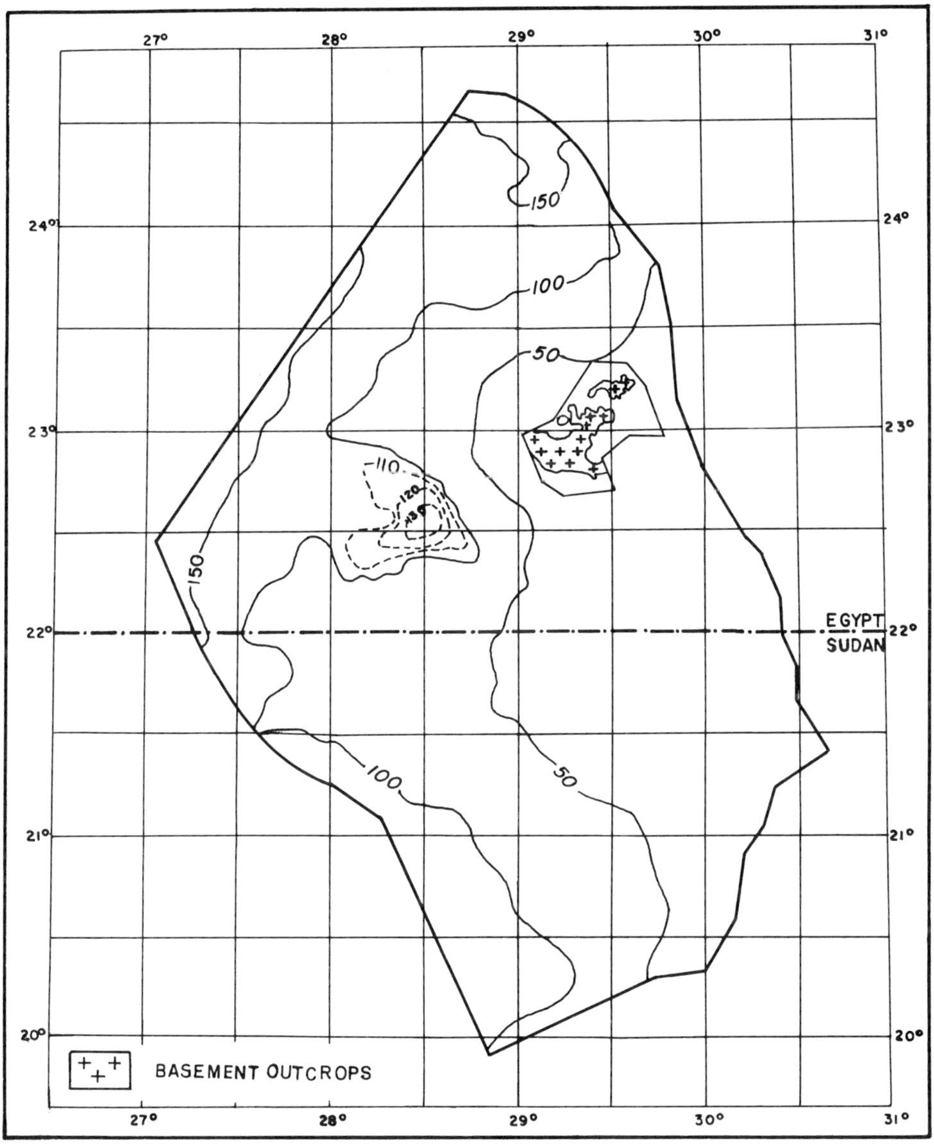

THE GENERAL PETROLEUM CO.
EAST OWEINAT AREA
DEPTH TO WATER TABLE
FROM GROUND SURFACE
EXTRACTION RATE $4.74 \times 10^6 m^3$/ day 100 year

FIGURE 4

water is calculated to be available for first-priority land development of some 189,000 feddans for a period of 100 years.

EVALUATION OF SOIL RESOURCES

A regional assessment was made of the whole area of about 7,000,000 feddans in 1982-83. The study was carried out by a team of specialists from the Desert Institute and the General Petroleum Company. The volume of work in this survey was comparable to that of 410 persons working for one month. A total of 4000 samples were collected and analyzed at the laboratories of the Land and Water Institute, Ministry of Agriculture. The results of the comprehensive study identify the varying tracts of arable land. These have been classified according to F.A.O. standards (1976).

The area studied was divided into thirteen zones, each comprised of about 500,000 feddans (Figure 5). Zones No. 2 through No. 13 were studied in the order of one profile per approximately 5,000 feddans, while Zone No. 1 was subjected to further detail with each profile representing 2,000 feddans. Land units were identified based on stereoscopic examination of aerial-photographs for the areas of appropriate scale (1:60,000) which were received from the Military Surveying Department of the Ministry of Defence. Soil classification and land capability maps produced as a result of this study show about 3.7 million feddans of promising sand soils -- quite an extensive area. About 1.5 million feddans of this acreage has a deep soil section (over 90 cm) and is ranked first-priority for development (Table 1). No doubt such results give this area good potential for rehabitation in Egypt.

Given the underground water and soil resources, one can be optimistic that the 1.5 million feddans of potentially fertile land in the presently uninhabited and barren desert can become productive by exploiting the enormous groundwater resources.

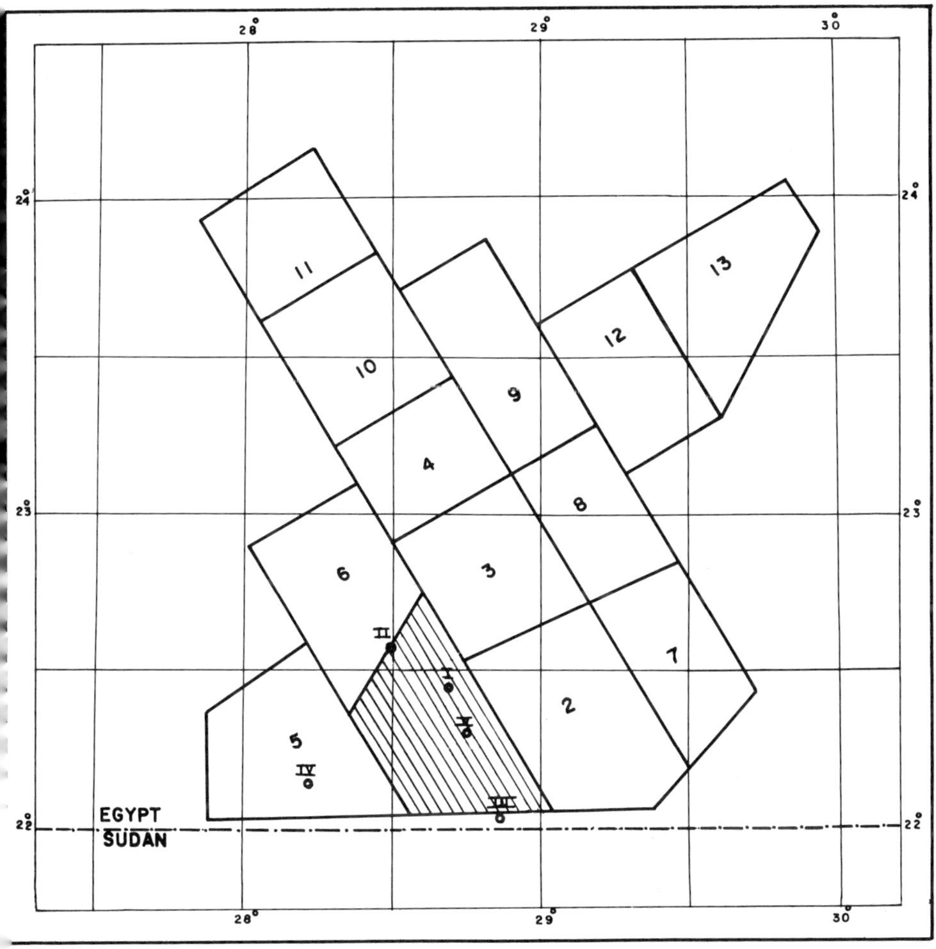

THE GENERAL PETROLEUM CO.
EAST OWEINAT AREA
SOIL CLASSIFICATION MAP
FOR 6 MILLION FADDAN

FIGURE 5

TABLE 1. Land capability classification of East Oweinat region

Approximate area in feddans

Classes Area	Total Area	Marginally suitable lands S3.1	S3.2	S3.3	S3.4	S3.5	S3.6	S3.7	Total	%
1	661900	190800	113700	37500	47600	34700	56900	14000	495200	75
2	807000	232400	119400	30300	42500	11500	21900	4000	462000	57
3	599600	227100	9500	12800	12300	21200	7000	-	289900	48
4	426900	31000	113300	1400	67400	6100	45900	29400	294500	69
5	657800	187700	102100	7100	3500	22400	12900	2300	338000	51
6	590900	379100	79500	4100	-	-	13100	-	475800	81
7	492100	46300	137900	-	19100	3600	21600	6800	235300	48
8	459300	33900	19600	63900	13700	9800	800	12100	153800	33
9	591800	68700	90200	20500	6600	125700	34600	-	346700	59
10	556100	18300	40700	45000	7700	74800	-	-	186400	34
11	470000	25300	14200	22000	29300	64000	-	18900	173700	37
12	538200	-	59000	1200	51000	400	53800	28600	194000	36
13	577900	3000	36300	8500	8500	9900	24700	6100	97000	17
Total	7429500	1443900	935400	254700	309200	394100	293200	122200	3742300	50.4

S 3.1 Deep sandy soils
S 3.2 Deep gravelly sandy soils (<40 gravels)
S 3.3 Moderately deep sandy soils
S 3.4 Moderately deep gravelly sandy soils (<40 gravels)
S 3.5 Shallow sandy soils
S 3.6 Deep gravelly sandy soils (>40 gravels)
S 3.7 Moderately deep gravelly sandy soils (>40 gravels)

TABLE 1: (Cont.)

Classes	Total	Not Suitable Lands			
Area	Area	N 1	N 2	Total	%
1	661900	109400	57300	166700	25
2	807000	232100	112900	345000	43
3	599600	294700	15000	309700	52
4	426900	106100	26300	132400	31
5	657800	155500	164300	319800	49
6	590900	63100	52000	115100	19
7	492100	16800	240000	256800	52
8	459300	134200	171300	305500	67
9	591800	123700	121400	245100	41
10	556100	91100	278600	369700	66
11	470000	180800	115500	295300	63
12	538200	89900	254300	344200	64
13	577900	126200	354100	480900	33
Total	7429500	1724200	1963000	3687200	49.6

N 1 Very shallow sandy soils and gravelly sandy soils
N 2 Rocky and sand dunes landscape

For such development, the region will need power, at an economic cost, primarily for lifting and distributing irrigation water. Primarily, it is lack of cheap energy that inhibits desert development in this area. Power for necessary supporting services and accommodations for the work force will also be needed. The conventional approach would be the use of diesel power, but this is neither realistic nor economical for the scale of operation.

Achieving a high level of local self-sufficiency will also be an essential requirement because East Oweinat is remote from the population centers of Egypt, and therefore the logistics of importing fuel, stores, and other provisions will be costly. This pioneering effort to cultivate a large area of presently barren desert will be of key importance to the Egyptian economy in the next century.

EVALUATION OF RENEWABLE ENERGY RESOURCES

Since 1979, measurements have been taken of sun insolation, windspeed and other meteorological data (Tables 2, 3, and 4). These proved encouraging for the testing of renewable energy resources.

The region is favored with exceptionally high average wind speeds. These, combined with solar irradiation, make the prospects for using wind and solar energy particularly promising. Seeking to exploit these resources is timely, as the wind and solar energy technologies that are necessary for such a scheme have reached acceptable levels of technical maturity. Also, their costs are declining in real terms due to technical improvements and the fact that the equipment is being manufactured at increasing scales of production.

TABLE 2. Monthly average hours of sun radiation in East Oweinat area (period 1980/1982)

Month	Hours	Month	Hours
January	7.7	July	9.9
February	8.8	August	9.6
March	9.8	September	9.6
April	30.3	October	9.2
May	10.2	November	7.9
June	10.0	December	7.3

TABLE 3. Monthly average hours of sun radiation Aswan area (1976-1979)

Month	Hours	Month	Hours
January	10.2	July	12.1
February	9.5	August	11.6
March	10.2	September	10.6
April	10.2	October	10.0
May	11.2	November	10.0
June	12.3	December	9.4

TABLE 4. East Oweinat average wind speed at 10m height in m/s

Month	M/S	Month	M/S
January	5.6	July	5.23
February	6.1	August	5.7
March	5.95	September	5.65
April	5.5	October	6.9
May	5.5	November	4.5
June	6.2	December	5.15

CURRENT IMPLEMENTATION

Based on data gathered on such insolation, Solarex Corporation and G.P.C. installed a photovoltaic pumping system with peak power and 21 kw at an experimental farm in the East Oweinat area to irrigate an area of ten feddans. This system is capable of delivering water at an average rate of $35m^3/h$ round the year for a daily duration of 10 hours in summer during July and August (a daily rate of $350m^3$/day as a peak value) and a daily rate of $200m^3$/day in winter as a minimum value. A dynamic head of 40mts and a surface water pressure of 3 atmospheres will be maintained the year round provided that the whole system is operated with a single well (Figure 6).

This system as a solar pumping system comprises the following:

- An array of Solarex PL-110 modules rated at 18.9kw peak power at 25°C.

- A Solarex ACR-6 system controller incorporating a programmable load controller.

- A 176 kwh battery bank consisting of EXIDE deep cycling E 120-17 cells.

- Two inverters, Abacus model 868-4-20.

- Two pumps, Berkeley model, one for lifting the water from the well and the other for distributing water against the back pressure of the irrigation system, a 92,000 gallon ($350m^3$) reservoir.

Concurrent with the 21 kw photovoltaic pumping system, the G.P.C. also operates a 3 kw photovoltaic pumping system. It was originally supplied by Lucas to the Military Technical College and then handed over to G.P.C. for operation under real working conditions where the value of photovoltaic powered irrigation systems could be

FIGURE 6. Simplified system block diagram

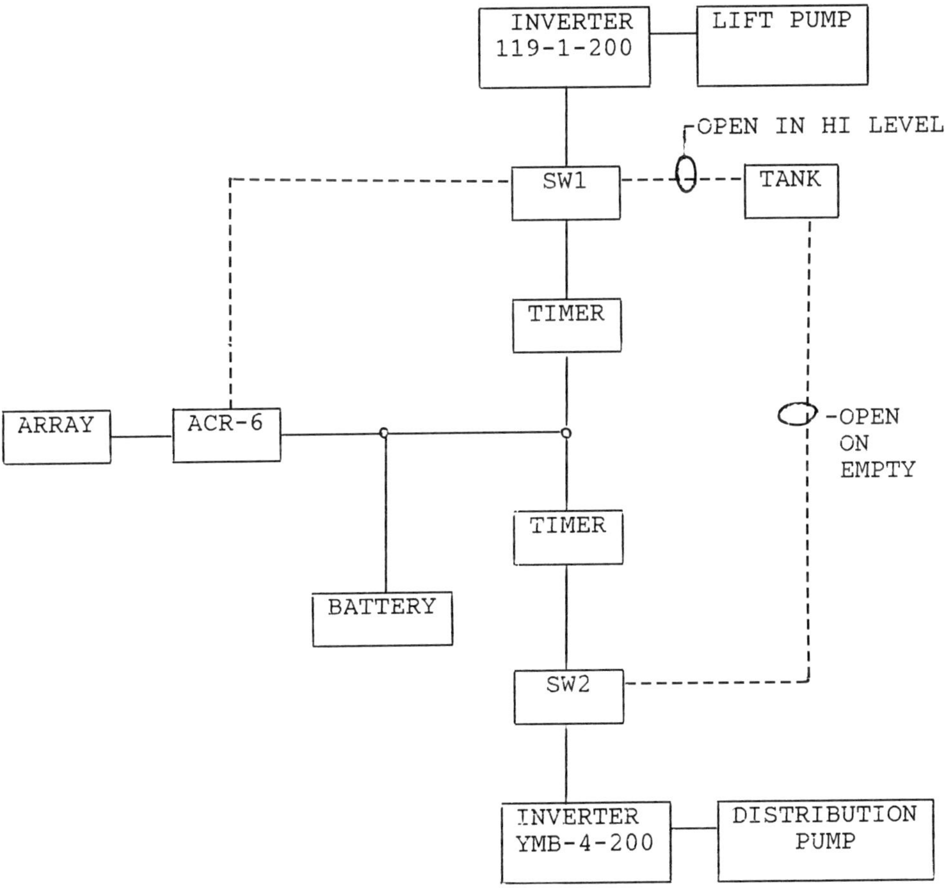

better evaluated. These systems have been used successfully for more than three years (1984-86), with only minor problems which have usually been solved within a reasonable time.

FUTURE IMPLEMENTATION

As a consequence of the United Nations Nairobi Conference held in 1981, the governments of Italy and Egypt reached an agreement to implement a Large Development Project (LDP) which aims to design, construct, and operate an agricultural settlement in Egypt. The project is called the Egyptian-Italian Renewable Energy Settlement (EIRES).

The main aims of the EIRES project are:

- To introduce modern agricultural and animal breeding techniques in the desert area, and to ensure comfortable conditions for a permanent human settlement.

- To make use of renewable energy resources such as solar (PV), wind, and biogas as a whole integrated system.

- To produce high-value human food.

- To produce a high level of self-sufficiency.

Once the settlement is a full-scale operation, the expected foods will be meats from beef and chicken, fresh milk, eggs and vegetables grown in greenhouses.

The annual consumption of electrical energy per operation is estimated to be as follows:

(a) irrigation: 42,000 kwh.

(b) dairy cows, greenhouse and buildings, cold storage and ancillary services: 106,000 kwh.

For this total annual load of 526,000 kwh there are three main local energy sources: wind generators, photovoltaics, and biogas - via a set of internal combustion engines on an experimental basis. The wind generators plus a 2000 kw peak PV would meet the electrical demand of the settlement according to the expected monthly energy production. Storage of energy is carried out electrically (daily storage) by nickel-cadmium batteries and hydraulically (weekly storage) by water basin. In the case of an emergency, energy is guaranteed by a diesel engine generator set. The energy required for the air conditioning of buildings is minimal because energy-conscious designs have been used, i.e. solar passive design and bioclimatic architecture.

THE STATUS OF THE EIRES PROJECT

The expected completion date of the EIRES project is the end of 1987. The basic design was completed at the end of 1984. At this time, all detailed engineering of civil works of the project has been completed. All the material and equipment to be procured from the Italian side as designated in the original basic design has been assigned; part has been delivered at the site and part is under delivery or construction.

The following construction activities are in progress or have been completed at the site:

(a) A water basin with maximum water storage of 40,000 m^3 is completely constructed.

(b) A pump house building has been completed.

(c) The foundation and steel structure for a PV system has been installed.

(d) Windbreaks have been planted around the entire farm.

(e) A complete net for a drip and sprinkler irrigation system has been installed for cultivating 200 feddans.

(f) The foundation for a power station and a battery room are being constructed.

(g) Erection of the wind energy system is planned for July 1987.

ARID LAND DEVELOPMENT IN PAKISTAN

ISHTIAQ AHMAD QAZI
Deputy Inspector General of Forests,
Government of Pakistan

Abstract Pakistan, a land of topo-climatic contrast, is predominantly comprised of arid and semi-arid tracts. The Indus River and its tributaries supply irrigation water. Over the years, agriculture has been developed on about 20 million hectares of the Indus basin through a network of canals. This paper discusses land use in the remaining area, which has been used historically as pasture land and wildlife habitats. With an increase in population, the area has been depleted and degraded to about one-tenth of its former capacity. Government policy emphasizes management of these wildlands for optimum utilization as pasture and wildlife areas. Development strategy envisages the achievement of goals through resource surveys, education and training programs, development of research, transfer of technology, participation of people and creation of institutions. Numerous programs have been launched on a limited scale in the fields of dry afforestation, range management, wildlife management, and soil and water conservation. Large-scale implementation has not been possible and is not envisaged in the foreseeable future due to numerous constraints and problems including lack of communication, the thinly-scattered and nomadic nature of the population, illiteracy, lack of a proper institutional framework and low priority.

ARID LAND DEVELOPMENT IN PAKISTAN

Pakistan, situated between 24° and 37° latitude and 61° and 75° longitude, is a country of topo-climatic contrast. It has a unique combination of deserts, alluvial plains and low and high hills. Except for a small hilly area in the northwest which lies in the monsoon zone, the rest receives less than 500 mm of precipitation. Climatic variations from region to region are pronounced. Areas in the south and near the coast have dry summers with temperatures rising as high as 45°C, while the northern areas are cool and pleasant at that time. Winters in western and northern parts are freezing while mild pleasant temperatures prevail in the south. About 90% of the area of Pakistan is either arid or semi-arid.[6] Pakistan is drained by the Indus River and its tributaries which, at the same time, supply irrigation water.

HISTORIC PERSPECTIVE

For centuries, the area has been known for its fertility, built up by the River Indus and its tributaries, and its rich resources of wild fauna and flora.[2] Human populations concentrated on flat areas along river banks, which received flood waters, and on areas where inundation canals could be dug. The fertility and the rich natural resources attracted migrants and invaders; thus a sedentary sector of the population were driven out. They shifted to comparatively infertile and far flung areas for seclusion and safety.

With an increase in population and the development of technology, the flat basin wildlands were converted into irrigated farmlands. The largest canal system in the world was established. The natural xerophatic vegetation was replaced by mesophatic species generally brought in from other areas. Thus dry-lands were converted into blooming farmlands of wheat, cotton, rice, sugarcane and many other crops. At this time, 20.34 million hectares out of a total area of 79.6 million hectares in Pakistan are used for productive agriculture.[1] The rest of the area continues to be used either by the nomadic population for cattle grazing and deferred cultivation or by wildlife. With increases in the populations of both human beings and livestock, arid land resources have undergone a process of depletion and degradation.

Many species of wild animals as well as plant communities have disappeared and many others are endangered.

CURRENT SITUATION

Although Pakistan is predominantly an agricultural country, and agriculture contributes 29% of the GDP and employs 55% of the labor force, it is practiced on only 26% of the total land area. About 42% of the area is wildlands used mainly for pasturing and wildlife management.[1] At the time that the country became independent in 1947, the percentage of wildlands was comparatively high. Much of this area has since been converted into irrigated farmland and forest plantations, such as in Thal in Southern Punjab and in Sind (Table 1).

TABLE 1. Current land use pattern[1]

Land use	Area in million ha	% of total area
Cultivation	20.34	26
Forests	3.10	4
Not available for cultivation	22.41	28
Wildlands	33.75	42
Total	79.60	100

The inhabitants of the wild tracts are predominantly pastoral and nomadic, moving from place to place with their cattle, camels, goats and sheep to exploit the range and water wherever it can be found. In remote areas where access to rivers and streams are not possible, the only way to make a place habitable is to dig a well or pond to get drinking water. When a well or a pond goes dry, the nomads move with their livestock, preferably to the fringes of irrigated areas where they stay until rains come.[5]

Life is extremely hard in the absence of means of communication, adequate resources and job opportunities. Their standard of living is low. They are usually underfed and most suffer from malnutrition. They live mainly on milk, millet and sorghum, supplemented by wheat, barley and wild fruits. Local trees like Salvadora, Zizyphus, Prospois, and acacia species are either lopped for fodder or are browsed. The arduous conditions have made the people hardy and somewhat immune to diseases.

The ranges have become unproductive and depleted because of a variety of adverse factors which have operated over the centuries. Large herds consisting mostly of goats, camels and sheep concentrate their grazing around the water points depleting the vegetation closest to the source. As a result of overgrazing, and in the absence of proper grazing management, deserts have become prone to wind erosion, causing sand movement, and dunes cover habitations and fields, blocking railways and roads. The devegetated and overgrazed arid mountains are exposed to erosion and, as a result, flash floods occur. The productivity of these lands is now about one-tenth of what it once was.[3]

POLICY FOR ARID LANDS DEVELOPMENT

At the time of independence, 14.3 million hectares were being cultivated which produced just enough grain for the population of 35 million. The remaining wildlands were utilized by wildlife and the livestock of the nomads. Even then, it was understood that the resources would be further depleted and their full benefits would be lost if they were not managed properly. Thus, soon after independence, the Government resolved to develop some of the arid lands for irrigated agriculture by extending canal networks. As a matter of policy, 10% of the irrigated land was to be developed into irrigated forest plantations, and other areas were to be stabilized by afforestation and fixing of sand dunes.

It was also resolved in 1960 to develop dry areas, unfit for crop production, for increased and improved livestock production. Scientifically developed pastures, animal husbandry programs and soil and water conservation measures were proposed.

The management of wildlands in accordance with their potential for recreation and wildlife was made an essential element of the 1980 Agricultural Policy. The promotion of wildlife conservation, consistent with other land uses and ensuring the preservation and enhancement of wildlife values, was a cornerstone of this policy.[4]

For the implementation of the policy for wild and arid land development, the following programs were conceived.

AFFORESTATION

The British left a legacy of over 40,000 hectares of irrigated forest plantations in 1947. Since then an additional 126,000 hectares have been afforested as in Table 2:

TABLE 2. Forest plantations in Pakistan

Project	Total project area	Area afforested
	Hectares	
Thal	631,578	37,426
Taunsa	695,546	15,342
Kotri	1,093,117	24,353
Guddu	187,854	20,540
Others	---	28,339
Total	3,608,095	12,600

The main objective for increasing the irrigated plantations was to produce fuelwood, timber and industrial wood. However, these plantations also helped immensely in moderating climatic conditions, ameliorating site conditions and providing a recreational resource base.

Dry afforestation, though on a smaller scale, has been and is still being done, mainly in the zones of inadequate and erratic rains. Objectives are to conserve soil, prevent sand dune movement, fill blanks in open depleted forests and revegetate eroded lands. Work on sand dune fixation has been carried out in uncommanded areas of irrigated plantations developed in the canal colonies and also in sandy desert areas as well as along the Makran coast of Baluchistan. *Calligonum polygonoides*, *Tamarix aphylla*, *Sachharum* spp. and *Arundo donax* were planted to bind the soil. *Acacia*, spp. *Parkinsonia aculeata*, *Prosopis cineraria*, *P. glandulosa*, *P. juliflora*, *Zizyphus maritiama* and *Eucalyptus* spp. were planted where the conditions were favorable, or after the initial sand fixation work was successful.

RANGE MANAGEMENT

Range lands occupy about 42% of the total area. Of that, about 90% fall in arid and semi-arid climatic zones. These differ greatly in their productivity and carrying capacity depending on precipitation, soil type, species, etc. Range lands are severely overgrazed. Forage production varies from 35 kg/ha to 960 kg/ha.[5] Large herds and unscientific grazing by large herds is the major cause of range depletion. In the late fifties, a range management project implemented on more than 47,000 hectares in the Maslakh area of Baluchistan Province established the feasibility of range land development on a scientific basis. However, vast areas of range lands are still used unscientifically by grazers.

About a dozen pilot projects were launched in the sixties in different ecological areas to demonstrate range rehabilitation technology which included reseeding of grasses, planting of fodder trees, and development of water points. Based on the experience gained from these pilot projects, operational projects were launched in Thal, Cholistan, and the D.G. Khan district in the Punjab, Khan district in NWFP, and Thar in Sind.

WILDLIFE MANAGEMENT

In centuries past, wildlife was abundant on the arid and semi-arid lands of Pakistan. The degradation and destruction of natural habitats is the major reason for a sharp decline in wildlife population, along with poaching, netting and indiscriminate shooting for fun, meat and trophies. The result is that now almost all of the species of large mammals and reptiles and many bird species are threatened with extinction. Chinkara in the Thal and Thar Desert, and urial in scrub forests are becoming rare. Black buck, once found abundantly in Cholistan, now need reintroduction. Cheetah, wild dogs and chowsinga have become extinct and 31 other species of mammals, 20 birds, and reptile species have become endangered.[7]

Expeditions of the World Wildlife Fund in 1966 and 1967 ascertained the wildlife situation in the country and suggested measures to arrest further decline. A Wildlife Enquiry Committee recommended measures in 1971 relating to game laws, game administration and conservation policy. Consequently, the National Council for the Conservation of Wildlife was established in 1974 in the Federal Government to formulate appropriate policies, coordinate the implementation of these policies with the provinces and to liaise with international agencies and non-government organizations. Consequently, seven national parks, 72 wildlife sanctuaries and 76 game reserves have been established in the country to preserve the country's major ecosystems and the endemic fauna and flora. Most of the national parks and protected areas are situated in the arid and semi-arid areas. Lal Sohara National Park is a reserve under the Man and Biosphere program of UNESCO and represents a desert ecosystem.

SOIL AND WATER CONSERVATION

Problems of water-borne soil erosion are common throughout the country and are severe in many parts, such as the arid mountain ranges of Sind and Baluchistan. Due to heavy seasonal torrents, the annual precipitation varies between 12 and 20 mm. Organic matter and productive upper soil may be totally washed away reducing the productivity of the land and making it unfit for agriculture and less productive for forestry, pasturing, and even for wildlife purposes.

High temperatures and low rainfall add to the severity of sand storms which have impoverished the soils by blowing away fine clay and silt particles and organic matter and leaving behind coarse sand in the form of big sand dunes. At present, 40% of the land area in Baluchistan, Sind and Punjab is subject to such active wind erosion. Soil and water conservation measures are extremely important for maintaining and improving the fertility of the tract. Soil erosion is checked by terracing, check damming, gully plugging, construction of spillways, flumes and regulators, cover cropping, and proper agronomical practices. Water conservation measures followed include contour ridging, trenching and pitting.

Soil and water conservation has been carried out on a limited scale. In Baluchistan, water conservation has been extremely beneficial for water recharge of aquifers used for irrigation. Water is also conserved by reducing evaporation.

DEVELOPMENT STRATEGY

The strategy for arid land development has to be multidimensional, comprehensive and dynamic. The strategy being followed in Pakistan comprises the following:

1. Resources are assessed through regular surveys.
2. Education and training is provided for development activities.
3. Development is supported by research findings.
4. Proven technology is transferred and popularized through demonstrations and operational projects.
5. Participation and involvement of people is encouraged.
6. Institutional arrangements are made for the implementation of the development programs.

RESOURCE SURVEY

The Soil Survey Department conducts soil classification and land-use capability surveys to provide a sound basis for land use planning and development activities. Most of the country, including representative areas in all ecological zones, has been surveyed.

The Zoological Survey Department and Wildlife Management branch of the Pakistan Forestry Institute are involved in surveying wildlife and habitats. The Range Management branch of the Pakistan Forestry Institute and provincial forest departments are responsible for range conditions and surveys of the carrying capacity of different range areas.

EDUCATION AND TRAINING

The Pakistan Forestry Institute for professionals and forestry schools for technicians cater to the training requirements for arid land resource development. These institutions have built up their capabilities to produce trained manpower in the fields of soil conservation, range management, wildlife management, dry afforestation and social forestry.

RESEARCH

Research is the primary responsibility of the Pakistan Forestry Institute and includes various aspects of arid zone use and development. Work has been done on dry afforestation, techniques, species, suitability, and water and soil conservation practices and range improvement techniques as well as other allied fields. The Pakistan Agricultural Research Council (PARC) has recently established an Arid Zone Research Institute (AZRI) at Quetta with stations at D.I. Khan (NWFP), Bahawalpur (Punjab), and Omarkot (Sind).

INFRASTRUCTURE

Infrastructure is considered in various disciplines concerning arid land development. The provinces of Punjab and Sind have set up range management circles and wildlife departments. The Government of NWFP has created a wildlife circle while the Baluchistan Forestry Department takes care of dry afforestation and water recharge programs.

POPULAR PARTICIPATION

Involvement of the people is low because they are

slow in adopting new technology. To enlist the peoples' cooperation in development activities, water points were developed at government expense in Cholistan. Similarly, all demonstration projects were funded by the Government. Saplings, seeds and technical advice are provided free of cost on request but an organized extension mechanism is not yet available in the undeveloped arid areas.

CONSTRAINTS AND PROBLEMS

In spite of the Government's commitment to develop arid lands, and numerous measures taken in that direction, achievements are few in light of the needs and requirements of the people. Factors contributing to this situation include:

i. Lack of Communication

The arid areas consist of either rugged mountain ranges or undulating deserts, where mechanized means of communication are not easy to develop. Long distances are covered on foot or on camel back. Thus, with poor communication and limited technical manpower, it is not possible to cover the total population and area in the tract. Moreover, the population keeps shifting in search of water and forage.

ii. Nomadic Nature of the Population

The population is sparse and spread over a very large area. Economically, it is difficult to justify investments.

iii. Illiteracy

The population, by and large, is illiterate. They are traditional and conservative by nature and quite resistant to new concepts, ideas, technology and changes.

iv. Mismanagement of resources

The populations of humans and livestock have increased tremendously resulting in the depletion of range lands and degeneration of livestock breeds. The situation is aggravated by the absence of veterinary services or advice about animal husbandry.

v. Lack of proper institutional framework

Although different nation-building departments have created technical manpower, they usually work in isolation. An integrated approach is lacking. In the absence of a proper institutional set up, efforts have been misdirected and unfruitful. Thus, meaningful programs cannot be effective.

vi. Low priority

The prevalent socio-economic conditions in the arid area are not conducive to obtaining political support. Consequently, the areas remain neglected and are given a low priority in the development program of the country.

NEW INITIATIVES

Learning from failures and shortcomings of the past, amends are being made to improve the development strategy. New institutions are being created for integrating approaches and supervising developmental activities in arid areas. Three such agencies recently created are:

i. Agency for Barani Area Development (ABAD)

Created for the rainfed districts of the Punjab, it is charged with the responsibility of coordinating all development activities in the rainfed areas.

ii. Cholistan Development Authority

The Cholistan Desert extends over about 0.28 million hectares but only one Civil Officer (non-technical) is in the Office of the Commissioner. Bahawalpur used to look after the affairs of Cholistan. Later on, nation-building departments initiated limited development activities pertaining to their sectors. Now the Government of the Punjab has established this autonomous body to organize development activities in the area.

iii. Sind Arid Zone Development Authority (SAZDA)

The Government of Sind has established this authority to implement well-coordinated development programs in arid zones including the hilly areas in the west and the desert in the southeast.

iv. Land Utilization Board

The Government of Punjab has formed a Land Utilization Board which has the responsibility to identify areas which are under public ownership and have not yet been allocated for specific land uses. Presently these lands are used for free pasturing. The surveys carried out by the Board will provide vital information, and the allocation of areas, coupled with change in land tenure, will provide a sound basis for the development of arid areas in the Punjab.

FUTURE PROSPECTS

The arid and semi-arid land resources in Pakistan are large, a positive factor if the land can be scientifically managed and harnessed on a sustained basis. Unfortunately, the land is depleted and continues to be further degraded despite efforts of the Government. These have reached only the demonstration stage. Scientific management for pasturing and wildlife is necessary, but is given low priority in the planning process of the country. Socio-economic conditions in the area and an improper institutional framework will continue to be constraints in the way of achieving the goals.

REFERENCES

1. Ministry of Food and Agriculture. 1984. Agriculture Statistics of Pakistan, Islamabad.

2. Ministry of Food and Agriculture, Forestry, Watershed, Range and Wildlife in Pakistan, Islamabad, 1986.

3. Ch. M. Anwar Khan. 1986. Range lands of Pakistan - a study. National Range Management Committee, Islamabad.

4. I.A. Qazi. 1986. Land use policy and forest legislation. Proceedings of Seminars on Social Forestry in Pakistan (PAK/83/021), FAO Rome.

5. M. Iqbal Sheikh. 1986. Arid zone forestry programmes - state of knowledge and experience in Pakistan. Pakistan Forest Institute, Peshawar.

6. M. Iqbal Sheikh. 1986. Afforestation of arid and semi-arid lands in Pakistan. Pakistan Forest Institute, Peshawar.

7. Ministry of Food and Agriculture. 1971. Wildlife Enquiry Committee Report, Islamabad.

EXPERIENCES IN DEVELOPMENT IN THE BAHAREYA OASIS OF EGYPT

S. NOUR EL DIN, S. ABDEL RAZIK,
M. EL-HARRAS, R. MARTELLO
Pilot Project for Agricultural Development
U.N.D.P., Zamalek, Egypt

Abstract The project aims toward producing food self-sufficiency for the 3000 inhabitants of the settlement associated with the iron ore mining operation in Bahareya Oasis in the Western Desert of Egypt. This goal is primarily approached by introducing workers of the mine to modern desert farming methods on a pilot farm. Animal protein self-sufficiency has been achieved at a level of about 50% through:

a- Chicken farming (broilers and layers)

b- A combined duck and fish farming operation employing the runoff waters from the iron ore mine;

c- An experimental implantation of French goats for milk production;

d- The introduction of rabbit farming.

Improvement and increase in the level of fresh vegetable production has been achieved through modern irrigation systems and farming practices. A training program, which introduces workers in the mine and the farmers of the surrounding areas, to the methods developed by the project, aims to propagate these methods throughout the six oases that make up the Bahareya Depression. A start along these lines has been initiated in the form of cooperative endeavors in the areas of water management, flour milling and various other schemes.

INTRODUCTION

Deep in the Western Desert of Egypt, near the borders of the Bahareya Oasis depression, four large iron ore deposits have been discovered. The development of this iron ore is making an important contribution to the industrial development of the country.

The start-up of production at the El Gedida iron ore mine increased production of steel by the Egyptian Iron and Steel Co. from 0.2 MT to 1.6 MT. It also led to the building of the first industrial community in the Western Desert of Egypt.

Until the 1970s, the Bahareya Oasis was one of the most remote and least-visited areas of Egypt. However, during the last few years, the area has been subjected to major changes and an accelerating rate of development. The key steps were the creation of an asphalt road and the mine.

LOCATION AND CLIMATE

The El Gedida mines are located 330 km southwest of Cairo and about 200 km west of the Minia Governorate. The northern entrance to the Bahareya Oasis depression is 12 km south of the mines area.

The Bahareya Oasis is an elliptical shape approximately 90 km by 45 km. It is surrounded by an escarpment about 100 m high with an altitude of 100 to 135 m. A few scattered villages are found inside the depression. The main village, El Bawiti, is 40 km from the mines. Its population of about 20,000 are mainly farmers.

The area is characterized by dry fair weather throughout most of the year. Winter nights are cold and sandstorms which last for a few hours occur in the spring.

GENERAL DEVELOPMENT

Infrastructure which makes the area more accessible began to be installed in the 1970s. An asphalt road extending 330 km was built from Cairo to the mine area in 1973. Later it was extended to the Bahareya depression and other oases in the Western Desert.

Other facilities include a railroad, wells, and electricity. A railway line was constructed from the mines to the steel facilities in Helwan (350 km) which is used mainly for transporting iron ore. Five deep water wells were sunk (more than 600 m) in the mine area to supply potable water and for industrial use. A sixth well was sunk in 1986 (1000 m deep). Two electric power lines of 132 kv each were constructed from the Aswan High Dam network through Samalout (El Minia) transformers to the mines. In 1986 the Oasis was supplied with electricity through two electric power lines of 33 kv from the mines' substation.

Habitation of the mine area was encouraged with the beginning of production at the mine in July 1973. During the next few years, the mine settlement was constructed as well as facilities such as a school complex, a hospital, bakery, cooperative, club, theater and cinema.

Greening of about 300 feddans of the township began at the end of 1974. The ground was leveled, suitable topsoil was transported to the area, and trees were planted for windbreaks and orchards. A private garden was provided for each worker. In 1976 an area was selected to be developed as a farm for producing fruits and vegetables.

In 1980 the Swedish Medical Aid Foundation donated $50,000 to plant 50,000 trees in the mine area. By the end of 1981, there were 120,000 trees planted in the area.

AGRICULTURAL DEVELOPMENT

A project to develop agriculture, irrigation systems, and animal husbandry in the Bahareya depression began in March 1983. Trials are conducted at the farms belonging to the mine and the successful results are passed on to the farmers in the Oasis. The Ministry of Industry represented by the Egyptian Iron and Steel Company is cooperating with the United Nations Development Program (UNDP) and the Food and Agriculture Organization (FAO) in the project.

The current status of the project is as follows:

HORTICULTURE

* The area farmed at the mines was increased from 70 feddans (in 1981) to 120 feddans. Two systems of irrigation, sprinkler and drip, are being used for the first time in the area.

* Different fruit trees have been planted from special selected seeds.

* New varieties of crops including jojoba, sisal, and some fodders are being tested.

* Small horticultural machinery is being introduced to save manpower and time.

ANIMAL HUSBANDRY

Domestic animals, poultry, fish, and bees are produced and cared for at the farm. The community is self-sufficient now with respect to eggs, chicken, ducks, and rabbits. These are produced in the following quantities:

Eggs: 1500 per day; to be increased to 3,000/day by June 1987

Chicken for meat: 3,000 per month; to be increased to 6,000/ mo by April 1987

Ducks: 1,000 per month

Rabbits: 1,500 per year

Dairy, sheep, and goats are also being raised. The herd consisting of fifty cows will be transferred to the farmers in the oasis at a later date. A flock of thirty Alpine goats was donated by the French Embassy. The desirability of crossing this breed with national goats will be studied.

A fish farm has been developed for the first time in this desert. At this time the farm contains mulet, tilapia, and carp and covers 7.5 feddans. The fish feed on the manure from the ducks. Besides the growing ponds, there are two hatcheries for tilapia. The ponds will be increased to cover 15 feddans by the end of 1987.

The water in the ponds is recycled. Initially, it is the used water from the crushing plant of the mine. From the fish pond, it is pumped to the fields to be used for irrigation.

Beekeeping is another endeavor that is being undertaken for the first time in the area. Ten hives are on the premises and this number will be increased to 40 this year.

One thousand pairs of pigeons complete the holdings of the mining farm.

DEVELOPMENT IN THE OASIS

From the beginning of the project, stress was laid on cooperating with the people of the Oasis and sharing activities and technical information. At first it was difficult to find the means by which to do so, but eventually local Community Development Associations provided an entree. Five such associations are active now in Bahareya villages under the auspices of the Ministry of Social Affairs.

The mine workers also transmit information to the villagers. Thirty-five percent of the workers in the mine (about 330) are from Bahareya Oasis. In 1973 when the mines were opened, these workers were completely unskilled. As a result of an internal training program, most of them are now introduced to complicated technical jobs in the mine. They are also able to inform the general population of the Oasis about other activities in the mining complex.

The project has also conducted tours for area farmers to teach them about activities at the mining farm and to show them similar activities in the Nile Valley.

Activities at the mines are being supported in the Oasis in the following ways:

a) In animal husbandry, several farmers in different villages are breeding laying hens and rabbits and have created their own fish ponds. Farmers have responded positively, especially after finding that these activities are profitable. During 1987, beekeeping and Alpine goats will be introduced to farmers. The challenge of the project is to insure that appropriate breeds and feed are available, as well as to provide ongoing technical advice.

b) For food and feed processing, a flour mill and threshing machines have been added to community facilities. One flour mill was built and began operation in 1986. A second is being built in another village and a third is planned. Half of the expenses for the mills are paid for from the agricultural development project, half from the community associations, and the mining company is to take care of maintenance and repairs.

One thresher was tested on wheat, beans, and maize in the Oasis from May to November 1986. The villagers were pleased because the machine saved time and effort and no crop was lost. Another thresher has been acquired and there are plans for getting a third. Farmers are able to rent the machines for 5 pounds per hour through the community associations. The accessibility of the threshing machines as well as the flour mill will probably encourage farmers to increase the amount of land devoted to these crops.

Other agricultural activities are about to be shared with farmers. The farmers asked the project to help with mechanization. In response, the project obtained a number of horticultural machines which will be introduced to the farmers through the community associations this year. It is planned that beekeeping and drip irrigation will also be introduced to farmers in the near future.

In the area of architecture, a model house is being built at the mining farm using local raw materials. The building project is a cooperative one with the French embassy. It is hoped that the results will be acceptable to the farmers.

At this time, a project to establish a pilot village on 600 feddans in the desert of the Oasis is being considered. The settlers would be pensioned miners, poor villagers, and young children of miners. Such a project may be a step toward encouraging immigration away from the crowded Nile Valley.

Finally, because there were no accommodations for visitors in the Oasis, the mining company prepared rest houses with a food service. About 200 visitors can be hosted now.

CONCLUSIONS

The Pilot Project for Development in the Bahareya area, supported by the facilities of the mining complex, has proven to be a good means for bringing new technology to the whole area. Such technology is not approved until trials have been made to introduce appropriate modifications and to make the technology as simple as possible.

In conclusion, it must be noted that there are a number of attractive possibilities for investments within the area. Achievements that have been made are only a beginning. With the experience that has been gained, future development can be more rapid.

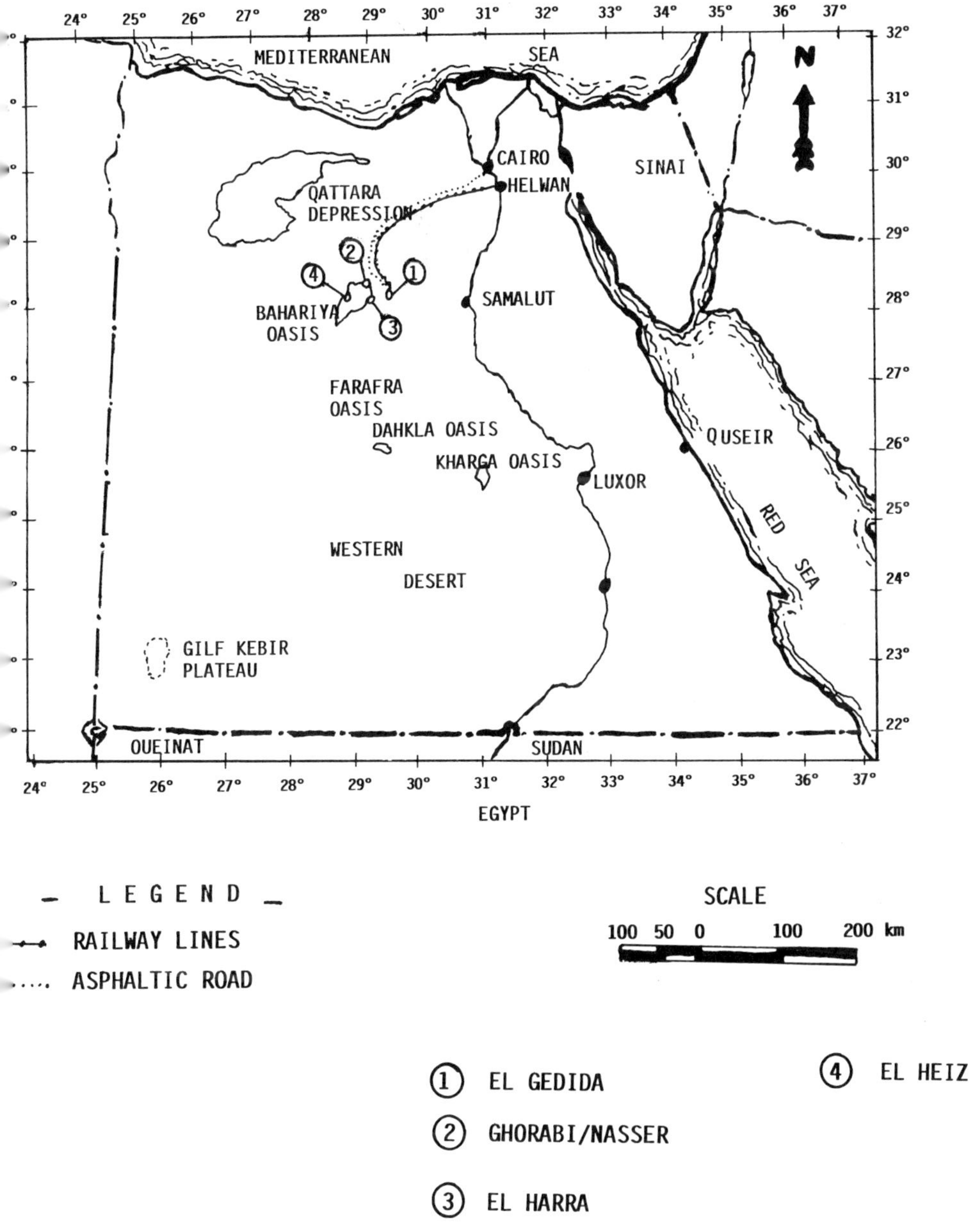
MEDITERRANEAN SEA
CAIRO
HELWAN
SINAI
QATTARA DEPRESSION
BAHARIYA OASIS
SAMALUT
FARAFRA OASIS
DAHKLA OASIS
KHARGA OASIS
LUXOR
QUSEIR
RED SEA
WESTERN DESERT
GILF KEBIR PLATEAU
OUEINAT
SUDAN
EGYPT
N
LEGEND
RAILWAY LINES
ASPHALTIC ROAD
SCALE
100 50 0 100 200 km
1 EL GEDIDA
2 GHORABI/NASSER
3 EL HARRA
4 EL HEIZ

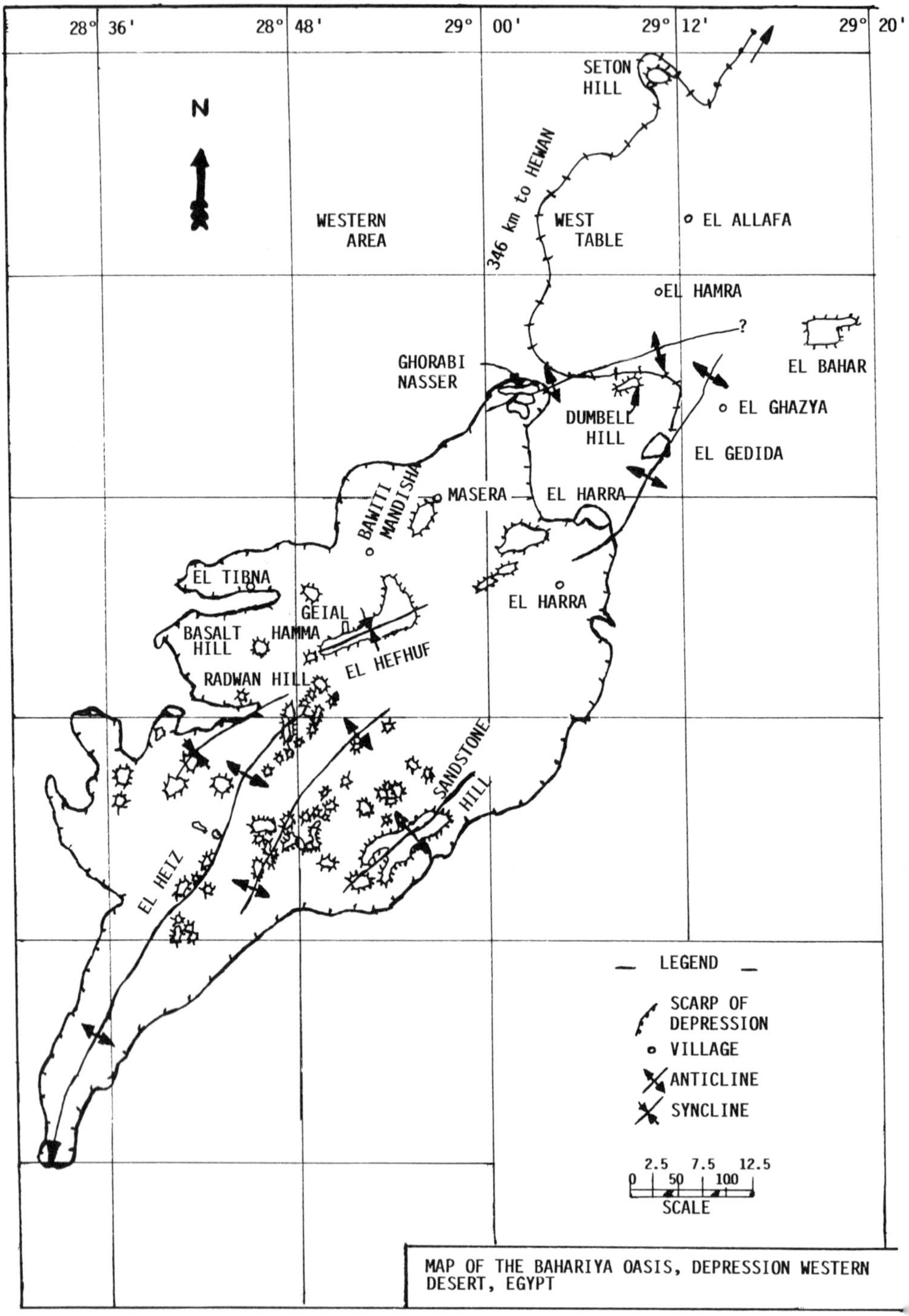

MAP OF THE BAHARIYA OASIS, DEPRESSION WESTERN DESERT, EGYPT

NEW AND RENEWABLE ENERGY STRATEGY FOR EGYPT (NRSE)

T. EL TABLAWI and A.I. HEGAZI
Renewable Energy Development Utilization Authority, Ministry of Electricity, Egypt

Abstract Egypt's development plan requires intensive programs in land reclamation, food production, industrialization and community development. To achieve these goals, the energy resources should satisfy about 49 M.M.T.O.E. by the year 2005. It will be difficult for existing resources and proven oil reserves to satisfy this level of energy needs, particularly since the national electric grid is not expected to cover remote areas.

In view of these facts, the Government, through the Supreme Council of NRSE and the Ministry of Electricity and Energy, has formulated a national strategy for the development and utilization of NRSE as well as for conservation of energy. In addition, intensive programs are being implemented for the promotion of NRSE use and adaptation to the Egyptian environment.

This paper discusses the goals of this "National Strategy" as well as some of the existing activities.

The overall goals are:

- To supply 5% of national commercial primary energy through the application of NRSE.

- To reduce national energy consumption by 10% through implementing energy conservation measures.

Existing activities cover a wide range of NRSE studies as well as demonstration and project implementation in the areas of solar, wind, biomass and mini-hydro.

INTRODUCTION

Egypt's development plan entails intensive programs in land reclamation, food production, industrialization and community development. To achieve these goals about 49 M.M.T.O.E. of commercial primary energy will be required by the year 2000, increasing to 60 M.M.T.O.E. by the year 2005. The present energy resources will fall short of satisfying this level of energy needs.

In view of these facts, the Government, through the Supreme Council for a New and Renewable Energy Strategy (SCNRSE) and the Ministry of Electricity and Energy (MOEE), has formulated a national strategy for the development and implementation of new and renewable energy as well as for energy conservation. This paper discusses the goals of this strategy and the existing MOEE activities in the field.

In order to achieve the objectives of the development plan, Egypt will have to face a number of constraints. Mainly these are: an increasing population, an extensive need for new agricultural areas and a high rate of industrialization. Therefore the demand for energy will increase sharply, while hydro and fossil fuel resources will fall short of domestic demand.

The MOEE has developed alternative scenarios for both domestic energy and primary commercial energy consumption. These alternatives assume three different gross national development rates: 6.3%, 5% and 4.3%, respectively, for high, medium and conservative assumptions. Meanwhile, NRSE and energy conservation strategies were developed as will be explained herein. Figures 1 and 2 show projections for primary energy production by source for the 1985 - 2005 period, based on the assumption of a conservative growth rate. These projections include: the introduction of nuclear power generation, a coal thermal power station and NRSE. These exhibits show that, if energy conservation measures are applied, the need for primary energy

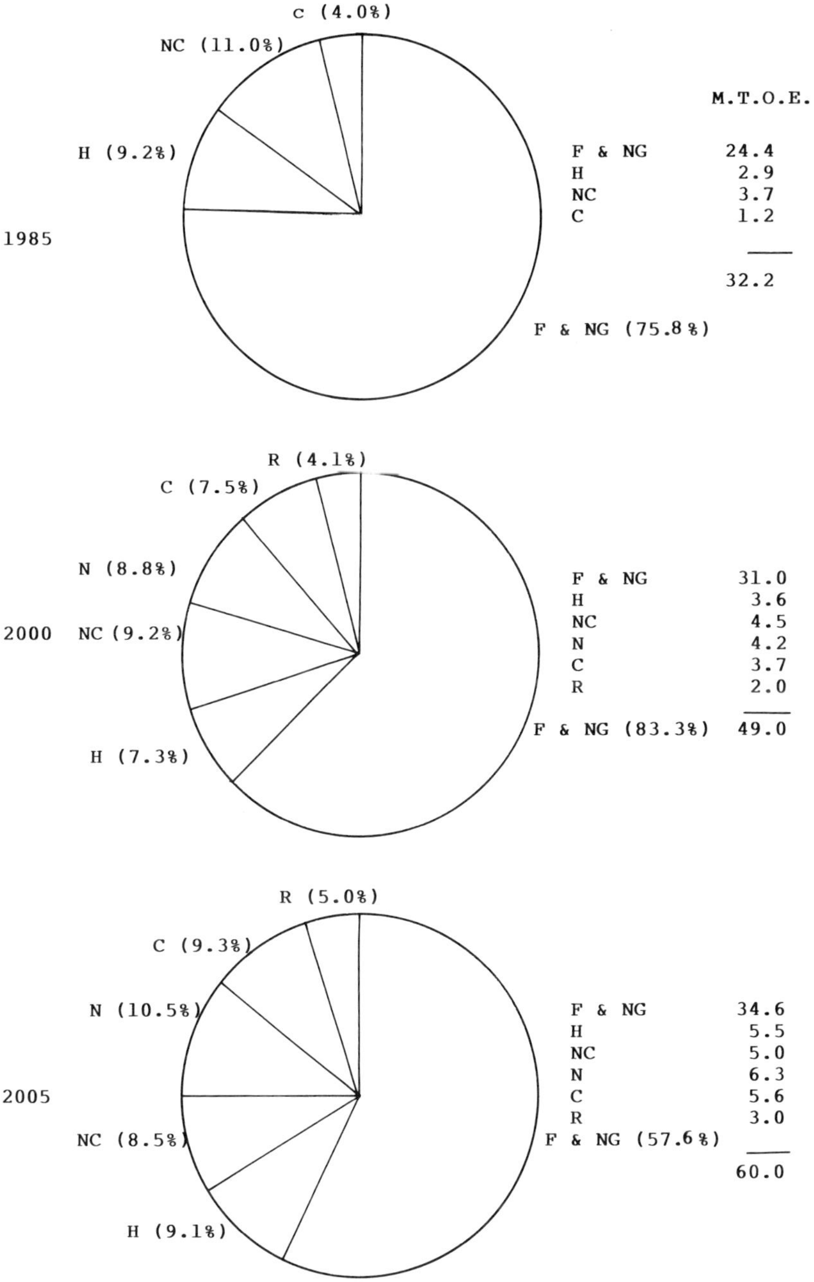

FIGURE 1: Total Primary Energy Production (1985 - 2005)

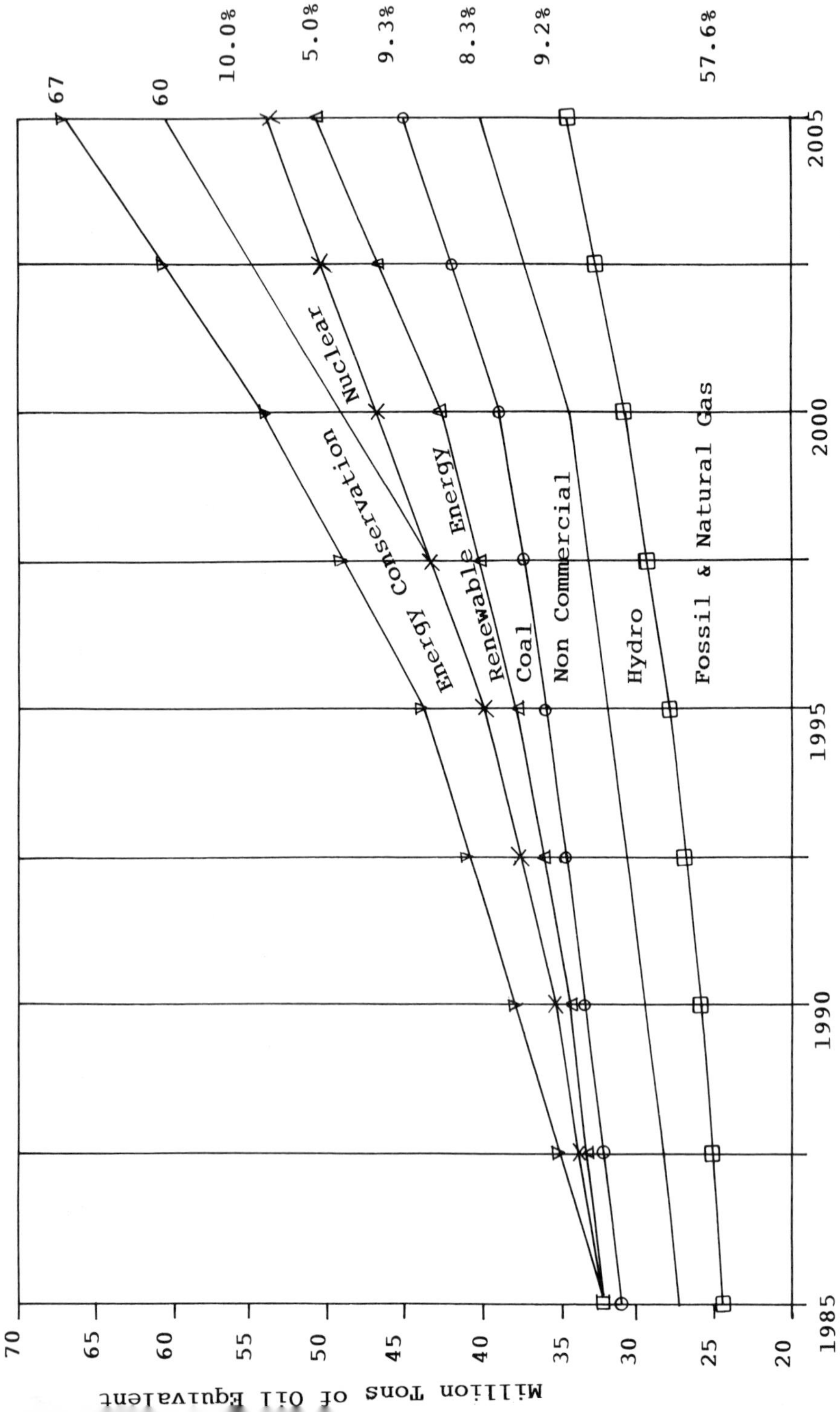

FIGURE 2. Projections of Primary Energy Production & Energy Conservation

production will be 49 M.M.T.O.E. by the year 2000 and 60 M.M.T.O.E. by the year 2005. An additional burden of almost 10% will be placed on fossil fuels if these measures are not appropriately applied.

NEW AND RENEWABLE ENERGY STRATEGY FOR EGYPT (NRSE)

Egypt enjoys high rates of solar radiation over all its territory (average of 6 $kwh/m^3/day$), as well as favorable wind resources in coastal areas, and the existence of intensive biomass resources including agriculture, animal and municipal solid wastes.

In addition, more than twenty locations for mini-hydro electric power generation, and a few low/medium temperature geothermal locations have been identified on the Sinai and Red Sea coasts. These have the potential of producing about 250 MWH/year.

In recognition of the above, as well as the lack of sufficient energy resources, the MOEE through SCNRSE, assisted by UNDP, has formulated a National Strategy for the Development and Utilization of New and Renewable Energy and Energy Conservation. This includes development of an institutional structure for NRSE activities to ensure project management and coordination of such activities within the country.

The overall goals of the New and Renewable Energy National Strategy are:

- To supply 5% of the national primary commercial energy needed by the year 2005.
- To reduce national energy consumption by 10% through implementing energy conservation measures.

Studies conducted by Egyptian experts regarding formulation of this strategy, as well as other feasibility and market studies performed by MOEE, have identified different applications for new and renewable energy derived from solar, wind, biomass and geothermal sources. Figure 3 shows projections for the potential application of NRSE from 1985 to 2005, where 5% of the total primary energy production is replaced by renewable energy, an equivalent of 3 M.M.T.O.E. annually.

Potential savings of conventional energy through the implementation of NRSE are expected to increase from 375 M.T.O.E. by the year 1990 to 2.1 M.M.T.O.E. by the year 2000 and 3 M.M.T.O.E. by 2005. The main contributor is expected to be solar water heating for domestic use and industrial processes, which will save about 1.0 M.M.T.O.E. by the year 2000 and over 1.5 M.M.T.O.E. by 2005.

It is estimated that solar photovoltaic applications for electrification of remote areas, water pumping and desalination will reach about 40 M.M.T.O.E. annually by the year 2005 with a total installed capacity of almost 60 MWP.

Favorable wind resources are available on the sea coasts and in East Oweinat. Wind power generation would be appropriate for desalination and pumping in these areas. The installed capacity is expected to be 360 MWP by the year 2005 if an East Oweinat land reclamation program is achieved. The annual energy savings will reach 159 M.M.T.O.E.

The proposed strategy for utilization of biomass calls for the development of public bakeries in rural areas as well as development of more efficient cooking facilities. The increased efficiencies of these bakeries and cookers could save about 175 M.M.T.O.E. annually. In addition, a promotional program for the use of biogas digestors is being considered to supply energy for 260,000 family units to be constructed in rural areas by the year 2005 as well as 150,000 units to be constructed in newly reclaimed areas. If these objectives are achieved by the year 2005, about 280 M.M.T.O.E. will be saved annually (Exhibit 2).

Furthermore, efforts are being made to utilize urban solid waste and sewage for power and fertilizer production. Technologies for

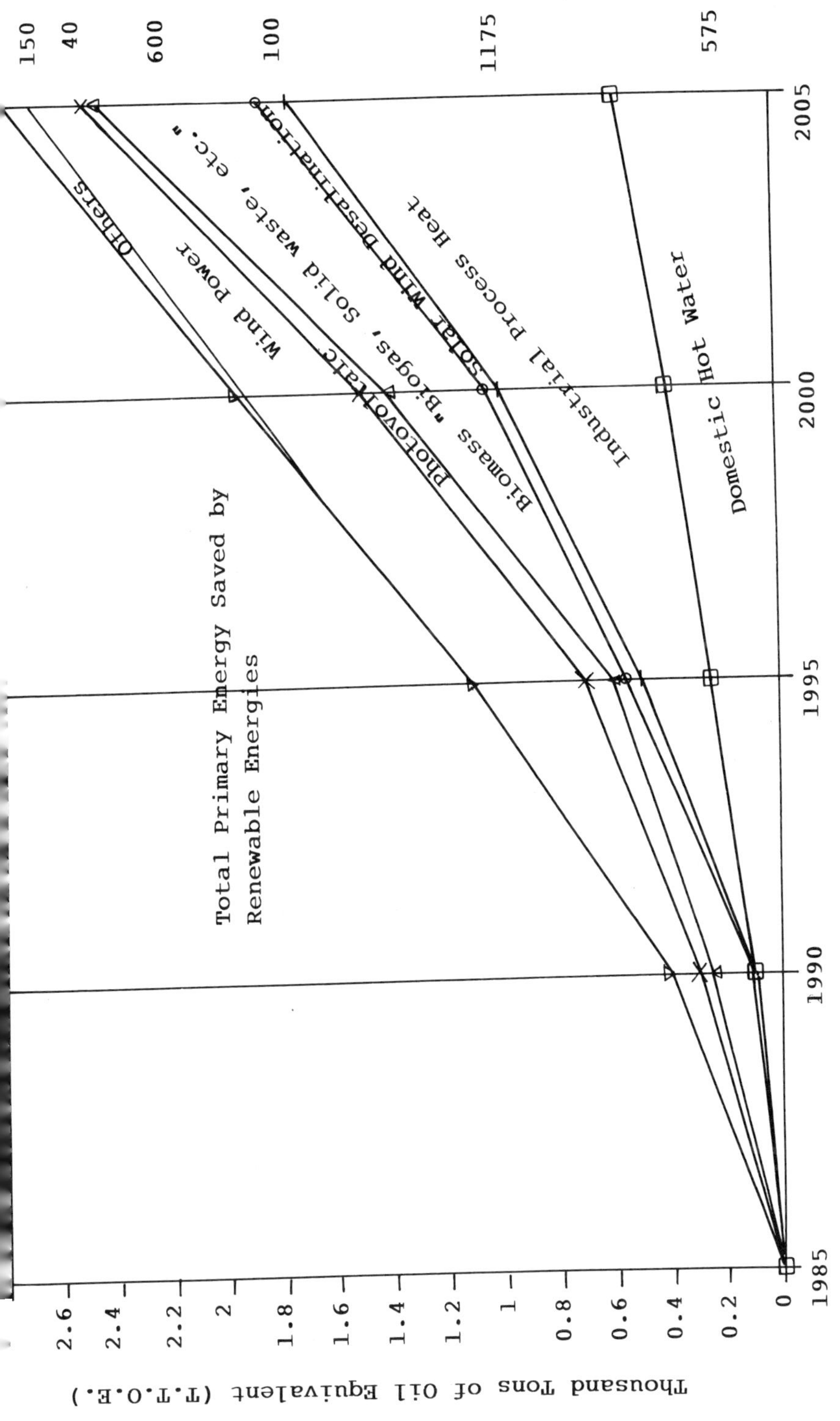

FIGURE 3: Projections for Renewable Energy Usage 1985 - 2005

utilization of such biomass can contribute effectively to NRSE goals.

Geothermal energy resources are very limited in Egypt. However, a number of potential sites have been identified in the Red Sea and Sinai areas.

INSTITUTIONAL STRUCTURE FOR NRSE

The SCNRSE is entrusted with coordinating the implementation of NRSE goals and activities within different national institutions. It is also responsible for the development of NRSE policies within the overall energy context of the country. Its technical secretariat has eleven specialized technical groups in different NRSE disciplines.

The Renewable Energy Development and Utilization Authority (REDUA) was established in 1986 to be the implementing agency for the SCNRSE in coordination with other concerned national authorities.

The main responsibilities of REDUA are:

- To identify and evaluate new and renewable energy resources and plan for their development and use within the frame work of the overall energy policies in Egypt.

- To carry out technical, economical and environmental studies and research needed for the development of renewable energy resources either alone or in cooperation with other technical institutions inside and outside Egypt.

- To identify, in coordination with relevant national authorities, fields where renewable energy sources can effectively replace conventional energy resources.

- The authority is the only organization authorized to issue necessary licenses in order to ensure technical standards.

- To implement NRSE projects, either on its own or in cooperation with others.

- To recommend standard specifications for renewable energy systems and equipment for Egypt and to perform standard performance tests for local and foreign produced systems and equipment under Egyptian environmental conditions, and then to issue certificates of performance.

- To carry out field testing of renewable energy technologies.

- To develop and implement the necessary training programs to disseminate renewable energy applications.

- To provide consultancy as needed in the field of renewable energy as well as the technical expertise needed for the development of a national industry of renewable energy equipment and its related programs.

- To implement all agreements in the field of renewable energy between the Egyptian government and local authorities or foreign governments and authorities.

Since 1978, the MOEE and its concerned authorities have been working actively to develop, promote and implement NRSE technologies in Egypt. These activities cover a wide spectrum of NRSE resources and applications which must be implemented in full coordination with concerned national authorities. Additionally, bilateral and multilateral cooperation agreements have been signed with several countries and international organizations, mainly France, FRG, Italy, Jordan, USAID, EEC and UNDP. The following is a summary of the completed and ongoing activities of REDUA. These activities are all directed towards the realization of Egypt's strategic goal of achieving 5% of total primary energy production by the year 2005 from new and renewable energy.

COMPLETED AND ONGOING ACTIVITIES

RESOURCE ASSESSMENT

One central station and six mobile units have been used since 1979 to assess solar radiation. In addition, wind measuring facilities have been installed in 19 sites on the coasts of Egypt.

SOLAR WATER HEATING

Programs of research, testing, demonstration and manufacture were conducted by MOEE and REDUA. These programs were supported by studies and regulatory activities for evaluation of the potential for Solar Water Heating applications in Egypt and promotion of their use. These activities included (but were not limited to):

- Demonstration of 35 different types and designs of Demonstration Solar Water Heaters (DSWH) at sites extending from Alexandria to Assiut.

- Renting 1000 DSWH to the public to encourage their marketing and to develop the required installation and maintenance capabilities.

- Demonstration of a number of collective DSWH systems at hospitals, social clubs and public utilities. Capacity of these DSWH is between 500 liters/day to 5000 liters/day.

- Development of laboratories for testing flat plate solar collectors.

- Assessment of the potential of the use of industrial process heat in Egypt, and development of a program for its promotion.

- Consideration of five industrial process heat projects for the use of poultry, textiles, food processing, metal treatment and food dehydration activities by REDUA in cooperation with USAID and other agencies.

- MOEE/REDUA has performed all of the studies and has taken action in coordination with the concerned Egyptian authorities resulting in the following:

  1- Issuance of government regulations encouraging the use of solar water heating in existing facilities while mandating its use in new communities.

  2- Preparation of Egyptian standard specifications for solar water heaters as well as the code of practice for its use.

DEVELOPMENT OF REMOTE COMMUNITIES

Priority is given to the use of NRSE in the development of remote communities, specifically for supplying of fresh water, food preservation, clinic refrigeration and village electrification. REDUA projects include:

Desalination

- Installation and operation of three photovoltaic desalination plants producing about 100 m^3/day at Abu Ghossoun, Hamrawain and the High Voltage Laboratory (HVL) near Giza.

- Consideration of Installation of a wind-driven desalination plant at Hurghada with a 80 m^3/day capacity and a 200 kw wind-energy conversion system.

- Installation of a unit to be located at Dahab in the Sinai with a capacity of 10 m^3/day powered by wind. In addition, a photovoltaic plant is under consideration for El Kasr village to produce 25 m^3/day of fresh water.

Cooling and Food Preservation

- A 10 kw fish-preservation unit powered by a

solar thermal system at Aswan.

- A combined photovoltaic/diesel-powered ice plant with a 6 ton/day capacity.

- An ice plant with a capacity of 3 ton/day using a wind/diesel hybrid system (55 kw) at Abu Ghossoun.

- A number of refrigerators have been installed in health clinics since 1979, and some others are currently under consideration.

Power Generation Projects

- A photovoltaic system (20 kwp) will be installed at the Narthin Oasis for electrification of El-Haiz village.

- A 400 kw wind farm at Ras Ghareb is to be connected to the local grid.

- A 25 kw wind/diesel system is to be installed at Matruh for village electrification.

- A wind farm at Hurghada consisting of 6 units, each of 30 kw rated output is to be connected to the local power station or grid.

- A research program for solar ponds is being developed in cooperation with Cairo University. Two ponds were built at Alexandria and a heat extraction system is under consideration.

Telecommunication and Navigation

- Photovoltaic navigation lights have been installed at the High Dam Lake since 1980.

- Telecommunication systems were installed at Abu Ghossoun, Hurghada, Siwa, and Mersa Matruh.

Irrigation and Agro-Industry

- A number of photovoltaic pumping stations with capacities varying between (2-25 Kwp) were installed at Mansoura and East Oweinat.
- A project is under consideration to install and test three photovoltaic pumps and one thermal pump at Wadi El Natroun.
- Underground water pumping at East Oweinat using 20 kw unit produced by MAN Company.
- Installation and testing of a simple solar dryer for agriculture products at Fayoum Governorate.

In addition to the activities listed above, a feasibility study of establishing an energy center of 1,000 acres at East Oweinat is currently under consideration.

Furthermore, a ministerial committee has been set up to perform a pre-feasibility study for manufacturing Wind Energy Conversion Systems (WECS) in Egypt.

DEVELOPMENT OF AN NRSE INFORMATION SYSTEM

Within the framework of a French-Egyptian agreement, a renewable energy documentation center was established. In addition, view of the rapidly expanding and diverse activities in the renewable energy field, an Egyptian/USAID contract agreement includes a separate task dedicated to the development of a computerized information system for renewable data bases. A wide variety of data sets and information will be collected, analyzed and integrated into this information system, and hence will be made available to interested parties in Egypt and possibly abroad. The information and data collected will cover a broad range of information on renewable energy resources including technical, engineering, social, financial, economic and market assessments and analyses.

This information system will be developed in three phases Phase I will comprise stand-alone personal microcomputers; phase II will comprise initial integrating in a local area network; and phase III will involve an integration of medium to large minicomputers. During phase II, the information system will be interfaced and integrated to other national information systems to make the information available to research centers, universities and public and private sector enterprises.

TRAINING AND INFORMATION DISSEMINATION

REDUA is planning a training program at all levels in cooperation with other national institutions, together with an information dissemination program for NRSE.

CONCLUSIONS

Sources of renewable energy are abundant in Egypt. Egypt offers, therefore, an excellent opportunity for renewable energy to be integrated in a National Energy Plan.

The 5% target for use of renewable energy by the year 2005 is based on a conservative National growth rate of 4.3%. An increased burden will be placed on fossil fuels should this goal of 5% not be reached, or should the development rate be higher than 4.3%. It is therefore imperative that all ongoing and planned activities undertaken under NRSE be carried out expeditiously and to a successful conclusion for renewable energy to attain market penetration.

SOLAR AND WIND ENERGY POTENTIALS IN JORDAN*

HANI EL-MULKI
Solar Energy Research Center,
Royal Scientific Society
P.O. Box 925819, Amman, Jordan

Abstract This paper presents briefly the methodology and the results of a study performed by the Solar Energy Research Center (SERC) of Jordan to determine the priority of research, development and demonstration (RD & D) activities to be conducted in Jordan in the fields of solar and wind energy. The determination process took into consideration the current energy situation and future trends, the current status of energy services in the different parts of Jordan, the available solar and wind resources, the current state of the art of solar and wind energy applications and technology.

INTRODUCTION

Jordan is one of the few countries in the world that lacks indigenous sources of energy which are readily available in the commercial sense. It has experienced great development in the energy sector in the last five years. At the same time, the energy demand has increased by approximately 9% annually.

To face this situation, it was essential that Jordan should direct some effort towards developing and utilizing its potential indigenous sources of energy in the most appropriate, efficient and accelerated manner. Among these indigenous sources of energy, solar and wind technology, especially for application in rural areas, ranks high.

---

* Data presented in this paper is that of 1981.

Since the scope of solar and wind energy applications and associate technologies is wide, it was necessary that one should carry out a comprehensive investigation to determine the priorities of RD & D activities in this field. It was important to also consider major factors which would affect future wide-spread applications of such energy in Jordan.

The paper hereafter briefly presents the broad objectives, the methodology followed, and the main results of the investigation.

OBJECTIVES

The broad objectives of the investigation carried out on the use of solar and wind energy resources in Jordan were as follows:

a. To determine solar and wind energy resource potential in Jordan.

b. To identify possible R & D projects in solar and wind energy.

c. To direct research and development in the area of solar and wind energy towards the benefit of Jordan.

The basic criteria for developing these broad objectives included:

a. Assessment of Jordan's current energy situation and its future trends.

b. Building a data base concerning the current status of services in Jordan for the purpose of identifying the basic energy needs.

c. Evaluating and compiling Jordan's available meteorological data for the purpose of assessing solar and wind energy resources.

d. Assessment of the state of the art of solar and wind energy applications and technology.

e. Assessment of socio-economic acceptance of solar and wind energy application in Jordan.

f. Assessment of Jordan's research and development capabilities.

METHODOLOGY

The general methodology that was followed in the investigation is that presented in Figure 1. This figure shows the interaction between the various activities of the investigation which ultimately led to decisions - that is, setting priorities for RD & D activities to be carried out in Jordan according to its present goals. In addition, the investigation considered the diversity of climate and topography of Jordan by dividing Jordan geographically into eight main regions as shown in Figure 2.

MAIN RESULTS

The main results of the investigation can be summarized as follows:

a. It was concluded from the investigation of the Jordan energy situation that:

Jordan has been totally dependent on one fuel oil supplied from one country, Saudi Arabia, through one system, the pipeline. The rate of growth of energy demand during the past decade has been in the range of 12 to 15% for crude oil and of 18-25% for electricity. This high rate is expected to continue for the coming few years, but perhaps at a slower tempo. In spite of the

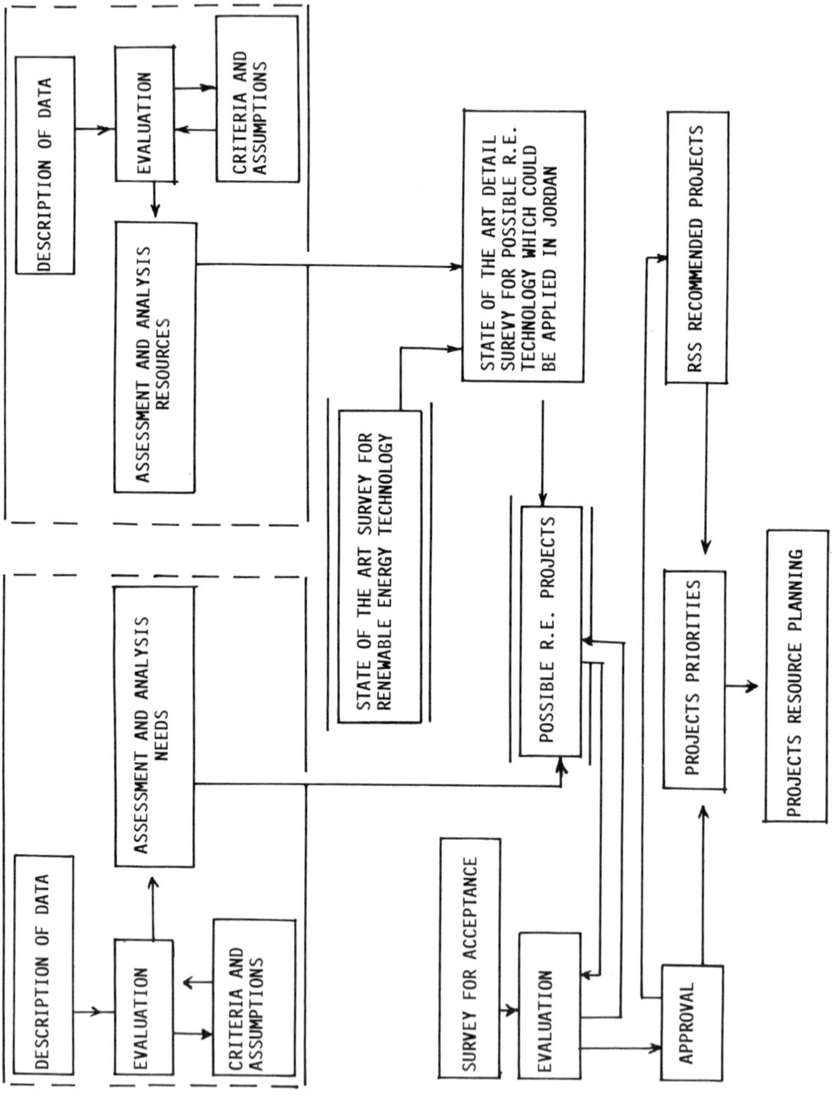

FIGURE 1: General outline of the methodology

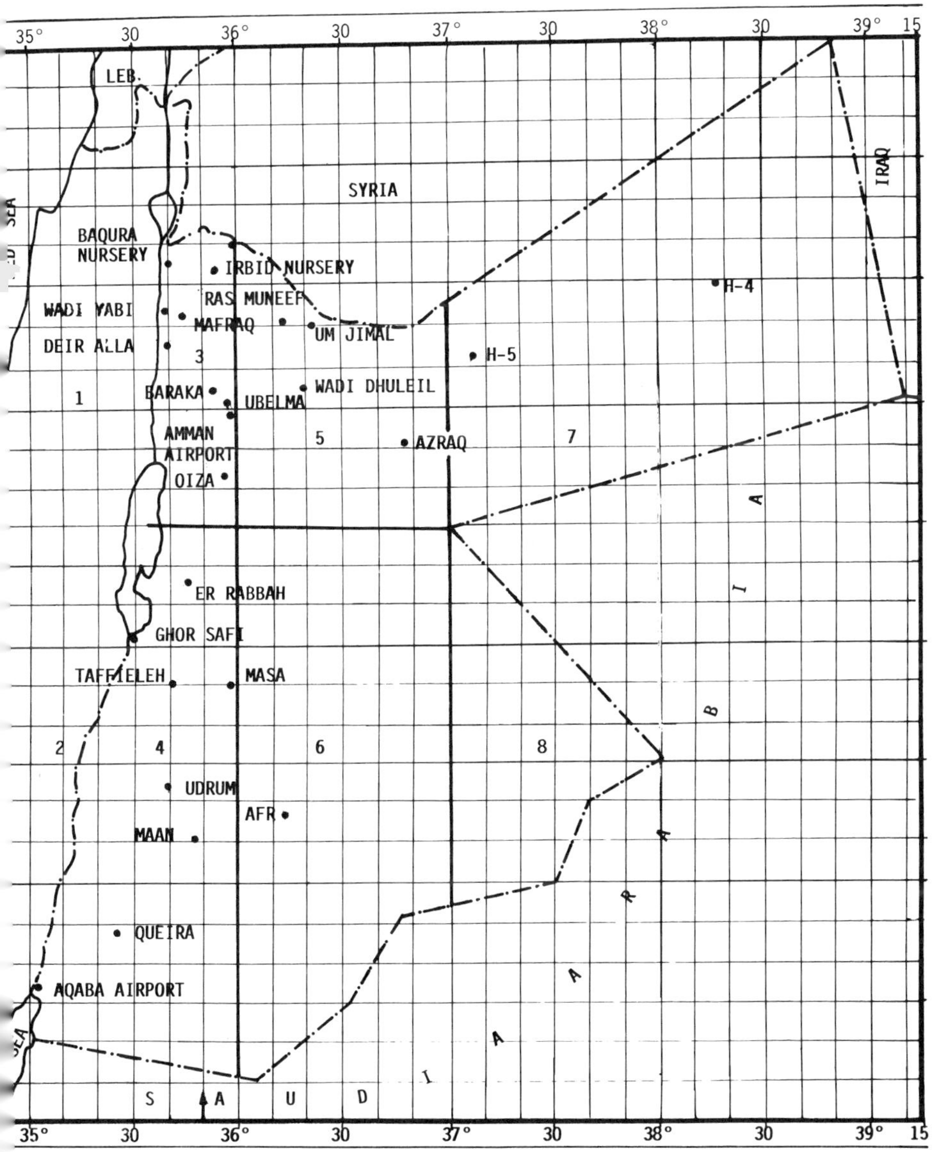

FIGURE 2: Jordan geographical regions

growth of demand, share per capita in Jordan is still 25% of that in developed countries.

The cost of energy is becoming a heavy burden to the Jordanian economy, to the extent that oil imports in 1982 exceeded 110% of exports and equalled nearly 13% of GNP. If the present state of affairs continues, it is expected that in the year 2000 the energy bill may exceed 20% of the GNP.

Potential indigenous energy resources, comprising shale oil, hydro-power, radio-active materials, solar energy, wind energy, biomass, geothermal, oil and gas, are not yet commercially developed. Exploration efforts for oil and gas have been intensified in the past three years. The results are positive and the government is quite keen to see the national oil exploration program implemented.

The demand for electrification of the countryside is increasing and efforts are being made to supply most villages and settlements with electricity by the turn of the decade. Yet, for economical reasons a number of small isolated, remote settlements may have no chance to be connected to the national grid system. Hence alternative sources, mainly solar and wind, can play a constructive role.

First, solar energy, and second, wind energy, can be considered to be technically and economically feasible alternative energy resources. Solar radiation in Jordan is quite high and sunshine days exceed 300 in most parts of the country.

Solar energy applications in Jordan cover a very wide spectrum: i.e. water heating, space heating, water pumping, electricity generation, greenhouses, crops drying, radio-telephone system and solar ponding for electricity generation and for potash recovery are possible.

b. It was concluded from the investigation of Jordan's basic energy needs (as of 1981) that:

The majority of Jordan's population (55%) are in or near Amman where services are available and even plentiful. The remaining 45% occupy four regions that constitute approximately 80% of Jordan's land. This indicates that migration towards major cities has been directly related to the availability of services and consequent job opportunities.

Many locations in Jordan lack certain services. Jordan has 54 locations without potable water, 755 locations without electricity, 557 locations without telephone, 577 locations with no water network and, finally, 270 locations with no services whatsoever.

Decentralized solar and wind energy systems are very suitable for satisfying the basic energy needs for remote areas in Jordan, and are potential candidates for the utilization of renewable resources.

Jordan has water wells possessing a variety of depths and yields. Looking at the appropriate technologies, it is clear that most of the Jordanian wells can be utilized in one way or another by solar or wind energy resources. However, real advancement in the state of art of those technologies would allow complete replacement of conventional energy with solar and wind energy in the near future.

The total basic energy required that may be supplied through solar and wind energy was estimated to be 2.8 GWh/d for heat energy (16% of total heat energy utilized in Jordan) and 325 MWh/d for electrical energy (4.5% of total installed energy in Jordan). In addition, assuming 25% market penetration, a 58 MWh/d of electrical energy would be needed for replacing conventional energy by solar and/or wind energy in water pumping.

As for the industrial sector, the RSS team felt that the amount of solar and wind energy that could be utilized at this time and which the available technology is marginal compared to other energy-using sectors. However, a comprehensive assessment could be performed through the flat-plate-collectors project since it is believed that such an application tops their priorities.

c. It was concluded from the investigation on availability of solar and wind resources, as shown in Table 1, that:

Jordan has excellent annual mean solar radiation. On a horizontal surface, solar radiation is 5.5-6 KWh/m^2 day, i.e. about 60% of that which falls on one square meter in Jordan.

Most of Jordan's regions, with the exception of the Jordan Valley, have moderate wind theoretical power of 150-250 watts per square meter of swept area. This wind power is well suited to water pumping operations. There is one region in the northern part of Jordan which has a wind theoretical power of 980 w/m^2 -- excellent for power production.

From the statements mentional above, one can conclude that research and development projects in the field of solar and wind energy are of utmost importance to Jordan since Jordan does not have any source of conventional energy.

The data presented in this study can be considered to be nucleus information for executing research and development of solar

TABLE 1. Summary of annual means on wind, solar and ambient temperature

Region No.	W. Speed (m/s)	Power (W/m^2)	Sunshine Duration (h/d)	Global Rad. (KWh/m253 d) (Estimated)	Av. Amb. Temp. (°C)	Mean Min. Temp. in Jan. (°C)	Mean temp. in July (°C)
1	3.0 4.0	40 90	8.5	4.5	23.5	7.2	38.5
2	3.0 7.5	40 400	9	5.5	23.5	10.5	40.1
3	5 10.5	150 950	9	5.5	16.25	1.0	33.0
4	5 6.5	150 250	9.5	6	16.25	1.4	34.0
5	5 5.5	150 250	9.0	6	17.5	1.7	34.7
6	6.0	210	9.5	6.5	17.5	-0.2	33.9
7	6.5	250	9.5	5.5	18.5	1.9	35.0
8	6.5	250	9.5	6	18.5	1.9	35.0

and wind energy projects; at the same time, it could determine the sites that are likely to be the best prospects.

In order to have an accurate assessment of the performance of solar and wind systems, several automatic weather stations that record data on a temporal and spatial basis will be needed. These stations would serve as good sources of information for statistical analysis and correlation among various stations.

d. It was concluded from the investigation of possible applications and priorities for solar and wind energy utilization that:

The most appropriate technology for water heating in all regions of Jordan is the flat plate collector. However, the price for the unit of energy delivered and, subsequently, the potential energy saving varies from one region to another depending on available solar radiation. The unit energy cost (JD/kwh thermal) varies from 0.006 to 0.0085.

For electrical pumping, wind converters were found to be suitable for most regions except Regions 1 and 2 where the solar pond was found to be more suitable. The price per unit of energy delivered (JD/kwh elect.) from wind converters varies from 0.014 to 0.087 depending on the availability of wind resources while that from solar ponds varies from 0.335 to 0.41.

For electrical power generation, wind converters were found to be suitable for most regions in Jordan except in Regions 1 and 2 where the solar pond was found more suitable. The price per unit of energy delivered (JD/kwh elect.) from wind converters varies from 0.009 to 0.057, while that from solar ponds for electrical power generation varies from 0.261 to 0.319.

For the remote locations of eight regions, the photovoltaic systems were the most appropriate technology to deliver the required power for the operation of emergency telephones, educational television, and refrigeration. The price for such applications differs from one region to another depending on the solar energy intensity.

Solar stills were found to be an appropriate technology for water desalination in Jordan.

Table 2 presents on overall view of the theoretical application of renewable energy as compared to conventional energy application. The total savings over the life of the equipment (15 years) show a favorable saving for all applications, if photovoltaic and solar pond technologies are excluded.

CONCLUSION

To conclude, priorities were determined as a result of this investigation. It is recommended that the following be used:

a. Flat plate collectors for domestic hot water applications with an anticipated energy savings contribution of 2,108,067 kwh/day.

b. Wind energy converting machines for electrical pumping with an anticipated energy savings contribution of 55,533 kwh/day.

c. Wind energy converting machines for power generation with an anticipated energy saving contribution of 407,139 kwh/day.

d. Solar stills for desalination with an anticipated energy savings contribution of

TABLE 2. Solar or wind energy technology suitable for Jordan, and the potential energy savings

Cost Items / Application	Potential Energy requirements KWh/day Elect.	Potential Energy requirements KWh/day Thermal	Total cost of renewable energy JD/day (a) Elect.	Total cost of renewable energy JD/day (a) Thermal	Total cost of conv. energy JD/day, (b) Elect.	Total cost of conv. energy JD/day, (b) Thermal	Saving JD/day (b)-(a)	Total saving JD/lifetime of 15 years
Domestic hot water - Flat plate Collector		2108067		14359		46377	+32018	+175298550
Electrical pumping	59645		5126		8171		+ 3045	+ 16671375
- Solar pond	4112		1567					
- Windmill	55533		3559					
Power generation	447822		31622		36274		+ 4652	+ 25469700
- Solar pond	40683		12656					
- Windmill	407139		18966					
Emergency Tel. - Photovoltaic	28		6.8		2.3		- 4.5	- 24638
Desalination - Solar stills		923327		5414		9233	+ 3819	+ 20909025
Educational TV - Photovoltaic	389		106		31.5		- 74.5	- 407887
Refrigeration - Photovoltaic	434		103		35		- 68	- 372300
Total Cost	508318	3031394	36964	19773	44514	55610	43387	237543825

923,327 kwh/day.

e. Photovoltaic for power generation with an anticipated energy savings contribution of 40,683 kwh/day.

f. Photovoltaic systems for emergency telephones, educational television and refrigeration with an anticipated energy savings contribution of 851 kwh/day.

AVAILABILITY OF WIND ENERGY IN A TYPICAL DESERT LOCATION IN EGYPT

AMIN MOBARAK
Department of Mechanical Engineering,
Cairo University

HEMMAT SAFWAT AND AMR ABDEL HAMID
Department of Engineering American University in Cairo

Abstract Wind speed at a typical Egyptian desert site was collected and analyzed for the period of one year extending from April 1985 to 1986. Measurements were accomplished using a three component anemometer. Data were recorded in a digital form using a data logger-computer system with optimum choice of scanning intervals and data storage size. Accurate estimates of measurements were assured, within practical constraints imposed by equipment capabilities. Measurements were analyzed on a weekly basis to yield wind speed and power density changes with time during the recording period. These summaries were, in turn, utilized to develop the wind-speed, wind power duration and diurnal curves. The paper presents these curves for various seasons of the year and for the whole year when applicable. The average wind speed at the site was found to be 4 m/s and this value is in conformance with a recently prepared wind speed map for Egypt. The wind speed in the winter season is relatively small with larger values exhibited in the spring. Daily peak wind speeds occur in the late afternoon and hence are not in phase with peak solar insolation occurring at noon time. The curves presented in this paper give the necessary wind information required for the selection and design of a hybrid wind/solar utilization system.

INTRODUCTION

The majority of Egypt's population (54 million) live in the Nile Valley and its Delta, in an area that amounts to less than 4% of the country's land area. Thus, 96% of the area is primarily desert. There is great pressure to reclaim the desert to meet the needs of a population which is increasing at a rate close to one million a year. In 1985, the American University in Cairo "AUC", established the Desert Development Center (DDC) which undertakes research and development programs to investigate how desert reclamation can best be achieved[1]. Because of the importance of adequate energy sources as part of the integrated development necessary to support new communities in the desert, and in view of the ever diminishing conventional (fossil/hydro) energy sources, the DDC has concentrated on investigating renewable energy sources. The harnessing of both solar and wind energy is being investigated.

In order to provide adequate energy to support a certain community, one must know the expected needs (quantities and use patterns). Also when considering utilization of renewable energy sources (solar or wind), reliable data is needed for these sources. For wind data, meteorological data have been collected in numerous stations in Egypt since the early 1930s. However, this type of data falls short of satisfying the needs of the wind utilization system (WUS) engineer for two reasons:

a) The data are not detailed enough.
b) The data may not be applicable for the site under consideration.

Thus, as a part of evaluation of a (WUS) at the Sadat City site (the location of the AUC's DDC), it was necessary to initiate an activity to collect and analyze wind availability at this site.

This paper deals with data recording and data processing in forms useful for the design of the WUS. After presenting these methods, the reduced data for a period of one year are included under the results section. Then, conclusions are drawn. It should be noted that the results serve as input to a separate and an accompanying paper found in the proceedings of this conference under the title "A Wind - Photovoltaic Hybrid System for Pumping Water in a Desert Environment".

Before proceeding with the sections which deal with the methodology of data gathering, it is appropriate to define the data sought and the various formats needed for subsequent usage.

The power available in the wind per unit frontal area of a rotor "$\wp$" (the power density) may be deduced from the wind speed according to the following equation:

$$\wp = \frac{\rho}{2} V^3 \quad W/m^2 \qquad (1)$$

V the wind speed m/s

ρ the density of the flowing air kg/m^3

Equation (1) can be easily derived from dimensionless analysis,[2]. Equation (1) indicates a strong dependence of power density on wind speed (the power three). To obtain the "energy available per unit area" over a period of one year, one needs to integrate the wind energy per unit area (energy density) available during infinitesimal time intervals. The energy density per infinitesimal time interval can be calculated from the instantaneous value of the power obtained from Equation (1).

The energy that reaches a wind turbine "E" over a period of time t, is given by:

$$E = \int_{o}^{t} \frac{\rho}{2} V^3 A \, dt \qquad (2)$$

Where A is the frontal area of the wind turbine

In practical terms, if the wind velocity is an average value over a period of Δt, Equation (2) can be rewritten using a summation notation as follows:

$$E = \sum_{n=1}^{N} \frac{\rho_n}{2} V_n^3 \, A \Delta t \quad (3)$$

Where t corresponds to $N\Delta t$. Equation (3), can be simplified if Δt is chosen to be constant, and an average value for the air density $\rho_{avg.}$ is assumed , reducing to:

$$E = \frac{\rho_{avg}}{2} A \Delta t \sum_{n=1}^{N} V_n^3 \quad (4)$$

It is important to note that, to obtain an accurate estimate of the available energy, one needs to consider small time intervals for Δt. This point is demonstrated in detail in [3].

To aid the designer of the WUS, the data should be presented in the following forms:

a) Wind speed frequency curve.

b) Wind duration curve.

c) Diurnal curve.

The first form represents a histogram form, specifically the relative frequency for various wind speeds. The second form gives the cumulative relative frequency of the data. This latter curve is particularly useful in estimating the energy available above various wind speeds. The diurnal form gives wind speed variation with the time of the day. This form is especially useful when considering storage requirements as well as hybrid (wind/solar) systems.

DATA ACQUISITION AND PROCESSING

From the previous section, it can be seen that the evaluation of wind energy data at a given site would ideally be reached using continuous measurements of wind speed (notice wind directionality).

Continuous measurements can be obtained using analog instruments. However, this results in large volumes of data that become quite difficult to analyze and interpret. The alternative is to monitor the required parameters periodically. The shorter the time between successive data collections, the more representative the data drawn from them. Typically, measurements are collected hourly but sometimes every three hours.

With recent advances in digital circuitry and computer technology, a new generation of data collection instruments has emerged. The "data loggers" can periodically sample a large number of input channels and store the information in memory for later analysis. The time span can now be shortened to seconds or minutes, thus approaching the continuous monitoring mode, and the large volume of the stored information now in digital form can be easily manipulated using computers.

Implementation of a computer-controlled data-logging procedure is dependent on a number of steps. All physical quantities to be measured must be sensed or converted to an electrical signal (voltage or current) to be sampled by the data logger. Transducer sensitivities must be accurately known and preferably be linear. Signals from the various transducers must be transferred to the location of the data logger without loss or attenuation. The data logger should be properly programmed to sample and manipulate the data for further storage in an identifiable form.

At the Sadat City site of the DDC, a microcomputer-based data logger has been operational since October 1984. The system,

Hewlett-Pckard 3054 DL, has been programmed to collect and process data without operator intervention for an uninterrupted period of one week.

To avoid the unnecessary storage of all raw data, initial analysis of the data and temporary storage of the results in the active computer memory are carried out in real time during the data sampling cycle. For the number of channels sampled and the amount of initial analysis carried out, a sampling cycle of 60 seconds was found to be adequate. During the sampling cycle, screen displays of the data, before and after this initial analysis, can be visually examined without interrupting the continuity of the data logging operation. Moreover, the system can be used to implement control strategies for different purposes by setting limit values for the directly measured or reduced quantities.

Every 60 minutes the information stored in the active memory of the computer is transferred to a file on a magnetic tape cartridge for permanent storage and further off-line analysis. At the end of one week, the tape cartridge is replaced and the stored data is extensively analysed on a compatible computer at the AUC. The results of this latter analysis are summarized in the form of hard copy plots and tables.

Reference [1], gives further details on the data acquisition/ processing systems used.

WIND RECORDS

The system described in the previous section constitutes a large system used for recording numerous measurements: solar, wind, operational, etc., that are monitored continuously at the DDC site in Sadat City. This paper is concerned with the availability of wind at this site.

An anemometer capable of measuring the three components of wind velocity has been installed at the site. This anemometer is the Gill UVW anemometer manufactured by the R.M. Young Company, Michigan, USA. Some of its details are shown in Figure 1. The wind energy sensors (wind turbines) outputs, are converted to DC power. The joints of the anemometer are kept under a pressure higher than atmospheric pressure through the small blower shown in Figure 1. The anemometer is 10 cm above the ground.

Figure 2 shows a sample of the weekly summaries of wind statistics after collection and processing. The plots in this figure show the relative frequency wind speed (with the average and standard deviation), the wind speed time of the day (with the average wind speed), the corresponding power density time of the day (with the integrated energy density for the week), and, the hourly wind speed for the whole week. The table found in Figure 2 states the average day wind speed for the seven days of the week.

RESULTS

In this section, the wind data recorded over the period from April 1985 to March 1986 will be summarized. As indicated in the introduction section, the summaries will be in the form of wind-frequency, wind duration and diurnal curves. Thus, using the collected/processed weekly summaries of the kind shown in Figure 2, these curves were prepared.

References [3 & 4] contain complete sets of these curves. For this paper, some of the results have been chosen and are discussed hereafter.

Figure 3 shows four wind-speed frequency curves for the months of April, July, October and January. These were selected to represent the four seasons. The data incorporate both day and night values (designated by "daily").

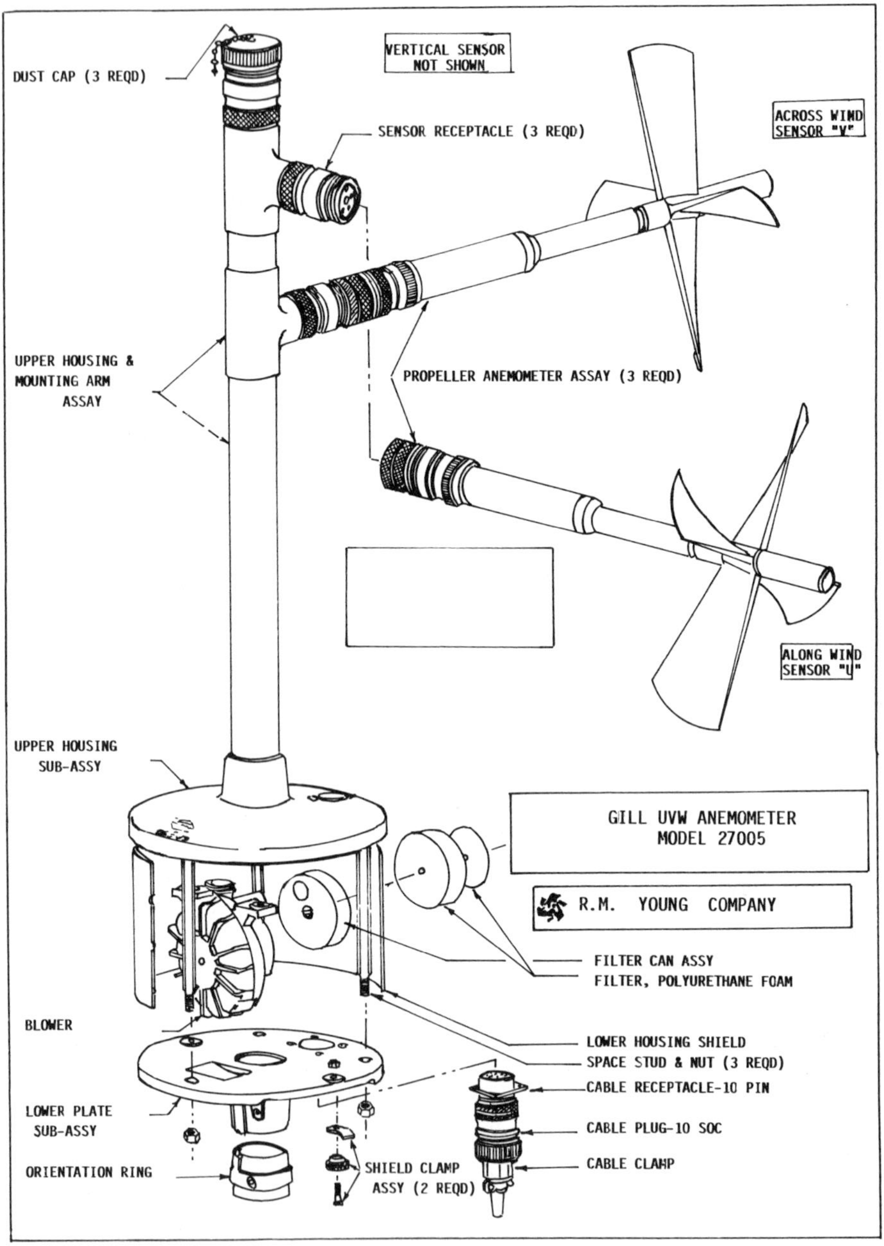

FIGURE 1: Major details of the UVW anemometer

Week: 22/ 9/85 to 29/ 9/85

Wind Statistics

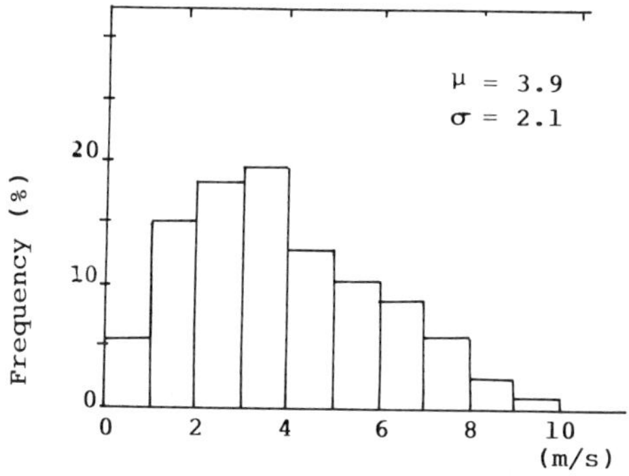

a) Wind speed freq. Total

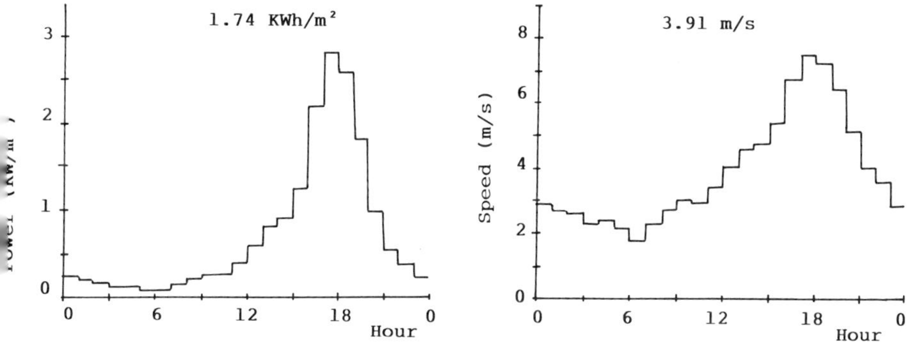

c) Wind Power

b) Wind Speed

Day	Avg. Speed
S	4.55
M	3.65
T	3.45
W	4.9
H	3.46
F	3.47
S	3.94
	3.92

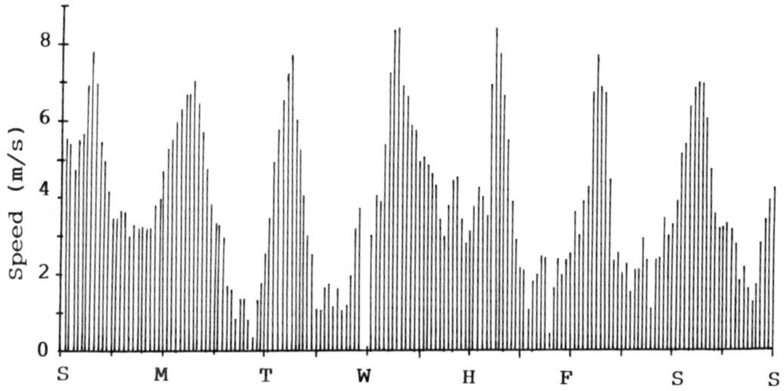

d) Wind Speed

FIGURE 2: Example representing typical summary of a week's wind statistics

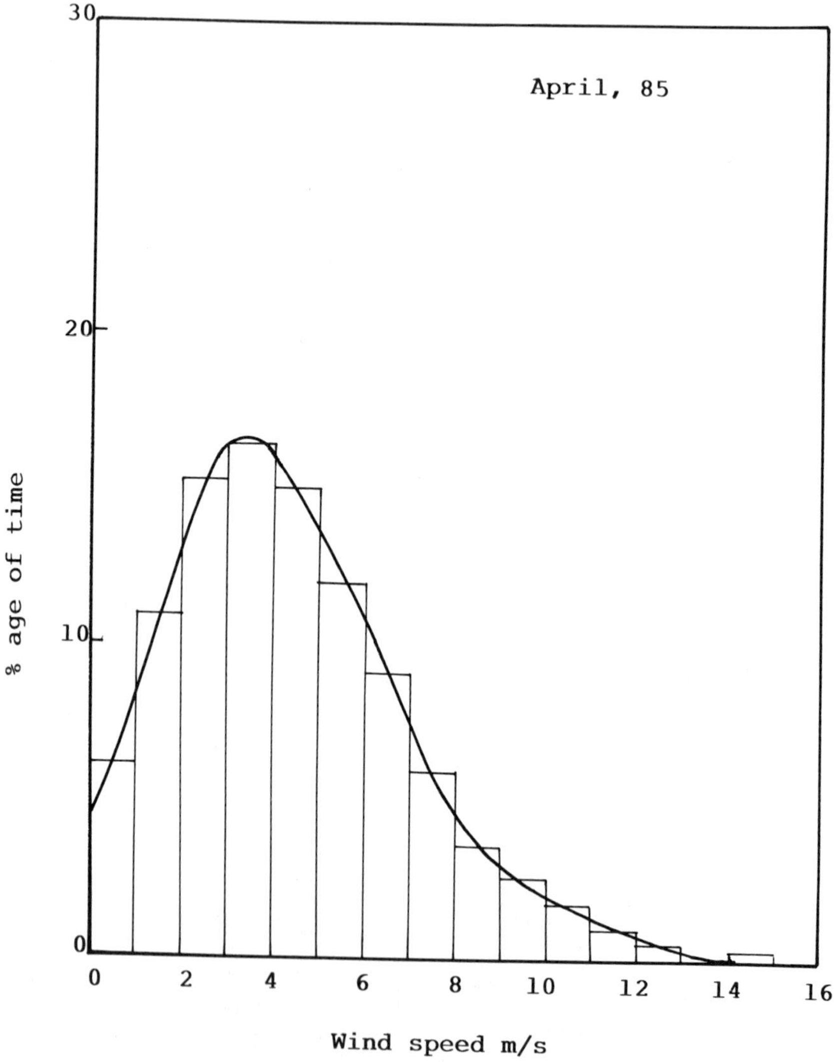

FIGURE 3a: Wind speed frequency curves (daily)

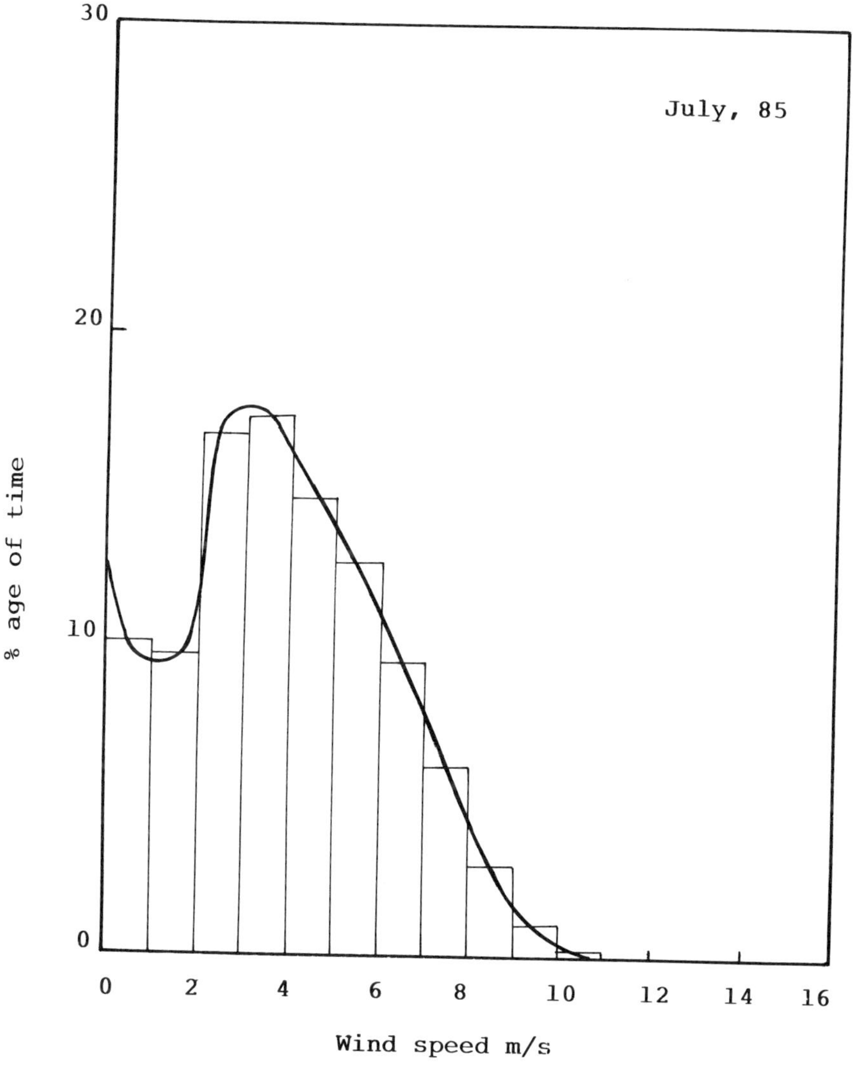

FIGURE 3b.

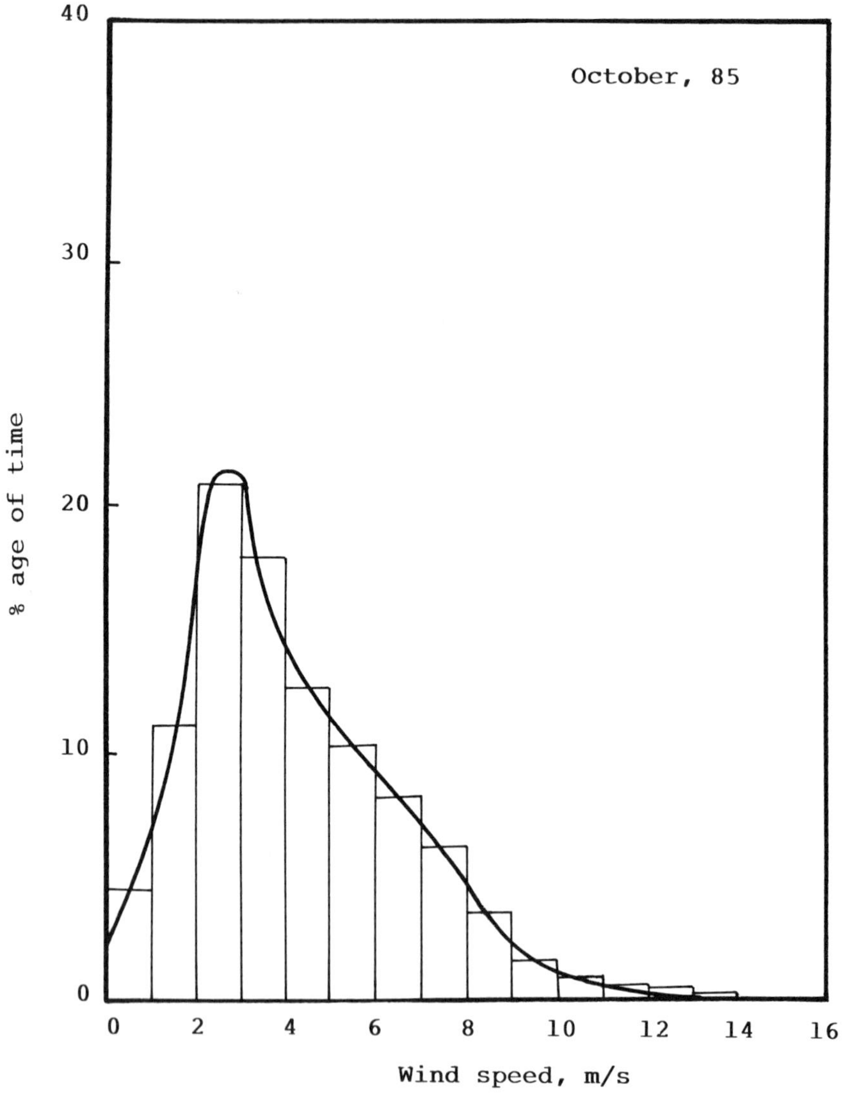
40
30
20
10
0
% age of time
October, 85
0
2
4
6
8
10
12
14
16
Wind speed, m/s

FIGURE 3c.

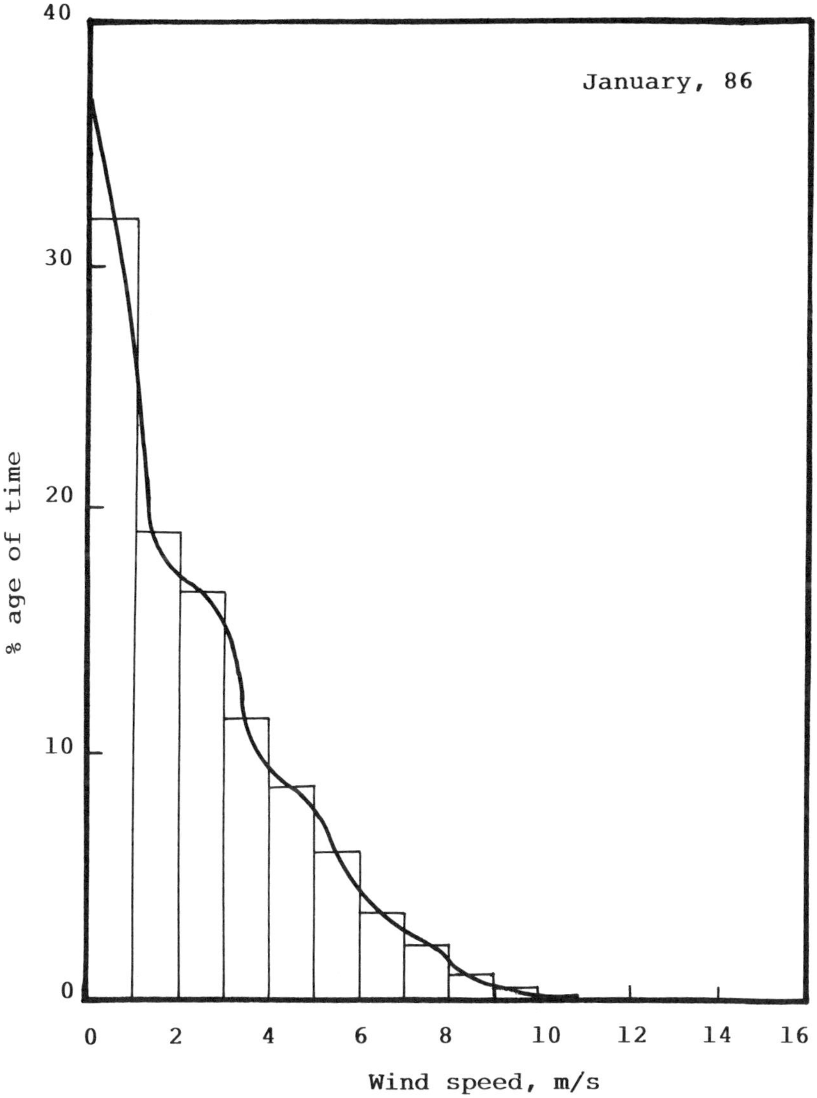
January, 86
% age of time
40
30
20
10
0
0
2
4
6
8
10
12
14
16
Wind speed, m/s

FIGURE 3d.

To aid in comparing the four seasons, the statistical properties of the four curves shown in Figure 3, are included in Table 1. From the plots in Figure 3 and the summary statistics of Table 1, one observes that, on the whole, the average mean speed is about 4 m/s and that velocity drops in January to 2.5 m/s. Higher velocities occur in April as can be detected from Figure 3 and Table 1.

TABLE 1: Summary of seasonal statistics (based on wind frequency data of Figure 4)

MONTH	MEAN m/s	MEDIAN m/s	STANDARD DEVIATION m/s
April	4.47	3.5	5.19
July	4.00	3.2	4.59
October	4.24	3.3	4.9
January	2.5	1.5	3.25

The wind speed frequency curve for the whole year (April 1985 - March 1986) is shown in Figure 4. The power duration curve for the same one-year-period is shown in Figure 5. The latter curve is of prime importance in terms of selection of size of the WUS. A comment is warranted with regards to the hump shown at the wind speed at 20 m/s in Figure 4. This wind speed appears to be excessive. Examination of the recorded data does not indicate malfunctioning of the instrumentation, however. Attention will be given to recordings during March 1987, and future years if necessary, to establish confidence in this hump. It is due to this hump, that the power duration curve has a substantially large value near the zero percent.

The diurnal curve for a certain month gives the average wind speed for the different hours of the day averaged over the whole month. Such presentation is important in the consideration of the hybrid (solar/wind) utilization systems. From Figure 6, it can be seen that the maximum wind speed usually occurs in the afternoon (16-18 hrs.). This pattern is typical for desert areas. In

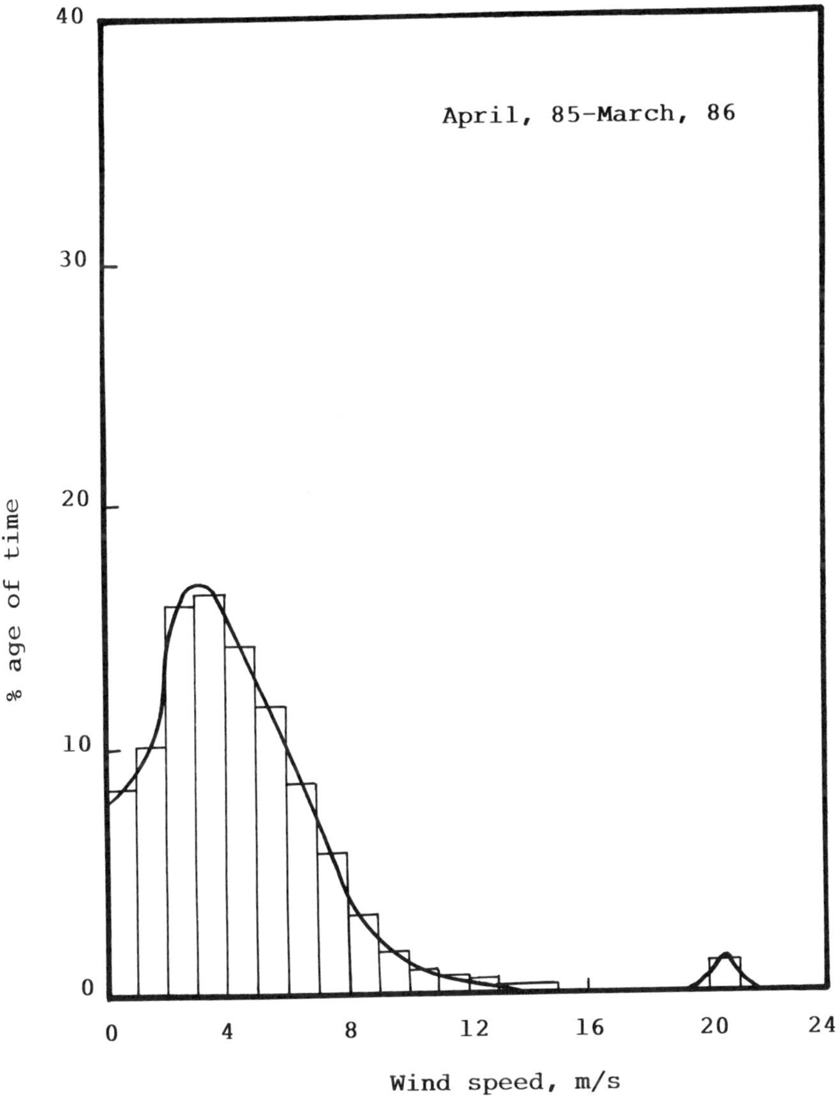

FIGURE 4: Wind speed frequency curve for one year

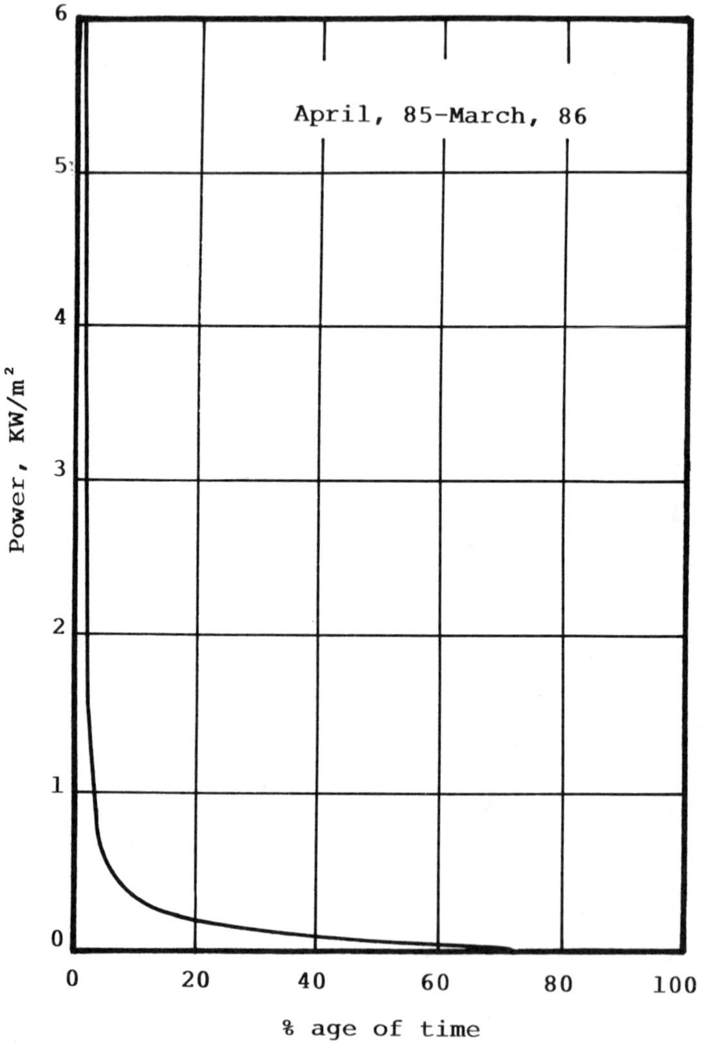

FIGURE 5: Wind power duration curve

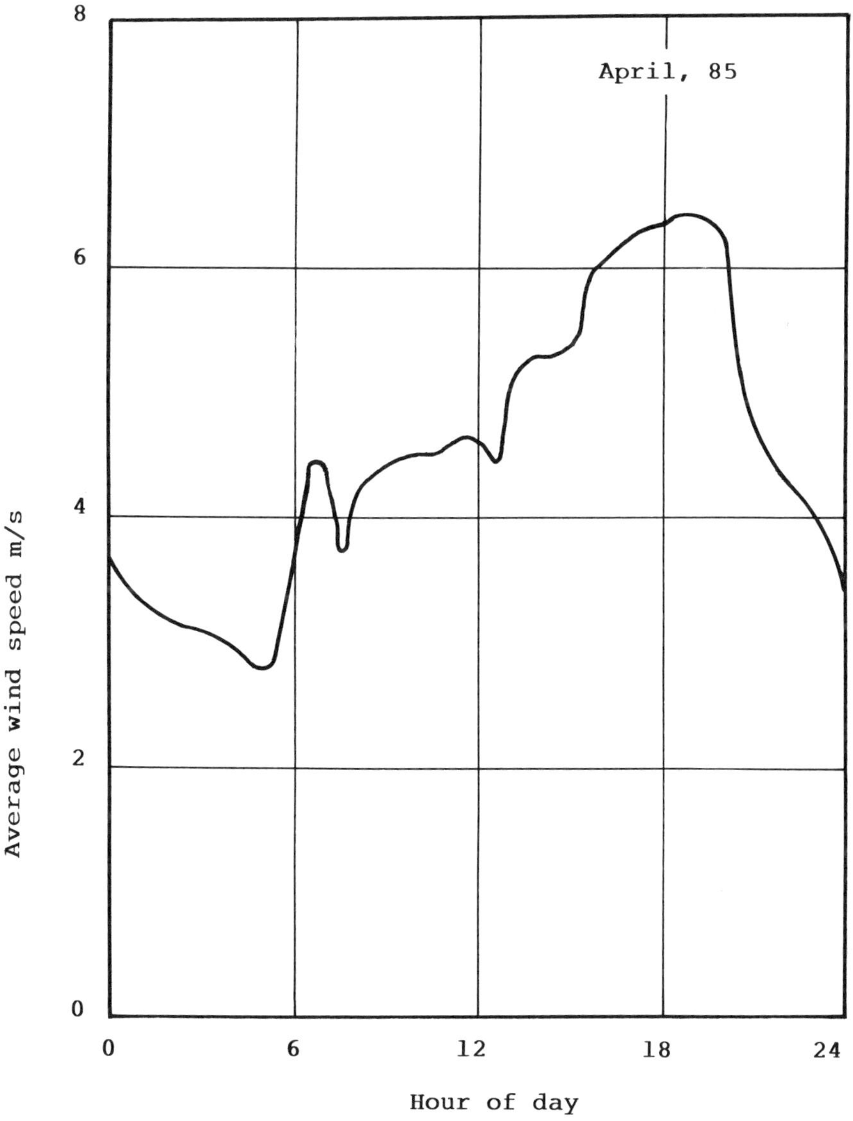

FIGURE 6a: Diurnal curves for various months

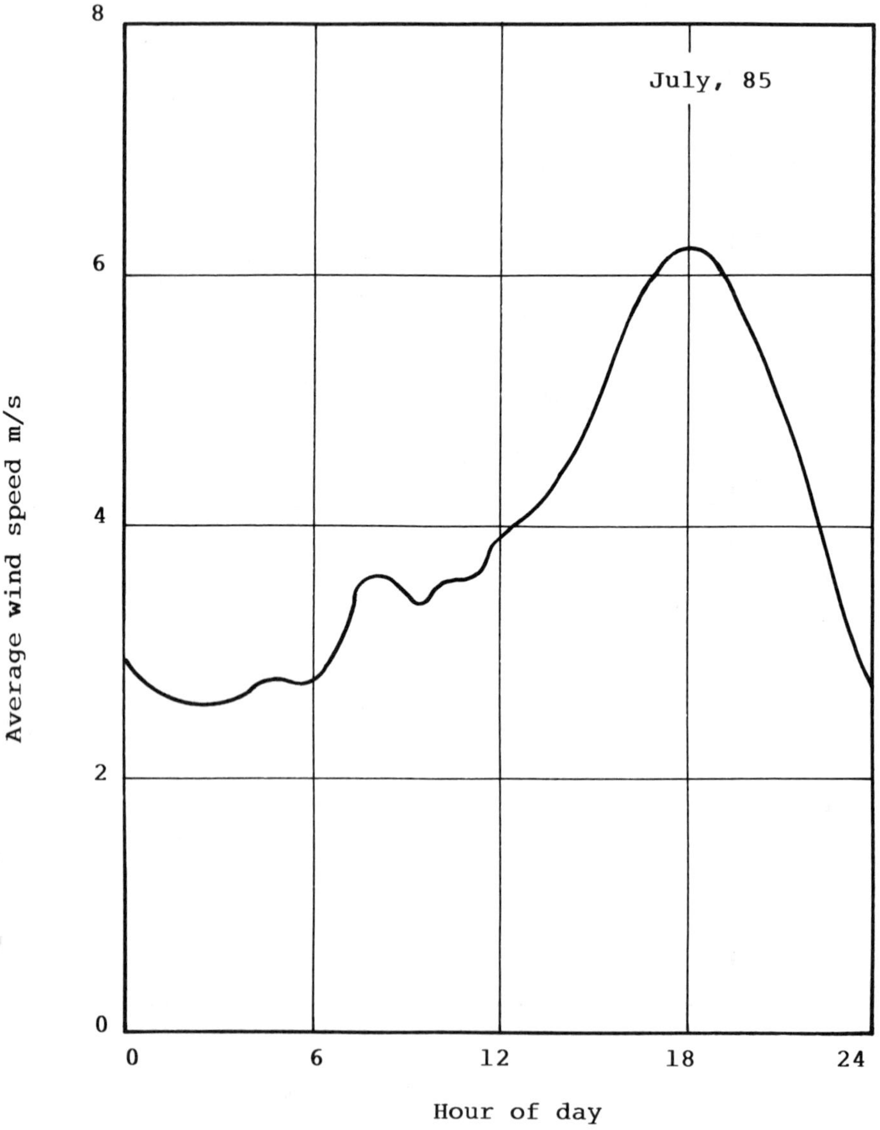

FIGURE 6b.

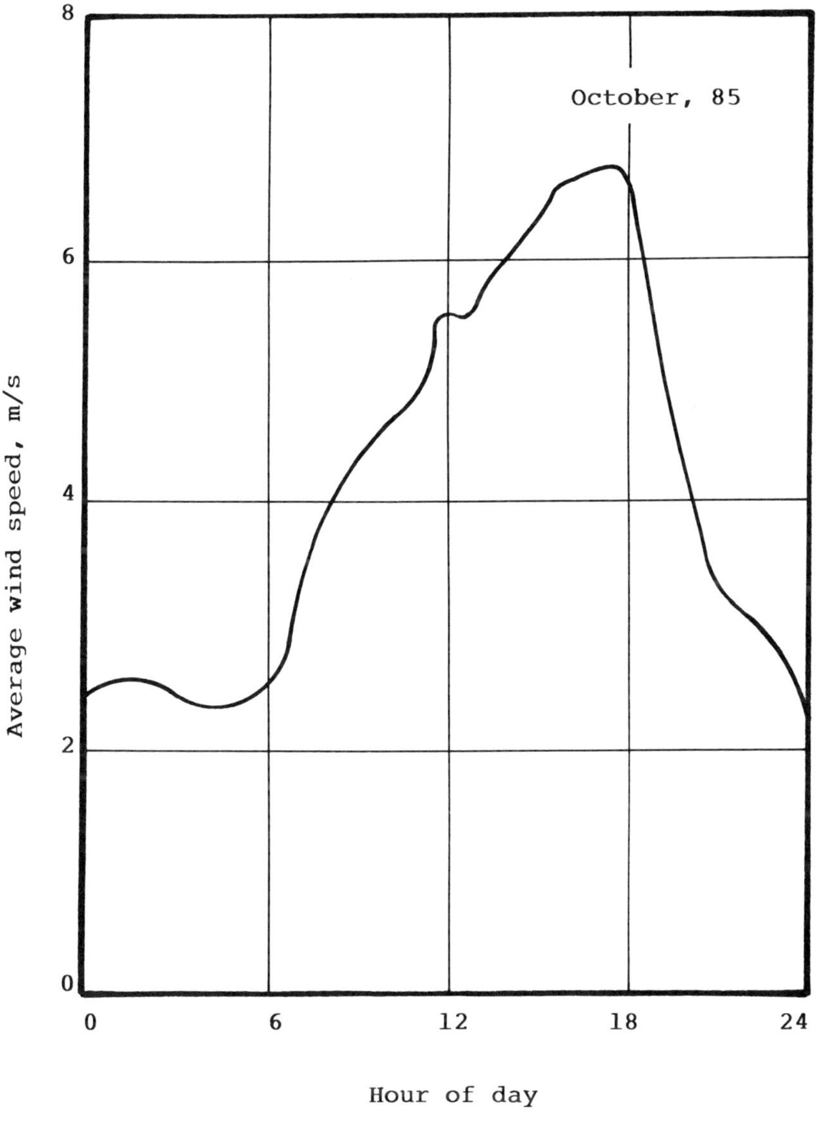

FIGURE 6c.

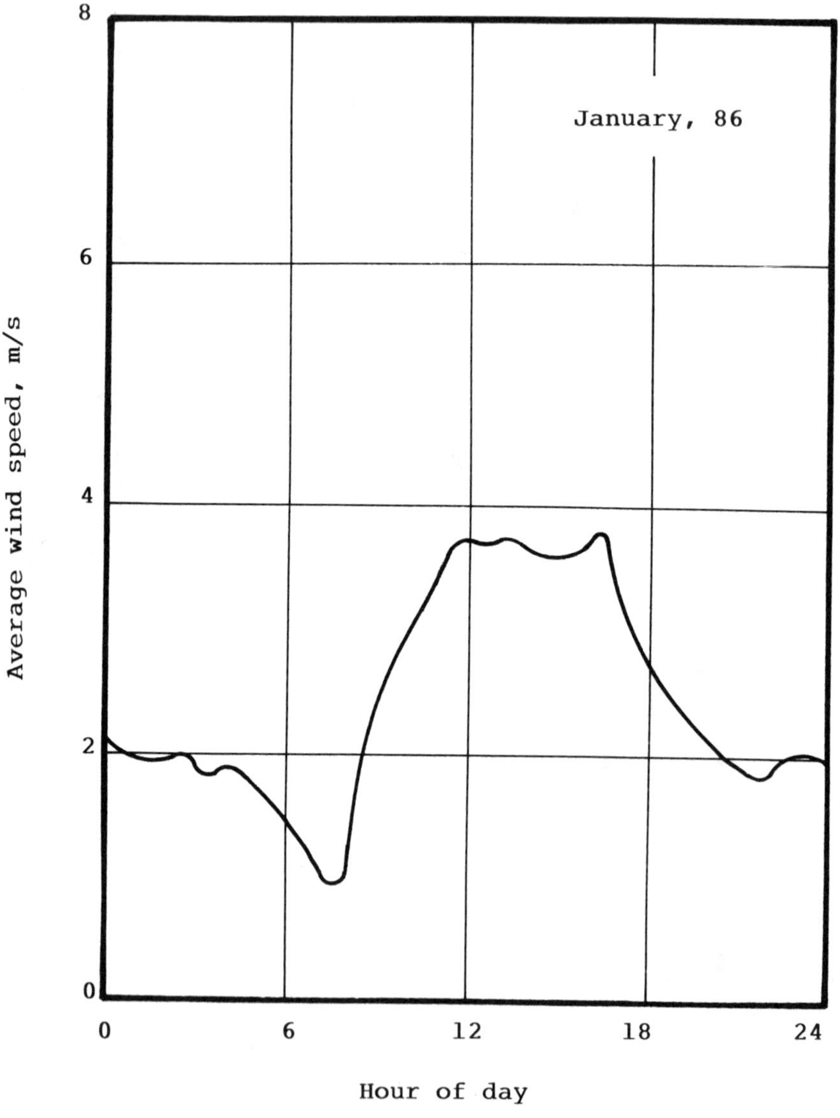

FIGURE 6d.

comparison, the maximum power of solar energy occurs at noon (12 hrs.). Thus, there is a period of 2 to 4 hours between the two peaks. Night time is marked with the minimum wind velocities. These observations regarding available renewable energy sources (for a hybrid system) must be viewed along with the required pattern of power usage to determine storage requirements.

It is appropriate to compare the results reported herein to the wind map for Egypt, recently completed and reported in reference [5]. The location of the DDC site is shown in Figure 7. From the data of that Figure, it appears that the average wind speed for zone 2, 4.4 - 5.1 m/s, is in close agreement with the average wind speed at Sadat City.

CONCLUSIONS

In this paper, the wind speed data at a typical location in Egypt's desert has been examined in detail. The average wind speed at the site was found to be in close agreement with the wind speed map for Egypt [5]. This average wind speed is of the order of 4 m/s.

The data have been presented in useful forms for utilization in selection and sizing of a wind/solar utilization system.

While the data do not cover many years, they cover a sufficient period to enable the sizing selection of a hybrid system. The recording of data continues and this will supplement the reported data and provide more information that is statistically reliable.

Egypt-Annual Average Wind Power

	10m	
Power Class	Wind Power (W/m^2)	Speed (m/s)
1	<100	<4.4
2	100-150	4.4-5.1
3	150-200	5.1-5.6
4	200-250	5.6-6.0
5	250-300	6.0-6.4
6	300-400	6.4-7.0
7	400-1000	7.0-9.4

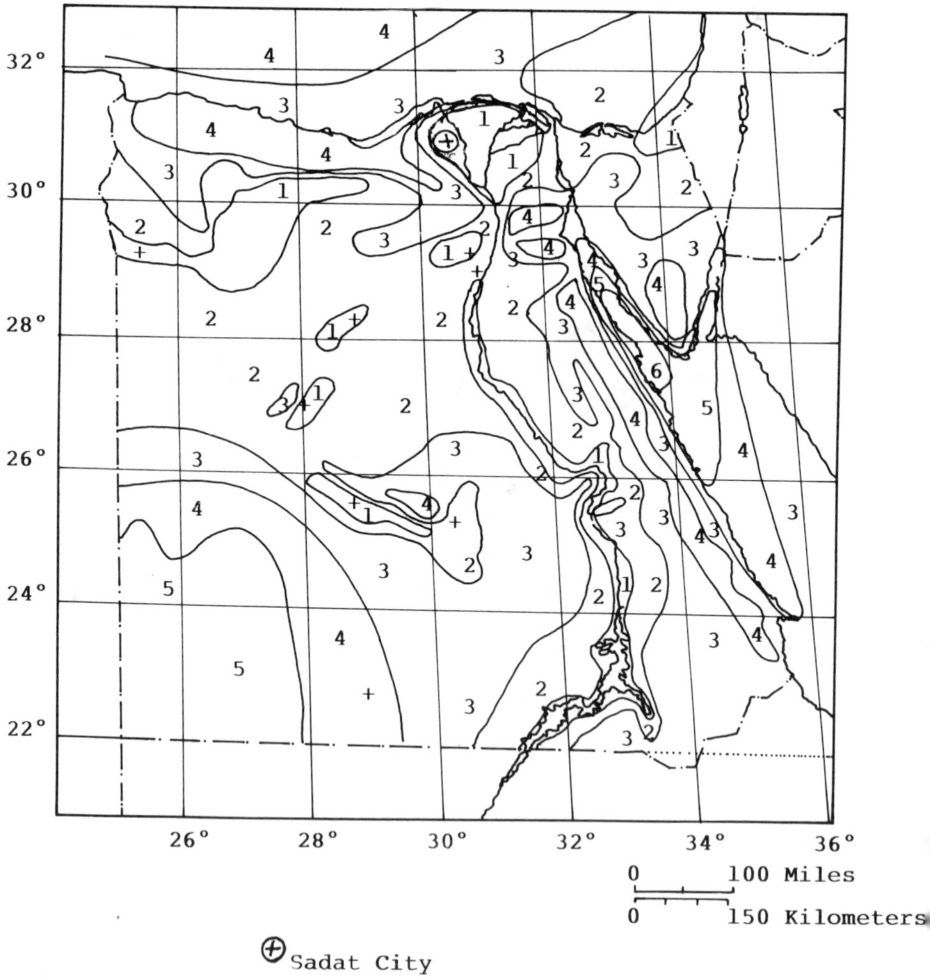

FIGURE 7: Location of AUC's DDC site

ACKNOWLEDGMENT:

The work presented in this paper was carried out under the sponsorship of the U.S. AID through the FRCU of the Egyptian Supreme Council of Universities (MAXI Project No. 842083.)

The authors wish to acknowledge efforts of other staff members in the AUC's Desert Development Center. The work done by Mr. Danyal Shoukry is particularly noted.

REFERENCES:

1) DDC 1985 Report.

2) Streeter, V.L., Wylie, E.B. Fluid Mechanics, McGraw Hill, 1981.

3) Desert Development Center and Alexandria University, "Applied Research, Demonstration and Training in Desert Development", Fourth Progress Report, June 1st-August 31st, 1985, submitted to FRCU, Egyptian Supreme Council of Universities.

4) Desert Development Center and Alexandria University, "Applied Research, Demonstration and Training in Desert Development", Eighth Progress Report, June 1st-August 31st, 1986, submitted to FRCU, Egyptian Supreme Council of Universities.

5) Mobarak, A., El-Mallah, A., Serag El-Din, M.A., "Wind Energy in Egypt", Mech. Power Dept., Cairo University, 1982.

NOMENCLATURE:

A	=	Frontal area of wind turbine	m^2
E	=	Energy reaching a wind turbine of frontal area "A" over a period "t"	KWh
P	=	Power density	KW/m^2
t	=	Time	Second
V	=	Wind speed	m/s
U,V,W	=	Three components of wind velocity	m/s
ρ	=	Mass density of air	Kg/m^3
t	=	Time increment	Second

Suffix

avg	=	average
n	=	Increment number "n"

SIMULATION OF THE PERFORMANCE OF A SMALL SIZE WIND TURBINE SYSTEM FOR HOUSEHOLD LOADS

M.A. SHALABY
Associate Professor, Dept. of Engineering, American University in Cairo and Cairo University, Cairo, Egypt

A.A. EL-MALLAH AND M. A. BADR
Mechanical Engineering Laboratory, National Research Institute, Cairo, Egypt

Abstract A small wind turbine system that serves a one-family household load is presented and simulated. The assumed system consists of a 1.5 KW turbine, battery based energy storage, and a domestic power load. Wind data and a power load curve are used as input to the simulation model, while performance measures such as percent of power shortage and average energy store level are used as output to the model. Wind data are collected and analysed for the "Borg El-Arab" site on the north coast. Statistical distributions are fitted, and they are found to follow Weibull and Gamma distributions for the various months. Based on a small sample, power demanded to support a small family is estimated as a stochastic variable that varies by the hour within a given day. The turbine is connected to a battery system to supply power at times of insufficient wind speeds. The average battery level is used as one of the performance measures to indicate how long the system can support the load during calm winds. Using a microcomputer, the simulation model is run for a period of 10 years during which hourly wind speeds and power demanded are randomly generated. Performance measures are reported and statistically analysed. It is concluded that a turbine of about 1.5 KW rating is required to support a small family at "Borg-El-Arab" with no more than an expected power shortage of 10%. The constructed simulation model may be used to analyse performance for other wind turbine ratings, wind, and load distributions, and at other sites as well.

INTRODUCTION

This paper is directed toward constructing and applying a simulation model to analyse the performance of a given wind turbine system. The system is to be located in Borg-El-Arab (North Coast of Egypt) to serve household loads for small families. The model estimates user-oriented measures as percent shortage, average storage level, and average output power. The system consists of a wind turbine, an electricity load, and an energy storage battery unit. The storage unit is essential since no back-up energy unit is assumed, and the hourly load pattern is different from the hourly power output pattern.

The literature on wind turbine performance is rich and diversified. Research is directed to evaluate site potential,[1,2] economics of wind turbine systems,[3] analysis of power output,[4] and reliability and failure analysis,[5] along with a variety of mechanical design problems. Of particular importance is the study of an integrated system of wind input, turbine configuration, and load requirements.[6] Such studies rely on algebraic analysis. They are confined to restrictive assumptions regarding wind speed and load distributions, and the types of control used in the realistic systems.

Simulation is a practical approach that can handle a large variety of wind speed and load patterns, and almost all types of controls used in reality. Haack[7] investigated a one-year simulation model that was updated using 3 hour increments. Galanis et al.[8] proposed a model to evaluate a wind-powered water heater. The model used deterministic wind-speed data tables. A system that supports air conditioners, water heaters, and other electric loads of an isolated user was simulated by Bandopadayay.[9] The study focused on the economic feasibility of this system. Cassale et al.[10] in a simulation model for an isolated user, used the number of shortage hours per year as a measure of performance. These models have the potential of providing practical and statistically analysed measures of performance.

In this study, a simulation model that uses raw wind data is presented. Wind speed data are automatically classified and fitted to either Weibull or Gamma distributions. Load is estimated through sampling for a small family. Wind-power transformations are based on a second order polynomial in the partial power range. Turbine rated power and maximum battery storage capacity are determined by an experimental simulation procedure that relies on user-defined criteria. Experimentation with the model is based on ten one-year simulation replications that provide confidence intervals for the selected measures of performance.

WIND SPEED ANALYSIS

Wind data were collected by the Egyptian Electricity Authority (EEA) for Borg-El-Arab in 1978. Daily speeds were recorded every two hours over a continuous period of one year.[11] Other wind data are available from the Egyptian Metrological Authority (EMA)[12] for some 69 sites, and from the Pacific Northwest Lab, USA,[13] for six selected coastal sites in Egypt.

Several distributions are applied to describe wind speeds at a site. Among these distributions are the Normal, Gamma,[14] Weibull,[15,16,17,18,19,20] and Rayleigh[2] distributions. Weibull is commonly used since it easily lends itself to analytical derivations. The Gamma can assume a variety of shapes including the normal, uniform, and skewed distributions. Both Weibull and Gamma are applied here with the following density functions:

$$\text{Gamma:} \quad f(v) = \left(\frac{v}{b}\right)^{c-1} [\exp(-v/b)]/b\,\Gamma(c), \quad v>0$$

$$\text{Weibull:} \quad f(v) = \left(\frac{c}{b}\right) v^{c-1} \exp\left(\frac{1}{b} v^{c}\right), \quad v>0,$$

where b is a scale parameter, and c is a shape parameter. Several methods are reported to estimate b and c parameters. Methods of percentiles and linear regression are used here for Weibull, and the method of matching moments is used for Gamma as detailed in [2]. Thus, for each month, 4 distributions (two Weibull and two Gamma) were fitted to the wind data using the different methods of parameter estimation. The Kolomogrov-Smirnov goodness-of-fit test[21] was used for each month. Table 1 shows the best fit distribution for each month for Borg-El-Arab data. A computer program that performs the classification, parameter estimation, fitting and testing wind data is in reference 11. Random variate generation from these distributions is found in reference[22].

WIND TURBINE POWER SYSTEM

The selected system is of the stand-alone type. The power generated is rectified to direct current, and stored in a battery storage unit. The required load is drawn from the storage through an inverter. The turbine is an electrical horizontal axis type which is widely produced now with different sizes. The main characteristics of a turbine are its rated power P_r, cut-in speed v_i, rated speed v_r, and cutout speed v_o. Batteries are chosen as the energy storage unit for this system. They are simple to maintain, operate, and are locally available. Although they have a low energy to volume ratio, they enjoy the highest efficiency among other capacitative or inductive storage types.[23] Acid and Nickel-Iron are most economical, nevertheless, Lead-Acid types are more reliable.[24]

A battery's main characteristic is its maximum storage capacity in ampere-hours (AH). The capacity must be a few times larger than the daily load if service continuity is sought.[10] Battery charging efficiency is almost constant (assumed as 75% here), while discharging efficiency depends on the state of the charge. Discharging efficiency can be taken from the manufacturer, or from

TABLE 1. Best fit distributions for Borg-El-Arab data

Month	Jan.	Feb.	Mar.	Apr.	May	June
Best fit*	G(4,.8)	G(12,.45)	G(5,.83)	G(12,.41)	G(21,.25)	G(19.,32)
Month	July	Aug.	Sep.	Oct.	Nov.	Dec.
Best fit*	G(23,.27)	G(19,.36)	G(4,.76)	G(15,.31)	G(5,.75)	G(4,1.15)

* G(...) is a symbol for Gamma (b,c), average wind speed =(b)(c)

capacity-rate curves.[23] The following relations are used here to determine the charging current I_{ch} (from turbine to battery), and the discharging current I_d (from battery to load);

$$I_{ch} = P\, \eta_r \eta_b / V \quad , \quad I_d = D/(V\, \eta_d \eta_I)$$

where, P is the power output from the turbine, V is the battery voltage, D is the load power demanded, and $\eta_r \eta_b$, η_I, η_d are the rectifier, charging, inverter, and discharging efficiency, respectively.

The power output of a wind turbine P is given as a function of the wind speed v m/sec, and other turbine characteristics;

$$P = 0 \qquad v < v_i$$

$$P = p(v) \quad v_i < v < v_r,$$

$$P = P_r \quad v_r < v < v_o,$$

$$P = 0 \qquad v_o < v$$

The power output in the partial power range p(v) can be approximated using either a linear,[25] a four line segment[7], or a second order[15] relationship. The second order approximation is used here where: $p(v) = a + bv + cv^2$, and a, b, c are determined using continuity conditions at v_r and v_i.[2]

POWER LOAD ESTIMATION FOR A FAMILY

Power load curves for domestic uses exhibit common characteristics. In the Netherlands,[6] the daily load pattern for 180 households was measured, while in the USA[26] the average hourly load for a typical resident was estimated. The peak load in both studies shifts with the sunset time. In Egypt, EEA all user load distribution statistics[27] exhibit two peaks due to the inclusion of domestic and industrial users. Other load curves for small villages in Egypt exhibit similar characteristics.[11] Domestic load curves are characterized by a steady low load from midnight to 6 A.M., a first peak between 6 to 9 A.M., and a second peak about one hour after sunset.

Domestic household electrical load is estimated here for a small family living in a residential shelter with four rooms. Appliances used are: TV, fan, radio, refrigerator, washing machine, and lighting fixtures. Relevant power for these appliances are found in references[28,29]. Appliance-usage times are normally obtained through statistical sampling and experimental monitoring.[6,7] Here, a sample of 52 families with an average standard of living was interviewed in Cairo. Information sought included the number of lighted rooms and appliances in use during each hour of a day. Based on this survey, the number of lighted rooms is assumed to follow a normal distribution; TV and radio are assumed to follow a Bernoulli distribution; and the outside light is used constantly from sunset to sunrise. The washing machine is scheduled during 8 to 11 A.M. and the fan is restricted to summer time only, following a Bernoulli distribution. Mean and standard deviation for these appliances are detailed in reference[11]. The year is divided into four seasons with adjusted distributions. Figure 1 presents the averages for summer and winter loads as estimated for Borg-El-Arab.

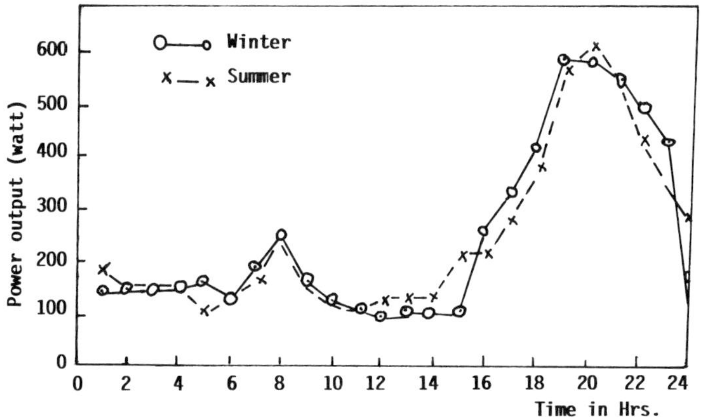

FIGURE 1: Estimated daily load power curve

THE SIMULATION MODEL

A computerized simulation model for the performance of a turbine would provide inexpensive information about its expected real behavior. It is a substitute for real time-experiments. To construct a simulation model, accurate representation of all relationships involved must be created, coded, and validated through a few pilot computer runs. The model presented here is a dynamic stochastic model with a fixed time increment clock that updates the system state every hour.[22] The model input may be divided into 5 items:

1. Daily wind speed observations in monthly groups, and hourly indices. An hourly index for a given hour in a given month is estimated as the percentage of the average hourly speed to the average monthly speed.
2. Load data: appliance rated power and probability of using an appliance during each hour. Four seasonal data groups are used.
3. Wind turbine specifications: rated power, cut-in, rated, and cutout speeds.
4. Battery, storage data: maximum capacity, and discharging capacity-rate curve.[23]
5. Efficiencies of electrical devices such as rectifier, inverter ....etc.

The computer simulation model is based on generating random daily speeds and hourly power loads according to the fitted distributions, and performing the necessary power transformations and battery charge updating, as presented earlier. While the simulation clock is being advanced, state-variables such as power output, shortage, battery level, and excess power are recorded. The model includes six modules: wind analysis and fitting, two random generators for Weibull and Gamma wind speeds, turbine output power calculations, hourly loads random generator, battery charging and discharging, and a statistical analysis module. Figure 2 presents a general flowchart for the computerized model.

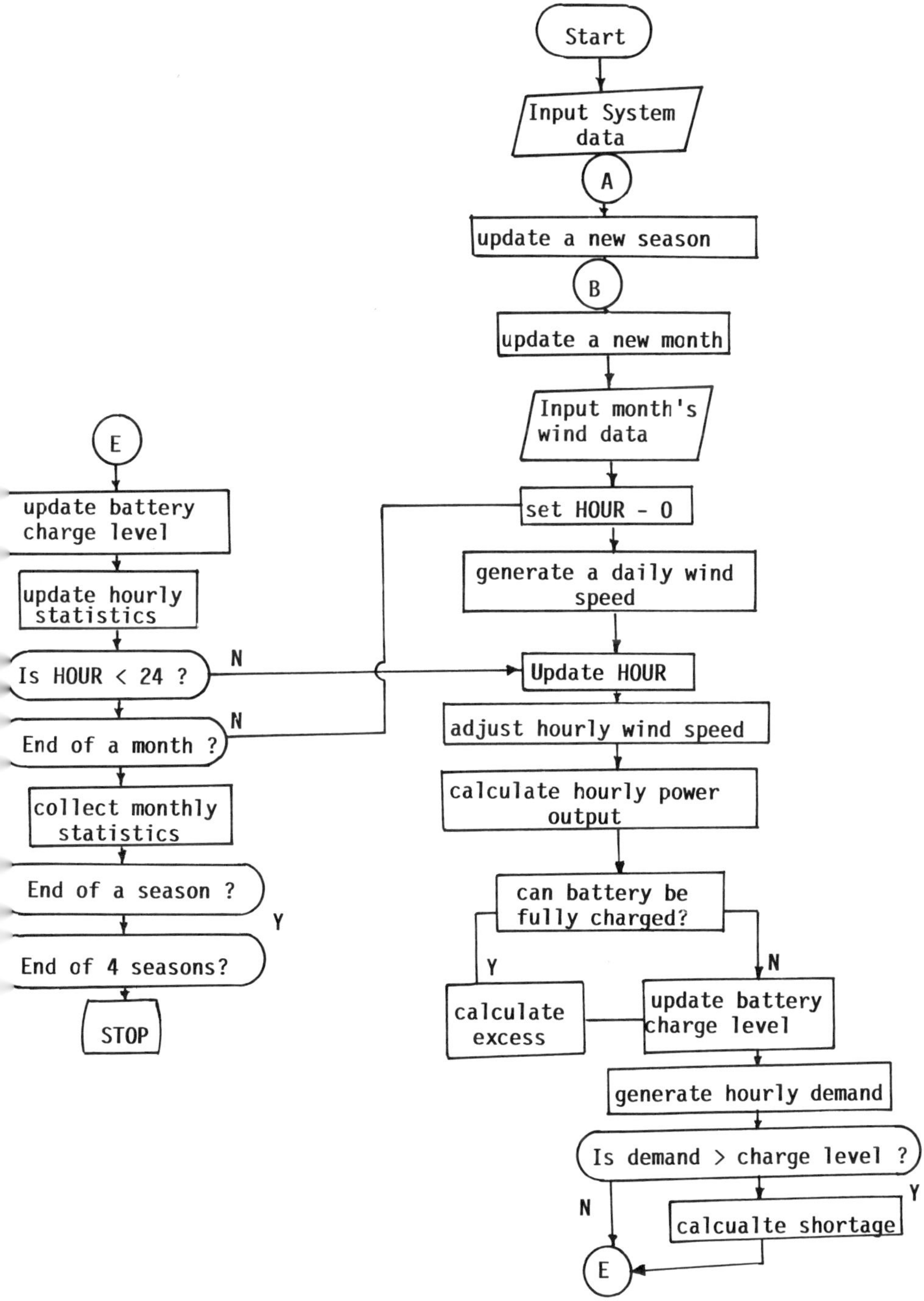

FIGURE 2: General flow chart for one year simulation

Model output measures of performance are obtained for daily and monthly periods. They are: average power output $\bar{P}$, average shortage $\bar{H}$, average battery storage level $\bar{S}$, and average excess power $\bar{E}$. Other measures of performance such as power coefficient, availability and capacity factor are not considered here. Let t be the time period, t=1,2....,T., and

P_t the power output at time t,
S_t the battery storage level at time t in AH (ampere hour)
D_t the power demand at time t,
V battery system voltage
H_t power shortage at time t,
C battery storage capacity (maximum) in AH,
E_t excess power at time t
η_b, η_d are the battery charging and discharging efficiencies

At the beginning of period t, an excess power situation may exist due to the finite battery capacity C;

$$E_t = \max\{ S_{t-1} + P_t\eta_b/V - C, 0\} V, \quad \bar{E} = \sum_{t=1}^{T} E_t/T$$

The battery storage is then increased by S = $\min\{C-S_{t-1}, P_t\eta_b/V\}$. As the load D_t is generated, a shortage situation may arise if $D_t > (S_{t-1} + \Delta S)\eta_d V$; $H_t = \max\{ 0, D_t - (S_{t-1} + \Delta S)\eta_d V\}$,

$$\bar{H} = \sum_{t-1}^{T} H_t/T.$$

The new battery storage level S_t is thus calculated as:

$$S_t = \max \{S_{t-1} + \Delta S - D_t/V \eta_d, 0\}, \quad \bar{S} = \sum_{t=1}^{T} S_t/T.$$

T is the length of the simulation run in hours,

$$\bar{P} = \sum_{t=1}^{T} P_t/T.$$

MODEL VALIDATION

Validation is intended to insure that the model is an accurate representation of reality, and that its output is statistically similar to real behavior. Subjective comparisons and statistical procedures are commonly used.[22] Wind-speed generation is validated here by comparing the generated wind speeds with their best-fitted distributions for two months. A X^2 test is used to test identity of the generated and the best-fitted distribution with 95% confidence level. Likewise, hourly load random generations are validated using a sample of five simulation runs (24 hours each). Hourly load is randomly generated from its estimated appliance usage distributions. For each hour the $\pm 1\sigma$ confidence limit for the load is calculated. The limits completely cover the previously estimated average load values for each hour. The power output is validated by comparing the analytically computed[2] average power with the 95% confidence limits that are obtained from simulation (using 160 hour simulation run). The computed average power output of a turbine with rated power P_r and wind distribution F(v), (see Figure 3), is given by:

$$\overline{P} = \int_{v_i}^{v_r} p(v)\, f(v)\, dv + P_r \int_{v_r}^{v_o} f(v)\, dv,$$

$$\overline{P} = \int_{v_i} (a + bv + cv^2) f(v)\, dv + P_r \int_{v_r}^{v_o} f(v)\, dv,$$

where f(v) is a Gamma or Weibull best-fitted distribution function. It is found that the computed average power is covered by the limits estimated from simulation.

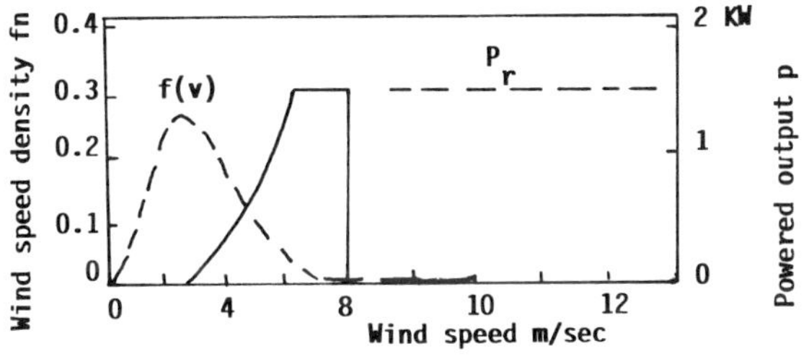

FIGURE 3: Turbine power output p(v), and wind speed density function f(v)

MODEL EXPERIMENTATION

The system, as introduced, has two parameters to be specified: the turbine rated power P_r and the battery capacity C. small turbines 1.2, 1.5, and 2.0 Kw rated power, and battery capacity of 40, 60.....300 AH are tried. To selected the two parameters, preliminary simulation experiments are conducted for a particularly low wind-speed month, September. For each rated power-battery capacity combination, model measures of performance are estimated. In these experiments, percentage excess is the total excess power (unstored) divided by the total turbine output, percentage shortage is total power shortage divided by the total power demanded, and percentage battery charge is the average battery charge level divided by the battery selected capacity. Figure 4 presents the relationship between percentage excess and battery capacity for three rated powers. Clearly, for a given load and rated power configuration there is an unavoidable percentage excess regardless of the battery capacity. These percentages are 2%, 3.5%, and 7% for the 1.2, 1.5, and 2.0 Kw rated powers. This also suggests that a too large battery capacity is not always beneficial.

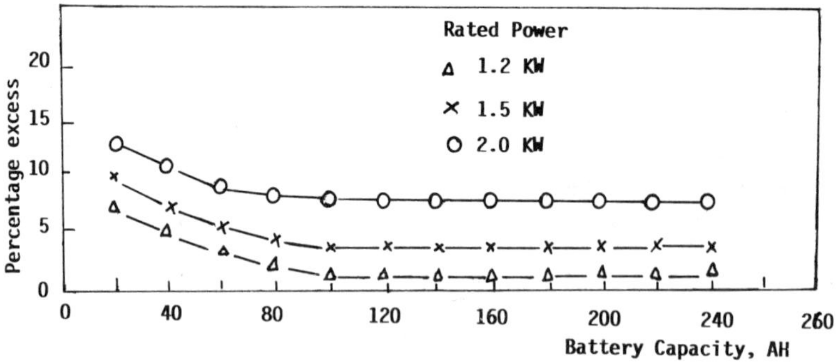

FIGURE 4: Percentage excess power and battery capacity

The percent average shortage-battery capacity relationship is demonstrated in Figure 5. A 0% shortage is possible through increasing the battery capacity. This indicates that the rated power range 1.2 - 2 Kw is suitable for the assumed household load. Figure 6 illustrates the average battery charge during September as a function of the battery capacity for the selected power range. Average battery charge is an indicator of how long the battery's own charge can support the load during calm wind situations.

To select a suitable rated power and battery capacity, a two-fold criterion that uses Figures 5 and 6 is applied. Performance of an average shortage of 5%, and an average battery charge of 135 AH is considered satisfactory as a system performance. The household load is estimated as 45 AH per day, and hence a battery with 135 AH average charge can support it for about three days.[3] Using Figures 5 and 6, it is found that a rated power of 1.2 Kw and a battery capacity of 200 AH is the least configuration which satisfies this criterion.

Detailed yearly simulation is conducted for the selected turbine. the following manufacturer specifications are assumed: P_r=1.2 Kw, C=200 AH, v_i=2.5 m/sec, v_r=6 m/sec, and v_o=15 m/sec. Ten full-year simulation runs are used. Yearly measures of performance are dealt with as independent observations. The classical method[30] of constructing 95% confidence limits for simulation replications is applied. Table 2 exhibits average and confidence limits for the monthly measures of performance. Figures 7 and 8 present these results. It is shown that the average power output, battery charge, and wind speed follow the same pattern, while percentage-shortage is inversely related to them. The yearly average shortage is less than 5% and the average battery charge is more than 135 AH.

FIGURE 5: Average percentage shortage and battery capacity relationship

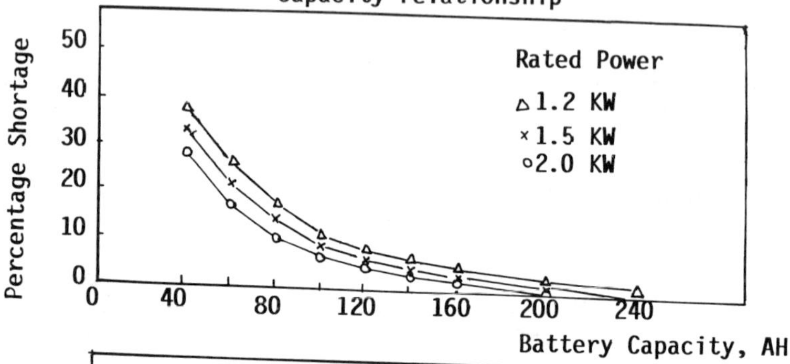

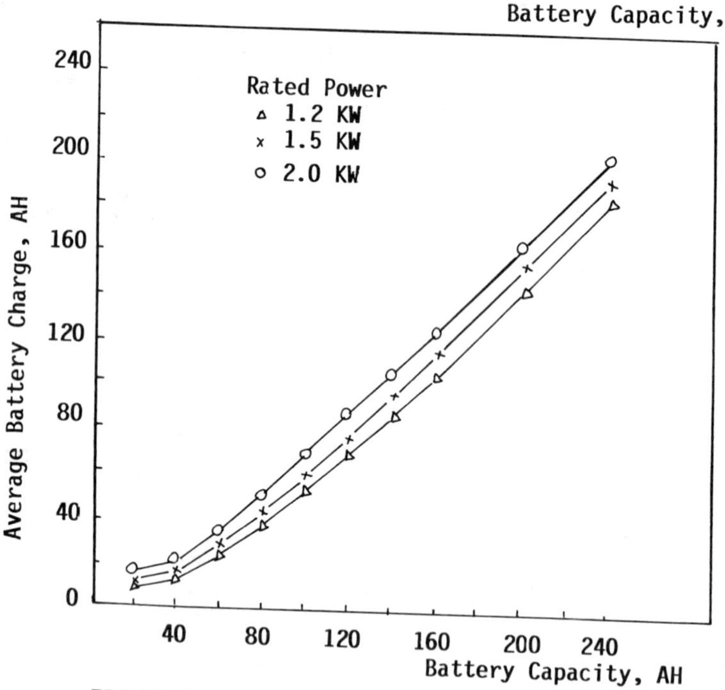

FIGURE 6: Average battery charge and battery capacity relationship

TABLE 2. Average and confidence limits for simulation results

Month	Wind speed v_t m/sec		Power P_t watt		Load D_t watt		Shortage% H_t		% Excess E_t		Battery S_t AH	
	V	CL*	P	CL*	D	CL*	H	CL*	F	CL*	S	CL*
Dec.	4.63	±.19	513	±34.7	232	±1.5	3.7	±3.6	8.4	±2.5	160	± 9.5
Jan.	3.39	±.18	323	±43	232	± .94	14.7	±7.8	2.5	±1.9	96	±22.7
Feb.	5.39	±.17	766	±47.4	232	±1.8	0	± 0	27.6	±4.5	188	± 3.1
Mar.	3.96	±.15	402	±34.1	232	±1.2	4.7	±4.8	4.0	±1.2	134	± 8.8
Apr.	4.91	±.18	637	±38.6	232	±1.2	0.1	± .18	19.0	±2.2	185	± 3.7
May	5.09	±.1	686	±25.3	232	±1.0	0	0	20.9	±4.5	194	± 1.2
June	5.88	±.24	882	±43.6	247	±1.9	0	0	32.5	±3.6	195	± .8
July	6.26	±.2	988	±36.	246	±1.2	0	0	46.4	±2.4	196	± .3
Aug.	6.70	±.16	1017	±27.2	246	±1.2	0	0	45.2	±1.5	198	± .3
Sep.	3.69	±.17	362	±39.2	233	±1.0	5.9	±6.2	1.5	± .5	134	±19.
Oct.	4.64	±.16	393	±40.5	234	±.75	2.0	±2.1	1.7	±1.0	126	±31
Nov.	3.49	±.24	302	±63.9	232	±1.4	13.4	±4.5	1.7	±2	91	±24.6

* Confidence Limits

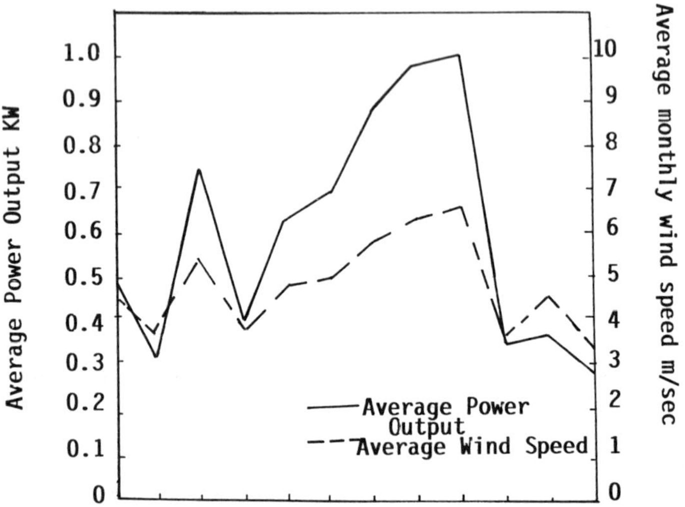

FIGURE 7: Monthly average wind speed and power output

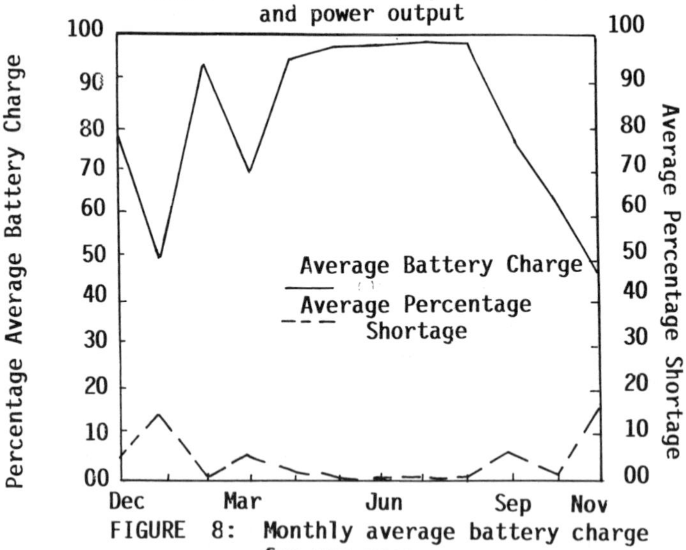

FIGURE 8: Monthly average battery charge for one year

CONCLUSIONS

The constructed simulation model is a valid approach to predicting the performance of a turbine system of a given configuration, at a given site, and to support an estimated power load. It provides means for selecting a turbine system that meets given user requirements concerning percentage-shortage and sufficient battery charge to support calm wind days. It has the advantage of handling analytically intractable wind and load distributions such as Gamma and Normal. In particular, it is shown that a system with a 1.2 Kw turbine, a 200 AH battery capacity, and a simple control charging / discharging system can support the estimated household load with less than 5% average shortage, and three days of calm wind. The approach presented can be applied readily to other wind turbine systems to evaluate or compare alternatives.

REFERENCES

1. R.B. Corotis. 1979. Simulation of Correlated Wind Speeds for Sites and Arrays, SUN II, Proc. Int. Solar Energy Soc. Congress, v2.

2. A. El-Mallah, M.A. Shalaby and M.A. Badr. 1986. A Model to Estimate Wind Power at Selected Sites in Egypt, Proc. Mech. Power Eng. 6th Conf., Monofia Univ., Egypt.

3. A.J. Unione, E.Y. Lim and A.S. McClymont. 1983. Estimation of Wind System Performance and Economics Using Unit Availability Criteria, Alter. Energy Sources, v(4).

4. L.H. Wegley, J. Ramsdell, M. Orgil and R. Drake. March1980. A Sitting Handbook for Small Wind Energy Conversion System, U.S. Dept. of Energy report.

5. W. Ramkumar, and R.G. Deshmut. 1979. Application of Probability to Evaluate Reliability of a Combined Electric and Conventional Generation System, SUN II, Proc., v(3).

6. P. Feron and E.A. Lysen. 1981. Effective Output and Availability of Wind Turbines for Household Loads, Wind Eng., v(5), no. 4.

7. B.N. Haack. 1980. A Simulation Model for Wind Electric System, Wind Eng., v(4), no. 4.

8. N. Galanis. 1980. Simulation of a Wind Powered Water Heater, Wind Eng., v(4), no. 4.

9. P.C. Bandopadayay. 1982. Total Energy Conversion System for Isolated User, Wind Eng., v(6), no. 2.

10. C. Casale and E. Sesto. May, 1985. Wind Energy Conversion Systems for Isolated Users, A special training workshop, Delphi.

11. M.A. Badr. 1986. A Computer Simulation Model to Analyse The Performance of a Wind Turbine System, Unpublished M.Sc. thesis, Fac. of Eng., Cairo Univ.

12. Egyptian Meteorological Authority. 1979. Meteorological Normals in Egypt, Cairo.

13. D.S. Renne and B.D. Holst. Nov. 1985. Characteristics of Wind Energy Use in Egypt, a USAID/EEA technical report, Cairo.

14. C.G. Justus. 1978. Wind and Wind Machines, Franklin Inst. Press.

15. C.G. Justus and A. Mikhail. 1979. Generic Power Performance Estimates for Wind Turbines, SUN II, Proc. Conf, v(2).

16. K.C. Chou and R.B. Corotis. 1981. Simulation of Hourly Wind Speeds and Array Wind Power, Solar Energy, v(26).

17. W.T. Pennel and H.L. Wegley. 1983. Effect of Site Wind Characteristics on Energy Production, Alter. Energy Sources, v(4).

18. J. Harper, D. Percival, T. Flaim and R. Sullivan. 1982. Wind System Value Analysis for Electric Utilities: A Comparison of Four Methods, Solar Energy Eng., v(104).

19. J. Manual and J. McGowan. 1979. A Design Procedure for Wind Power Heating Systems, SUN II, Proc. Conf.

20. F.R. Eldridge. 1980. Wind Machines, Mitre Corp.

21. W.J. Conover. 1971. Practical Non Parametric Statistics, John Wiley.

22. J. Banks and J. Carson. 1984. Discrete Event Simulation, Prentice Hall.

23. C.L. Mantell. 1979. Batteries and Energy Systems, McGraw Hill.

24. M. Conte. 1986. Personal Communications, The Italian Nuclear Energy Centre.

25. C.G. Justis and W. Hargraves. 1977. Wind Energy Statistics for Large Array of Wind Turbines Solar Energy, v(20).

26. G. Saliem. 1980. An Assessment of Energy Storage Systems Suitable for Wind & Solar Energy Applications in Egypt, Cairo Univ./MIT Center report, Cairo Univ.

27. Ministry of Electricity and Energy. 1980. Annual Report of Electricity Statistics, Government of Egypt

28. C.G. Justis. 1979. Is Wind a Practical Source of Energy For You?, A special public., DOE Rocky Flats Wind Systems Program, USA.

29. J. Park. 1981. The Wind Power Book, Cheshire Books, England.

30. M.A. Shalaby. 1984. Using Confidence Intervals in Queueing System Simulation, J. Egyptian Stat. Soc., v(28).

WATER YIELD FROM WIND PUMPS FOR DESERT DEVELOPMENT IN EGYPT

A. EL-MALLAH and A.M. SOLTAN
Mechanical Engineering Laboratory,
National Research Center, Cairo, Egypt

Abstract A procedure is presented for selecting appropriate wind turbines for water pumping on the northwestern coast of Egypt. Wind speed data from different locations have been collected and classified and then fitted to the Weibull probability distribution function. The fitted data was corrected to the standard 10-m anemometer height. A survey of the water resources on the coastal area is also presented. Wells are classified with respect to water depth and safe-withdrawal rate of discharge. Mathematical formulas are established and a Nomogram is constructed to help in easy selection of a candidate wind turbine. A comparison of the performance of six machines with different characteristics is given for each of the sites. Practical examples are given to show how to replace existing pumping methods with suitable wind pumping systems.

INTRODUCTION

Water is necessary for desert development, to make settlement of nomadic populations possible and to help in creating new communities. The northwestern coast of Egypt is a target area for development that is attracting great interest at present. The coastal zone from Sallum to Alexandria possesses both water and wind resources which make it possible to use wind energy for pumping water.

Local water resources originate mainly from rainfall which feeds underground water through sheet runoff and wadi run-offs.[1,2,3,4] At present, underground water is delivered to the users by electric and diesel-driven pumps from boreholes or galleries.

Small-size wind pumping units were installed in the early sixties along the coast to pump water for irrigation. About 1500 dug wells were equipped with such machines. Most of these units are now in poor condition due to lack of maintenance and repair. Some of the owners suffered from an insufficient water supply and others complained that excessive pumping exceeded the safe withdrawal rate of the well and quickly led to deterioration of the water quality and silt-up of the wells. Moreover, with the introduction of electricity and Nile water through canals and piping systems, wind turbines were replaced by electric and diesel-driven pumps.

Although the expected lifetime of the wind pumping units has expired, some of them are still in good condition and working very efficiently. Therefore, it is worthwhile to replace the old machines by new, reseized, and well-designed Wind Energy Conversion Systems, WECS, for pumping water. It is necessary to take into account the water needs, the safe withdrawal of the water sources, and the wind characteristics of the site.

The procedure described in the present work is suggested as a way to select the appropriate WECS for different locations. Emphasis is on the northwest coast of Egypt.

SURFACE AND UNDERGROUND WATER SOURCES

Water for drinking and irrigation could be obtained in different ways from both surface and underground sources. Dykes, cisterns, dug and drilled wells, and collected galleries[2] are possible means.

Cisterns, dating back to the Roman period, have been excavated in the rock. Their capacities vary from 100 to $1000 m^3$. Most of them were silted up. In 1979, 2200 cisterns were excavated to provide $356{,}800\ m^3$ of water. The power needed to drive water 500 m from a cistern, at a pressure head of 4, is estimated to be 2 to 5 hp.[3]

Besides cisterns, there are approximately 1500 dug wells equipped with windmills. About two-thirds of these wells are in Burg El-Arab, Dabaa, Mersa Matruh and Sidi-Baranni.[2] About 30% of them are still in use.[4] The dug wells range from 1 m to 4 m in diameter and the depth of water varies from a few meters to more than 30m.[4] The average rate of discharge per windmill varies from 1 to 5 m^3/day.

Drilled wells are found in some places on the coast. The wells at Fuka (near El Dabaa) are capable of producing 20 m^3/hr for five hours daily using electric driven pumps rated at 5 hp.[2]

The most extensive gallery development is at El-Qasr (Mersa Matruh), where there are ten collecting galleries with a total length of 1.5 km. The estimated output of the galleries is found to be 0.056-0.071 m^3 per meter length of gallery per day. This water is used for irrigation far from the galleries using diesel driven pumps of 6-12 hp. The daily discharge of a pumping unit installed on the gallery ranges between 60 and 120 m^3/day.

WIND RESOURCES

The annual average wind speed along the northwest coast fluctuates between 4 and 5.5 m/s.[5] These values were confirmed by recent measurements.[6] The data were obtained from the Egyptian Meteorological Authority[5] after which it was processed and analyzed.[7,8,9]

Suitable probability density functions, PDF, can also be used to describe the wind speed distribution mathematically.[10] The favorable PDF that can be used is the Weibull distribution which takes the form:

$$f(v) = \frac{k}{c}\left(\frac{v}{c}\right)^{k-1} \exp.\left[-\left(\frac{v}{c}\right)^{k}\right] \qquad (1)$$

where:

f(v) is the wind speed Weibull distribution function, k is the shape parameter and c is the scale parameter. The meteorological data mentioned above was fitted to the appropriate Weibull PDF.[11,12] A method of linear regression[12] was adopted in the present work to compute the Weibull parameters k and c for each of four meteorological stations along the coast (Sallum, Mersa Matruh, Dabaa, and Dekheila), month by month at a reference height of 10 m above ground level. The computed results were used to estimate the power output of wind turbines installed at these sites.

AVERAGE POWER OF A WECS

The power output of a WECS is generally divided into two parts. The first part is measured from a cut-in wind speed v_i to a rated wind speed v_r. The second part is the rated power range which begins from the v_r under speed to a cutoff wind speed v_o. Practical solutions of the partial power range have been discussed earlier[12] by considering quadratic and exponential relations as follows:

$$p\ (v)_{vi-vr} = A + Bv\ Cv^2, \text{ or} \tag{2}$$

$$p\ (v)_{vi-vr} = a + b\ v^k \tag{3}$$

where:

A, B, C, a and b are constants and could be evaluated using WECS characteristics, and k is the Weibull shape parameter.

A comparison between the two formulas was carried out using the actual data given by manufacturers (in different manufacturers' brochures) of ten different machines, as shown in Figure 1, using v/v_r as abscissa and p/p_r as ordinate. Both relations were drawn for v_i/v_r = 0.27 and 0.36. These values represent the lower and higher limits of the checked machines. The exponential formula was further computed for three different k-values (1.5, 2.0, and 2.5). The figure shows that the quadratic relation generally underestimates the output, and is insensitive to variations of the cut-in wind speed. On the contrary, the exponential relationship is more realistic, except in the case where k is less than 1.5. It was found that the majority of the k-values for the above-mentioned sites lie between 1.6 and 2.2; hence the exponential relationship is found to be more practical and will be used for further calculations.

The average power output, p, of the WECS in the whole range from v_i to v_o will be given by:

$$\bar{p} = \int_{v_i}^{v_o} f(v)\, p(v)\, dv \qquad (4)$$

where:

f(v) is the PDF of the wind speed at the site for the pertinent period of time given by Eq. 1, p(v) is the output power as given by Eq. 3 for $v_i<v<v_r$ and $p(v) = p_r$ for $v_r<v<v_o$. Equation 4 is computed for six different categories of WECS assumed to be used under the wind conditions at the four sites. The characteristics of these machines are given in Figure 2, and are based on manufacturers' data,[13] and some practical experience. It may be noted that WECS I and II differ from each other in v_o, systems III and IV vary in v_i and v_r whereas systems V and VI have different v_i only.

The data of the monthly PDFs of the four sites (k and c values) and the characteristics of the WECSs are used in Eq. 4 to compute the capacity factor $\bar{p}/p_r$ for each WECS. The results are given in Figures 3 (a-d). The figures show that the capacity factor varies monthly depending on both site and machine characteristics. In some cases, e.g. for site 1, systems I and II have almost identical values despite the difference in v_o This phenomenon is particular to that site and can be attributed to the fact that the average wind speed is relatively low. As the cut-in wind speed decreases, the capacity factor increases (See systems V and VI). It is further increased by using WECS with low-rated wind speeds for all sites.

SELECTION OF SUITABLE WECS

A suitable WECS may be selected for a certain water requirement after considering the interaction of site and machine characteristics and the characteristics of the water source, as mentioned above. To facilitate the selection of an appropriate WECS to replace the existing pumping methods on the northwest coast, a chart is constructed in the form of a Nomogram, Figures 4 (a,b). The variables represented on the Nomogram are the average water discharge rate Q, the overall pumping head H, the overall pumping efficiency η the rated power of the WECS and the capacity factor. To increase the accuracy of estimation, two Nomograms with different scales are submitted. The first one is suggested for a low discharge rate and for WECS with low rated power that can be used for dug wells with limited capacity. The second one is for a relatively higher rate of discharge that can be used for boreholes and collecting galleries.

The Nomograms can be used in conjunction with Figure 3 in the following way:

1. Define a monthly average discharge rate needed to satisfy certain requirements based on monthly average values, point A on the far left scale, Figure 4.

2. Define the overall pumping head of the system, point B.

3. Join A B and extend it to C.

4. Draw a ray from O' representing the average overall efficiency of the system on the η

scale; a horizontal line from C meets this ray in D.

5. Drop a vertical line from D.

6. Draw a horizontal line representing the appropriate capacity factor of a suitable WECS category using Figure 3, point E.

7. The horizontal and vertical lines from E and D, respectively, will meet in F.

8. The line OF will define the required rated power on the p_r scale.

CONCLUSIONS

Water is supplied on the northwest coast from dug wells which have a limited daily output of between 1 and 5 m^3 and from boreholes and collecting galleries with a capacity of around 100 m^3/day. Each of these sources can be equipped with suitable WECS according to the wind characteristics at the site. The specifications of the WECS can be selected with the aid of the accompanying Nomograms and the capacity factor chart for each of the four mentioned sites along the coastal area from Sallum in the west to Alexandria in the east. Replacement of the existing wind pumping units with new well-designed and sized units is essential to avoid problems due to mismatching between the WECS and well specifications. The present method can be adopted wherever wind turbines are used in pumping water.

ABBREVIATIONS

c Weibull scale parameter, ms^{-1}.

H Pumping head, m.

k Weibull shape parameter.

p Output power, W.

$\overline{p}$ Average power, W.

p_r Rated power, W.

Q Rate of discharge, $m^3\ s^{-1}$.

$\overline{Q}$ Average rate of discharge, $m^3\ s^{-1}$.

v Wind speed, $m\ s^{-1}$.

v^i Cut-in wind speed, $m\ s^{-1}$.

v^o Cutoff wind speed, $m\ s^{-1}$.

v_r Rated wind speed, $m\ s^{-1}$

η Efficiency, %.

REFERENCES

1. H. Ismail. August 1976. General plan for the coastal zone, ARE. Technical Report, Ministry of Housing and Construction.

2. FAO. 1970. Preinvestement survey of the NW-coastal region, UAR, Physical conditions and water resources. Tech. Report 2, ESE:SF/UAR 49.

3. Abou Khalid. 1984. Improvement of irrigation practices and methods in NE-coastal zone of ARE. Consulting Report, EG/82/015, FAO.

4. Agricultural development in NW-coastal zone, Egypt. 1985. Project AG. TCP/EGY/2205, FAO.

5. Egyptian Meteorological Authority, Cairo. 1979. Meteorological Normals of the Arab Republic of Egypt.

6. P.S. Renne and B.D. Holst. November 1975. Wind characteristics of wind use in Egypt. Report prepared by Battle PNL, USAID contract NEB-0123-C-00-4024-00.

7. M.S. Harb. November 1975. Characteristic features of wind field in windy regions in Egypt. Egyptian Energy Resources Study Project, Final Report, supported by NSF, Grant No. GF-38938.

8. A. El-Mallah. 1981. Wind water pumping for selected sites in Egypt. Fourth General Conference of the Association for the Advancement of Agricultural Science in Africa, AAASA, Cairo.

9. M.A. Mobarak, A. El-Mallah, and Sirag El Din. 1982. Wind energy in Egypt. The Fourth International Conference for Mech. Power Eng. ICMPE 4, Cairo.

10. C.G. Justus. 1978. Wind Energy Statistics for Large Arrays of Wind Turbines (New England and Central U.S. Regions), Solar Energy 20.

11. A.M. Harranein and A.O. Darwish. 1985. On the evaluation of available wind energy using climatological wind data. Alternative Energy Sources and Technology Conf. AESTC 1, Mansoura.

12. A. El-Mallah, M.A. Shalaby and M.A. Badr. 1986. A model to estimate wind power at selected sites in Egypt. The Sixth Int. Conf. for Mech. Power Eng., ICPME 6, Cairo.

13. Rockwell International. June 1978. Commercially available small wind machines. A Checklist of Systems, Manufacturers and Distributers.

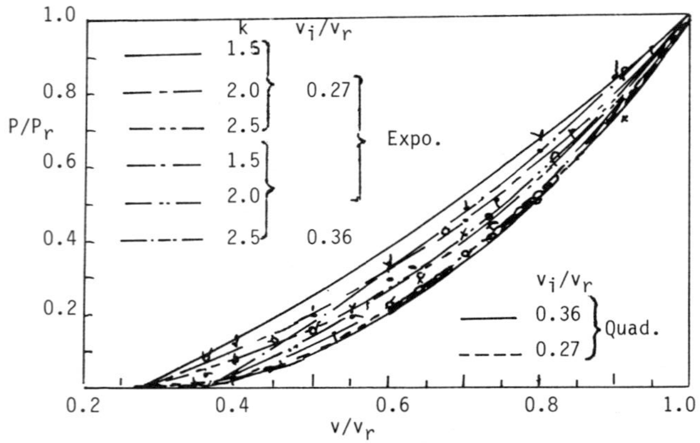

Type	P_r	v_i/v_r
Cencenbaugh	0.5	0.36
"	1.0	0.33
Winco	1.1	0.35
"	0.9	0.35
"	1.3	0.30
"	2.5	0.27
Jacobs	20.0	0.32
Elektro	10.0	0.27
"	6.0	0.30
Enertech	40.0	0.31

FIGURE 1: Comparison between power output and calculated output of different wind turbines

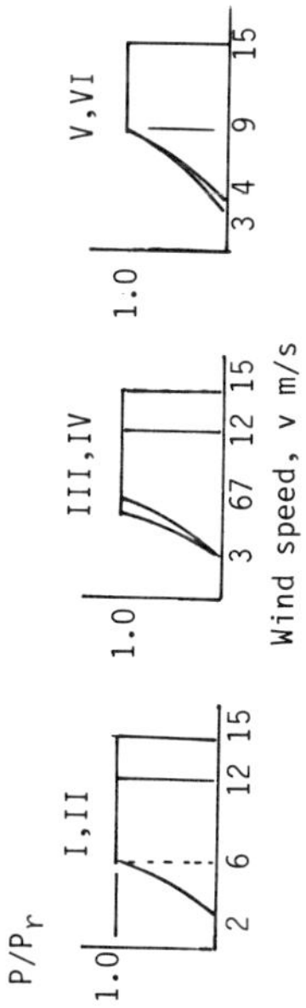

WECS No.	v_i	v_r	v_o
I	2	6	12
II	2	6	15
III	3	6	12
IV	3	7	15
V	3	9	15
VI	4	9	15

FIGURE 2: Comparison between different WECS characteristics.

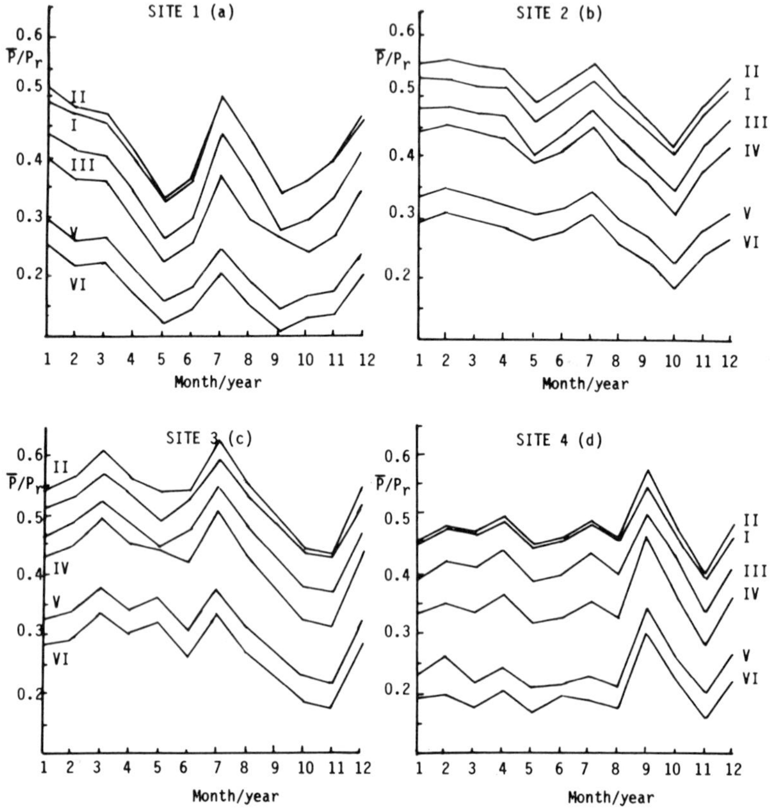

FIGURE 3: Monthly capacity factor of different sites and WECSs.

Site 1: Sallum Site 2: Mersa Matruh,
Site 3: Dabaa, Site 4: Dekhiela.

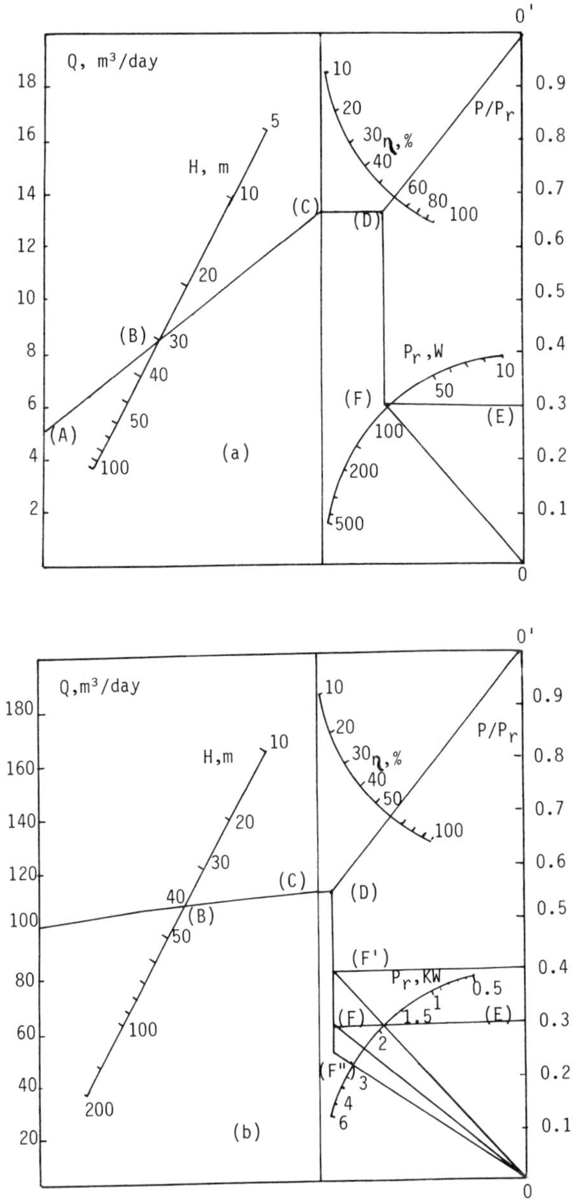

FIGURE 4 (a,b): A nomogram for estimating the rated power of wind pumping systems.

A BIOENERGY PROJECT TO REDUCE DEFORESTATION IN SUDAN

EL TAYEB IDRIS EISA
Sudan Renewable Energy Project

Abstract In Sudan, the primary energy sources are traditional fuels, charcoal and fuelwood. Sustainable supplies for wood fuels are declining while the demand for these fuels continues to rise. The solution to the problem requires a strategy that (1) uses available biomass resources, (2) increases the fuelwood/forestry supply and (3) improves consumption.

INTRODUCTION

The primary energy source is Sudan is traditional fuel, that is, charcoal and fuelwood. Of the total energy consumed in 1980, about 78% was fuelwood and charcoal, and of that 78%, 95% was used for home cooking. Projections for future energy consumption show that charcoal and fuelwood will continue to be the predominant energy source for many years.

Sustainable supplies of wood fuels are declining while the demand for these fuels continues to rise. The numbers are startling: the National Energy Administration predicts that by 1987 the fuelwood resource in the central regions, the major suppliers for the country, will be depleted. By 1990, fuelwood in southern Kordofan will have vanished and southern Darfur will run out of fuelwood as we approach the year 2000.

While charcoal and fuelwood are the primary sources of energy, this consumption is not the major cause of deforestation. The major cause is land clearing for mechanized farming. In the central regions where mechanized farming is expanding the agricultural production, thousands of feddans of forest are cleared and burned each year. This fuelwood resource could be used to produce charcoal, and thereby relieve the pressure to deforest lands in other parts of Sudan. However, this utilization of trees from cleared agricultural lands is not enough.

The solution to the problem of deforestation or desertification requires a threefold strategy. Sudan must:

- Use all available biomass resources and not lose resources through wasteful burning;
- Increase the fuelwood/forestry supply through increased tree planting;
- Introduce more efficient methods of fuelwood and charcoal consumption.

SREP PROGRAMS

The Energy Research Council (ERC) along with other governmental and private organizations, has been working to reduce the woody biomass energy shortage. The Sudan Renewable Energy Project (SREP), a project funded jointly by USAID and the Government of Sudan and administered by ERC, is implementing pilot and demonstration projects. These are:

- to design, test and promote fuel-saving charcoal stoves;
- to study charcoal production efficiency to determine possible improvements;
- to promote various agro-forestry initiatives to increase fuelwood production;
- to study commercial use of fuelwood to improve efficiencies;
- to introduce briquetting of agricultural residues as a possible fuel substitute.

CHARCOAL STOVES

As stated previously, 95% of the charcoal and wood used in Sudan is for domestic cooking. Therefore, minor improvements in the combustion efficiency of traditional charcoal stoves will produce significant charcoal savings nationally.

In the Greater Khartoum area, it is estimated that 360,000 stoves are sold annually (based on a report by SREP marketing consultant, Arthur Brown, October 1985). Potential savings of charcoal using the improved charcoal stove now being promoted by SREP is estimated at 281 kgs per year per stove. At this time, stove production and sales of the improved stoves as a result of SREP dissemination activities are estimated at 4 to 5% of the market share. Charcoal savings from these stove sales are about 51,000 sacks of charcoal, valued at L.S. 869,550 per year.

Savings in forest plantings to replace the charcoal used is estimated by dividing the 51,000 sacks saved by 30 (the number of sacks of charcoal produced per feddan), which equals 1,700 feddans. SREP believes that with improved production and marketing, 20 to 30% of the stove users in the Khartoum area could be reached, thereby increasing the impact five to six times. In other words, over 10,000 feddans per year could be saved.

AGRO-FORESTRY

SREP has promoted several types of fuelwood/ forestry initiatives. Among these are:

1. Shelterbelts for individual farms.

2. Village forestry projects.

3. Community nurseries.

4. Shelterbelts, canal-side planting, and fuelwood lots for agricultural schemes.

If implemented on a national scale, all of these initiatives would contribute to reduced deforestation, greater agricultural production and an improved environment. The potential impact in financial terms is not easy to calculate. However, let us consider Sudan's irrigated scheme areas. If only 5% of the total arable land (not the 10 to 15% required by law), were placed under forestry this would mean 275,000 feddans of forest. These lands would produce 275,000 fd X 6.25 m^3/fd/yr = 1,718,750 m^3 per year of usable fuelwood. If this production is used for fuelwood and charcoal only, it would provide 35% of the demand in the Gezira, Blue Nile, and Khartoum provinces (based on 1983 National Energy Plan demand estimates).

CHARCOAL PRODUCTION ON MECHANIZED FARMS

SREP has initiated a new project on mechanized farms in rain-fed areas. The two groups cooperating in the project (Arab-Sudanese and Damazine Agricultural), state that the potential production from the trees to be cleared in 1985-86 is 1,305,000 sacks of charcoal, with a value of L.S. 3,900,000. In addition, the project is assisting in planting shelterbelts to provide fuelwood resources for the future.

FUELWOOD SUBSTITUTION AND THE USE OF ALTERNATIVE ENERGY RESOURCES

Other fuels and energy sources can be used instead of charcoal and wood. Cotton stalks are being considered as a potential household fuel. Preliminary tests have shown that it is possible to char the cotton stalks and then briquet them to produce a more transportable and uniform-burning fuel. The potential for charcoal production, considering 75% availability of total stalk production and a 25% conversion efficiency, is 22,780 tones per year from the Rahad Scheme and 112,500 tones per year from the Gezira Scheme.

Another measure that could contribute to reduced desertification would be the use of renewable energy sources for water pumping and lifting water. Included would be wind pumping and photovoltaic devices for rural areas. These renewable energy technologies could provide reliable power sources for water pumping in areas which now depend on conventional sources of energy.

CONCLUSIONS

Immediate action with the enlistment of the national government, local communities, and individuals is necessary. Despite the fact that it will take massive mobilization of human and material resources, the projects should be small and numerous instead of large and few. Too often, excessive organizational requirements prohibit actual implementation of projects. I would suggest initiating modest projects and developing infrastructure as the number of small field programs grows.

Specific actions recommended are:

1. More efficient utilization of the forests that are cleared for mechanized farming schemes for agricultural production;
2. Increased production of wood fuels in an economical way using methods that have been successfully demonstrated.
3. Consideration of fuel substitutes to relieve the demand for woody biomass.

To realize these goals, it is essential that cooperation and goodwill exist among the agencies, communities, and governments who share this concern for Sudan's energy needs.

WIND-PHOTOVOLTAIC HYBRID SYSTEM FOR PUMPING WATER IN A DESERT ENVIRONMENT

AMR ABDEL HAMID AND HEMMAT SAFWAT
Department of Engineering
American University in Cairo

AMIN MOBARAK
Department of Mechanical Engineering
Cairo University

Abstract The use of wind and solar energies to drive a water pumping system at a typical desert site has been investigated. The complete system is conceived to be composed of a wind turbine, photovoltaic panels, electrical battery storage, water pumps and a water storage tank. Actual measured solar and wind energy parameters at the site have been collected over a period of twelve months and were used as an input in the investigation. Performance data of various types of commercially available system components were analyzed and the appropriate types of components were selected for the system. A computer model for the detailed performance characteristics of the system was generated. An optimization study was then carried out to determine the most suitable number of solar panels, wind turbines, batteries and pumping units compatible with the site. Based on these results, the system components have been ordered and are now being assembled. The system simulation model is quite versatile and provides a basis for the design and development of energy management strategies for general wind-solar hybrid energy systems with loads other than water pumping.

INTRODUCTION

Utilization of renewable energy sources to pump water in remote areas has received considerable attention in recent years. Of the many systems considered, for areas where conventional electrical power is not available, those which incorporate wind or solar energies seem to dominate the market. Several advanced systems are designed to operate in a hybrid mode in order to optimize the use of the available renewable energy resources at a given site.

Appropriate engineering of a hybrid renewable-energy water pumping system depends on the availability of accurate information on the characteristics of the renewable energy sources at the proposed site, the characteristics of the various components involved in the envisaged system and the development of an energy management strategy.

The appropriate forms of renewable energy characterization have been discussed in another publication[1]. For a hybrid energy system, the required information is the variation of the available energy with time of day for different days of the year.

As part of a major program currently in progress at the Desert Development Center (DDC) of the American University in Cairo (AUC), it was decided to put together a wind-photovoltaic hybrid system to pump water at the Sadat City site of the DDC. This paper describes the major components of the system and discusses the results obtained from the computer simulation for system operation.

TYPICAL CHARACTERISTICS OF RENEWABLE ENERGY SOURCES AT THE AUC DDC SADAT CITY SITE

Solar and wind energy parameters have been monitored at the site for more than twelve months starting in April of 1985. The detailed results were presented previously[1]. For the purpose of demonstration, the renewable energy data corresponding to July and December are shown here. These months correspond to maximum and minimum total energy availability, respectively. Figures 1 and 2 show

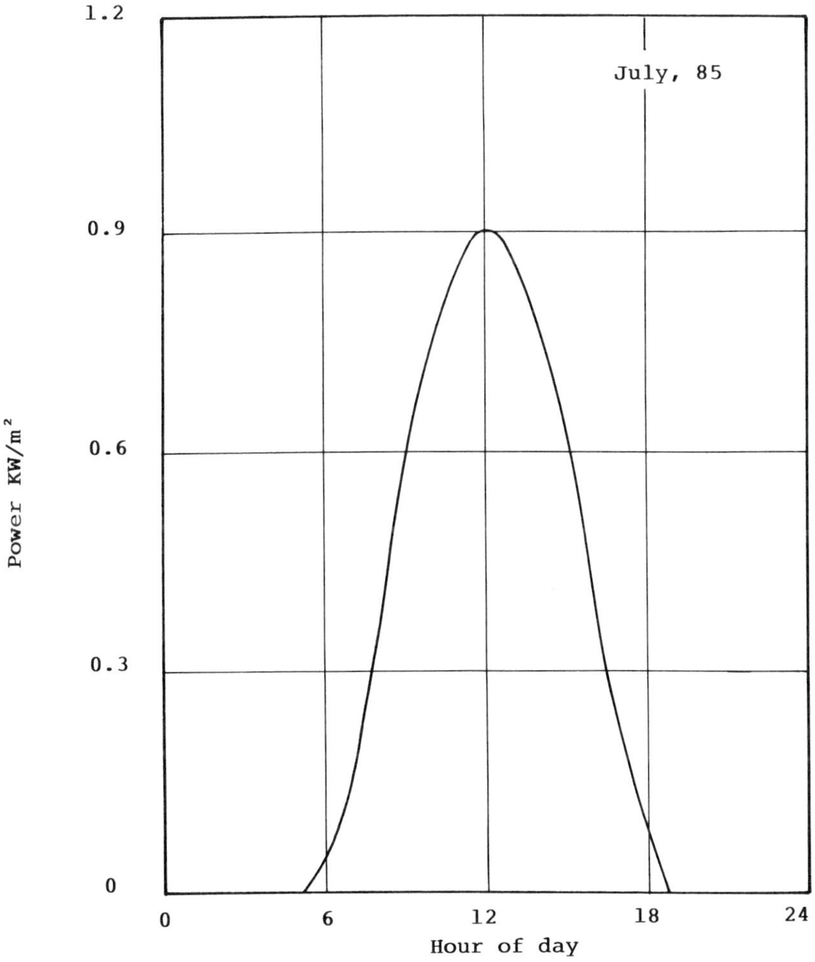

FIGURE 1: Typical variation of the solar intensity with time of day during the month of July, 1985 at DDC Sadat City site

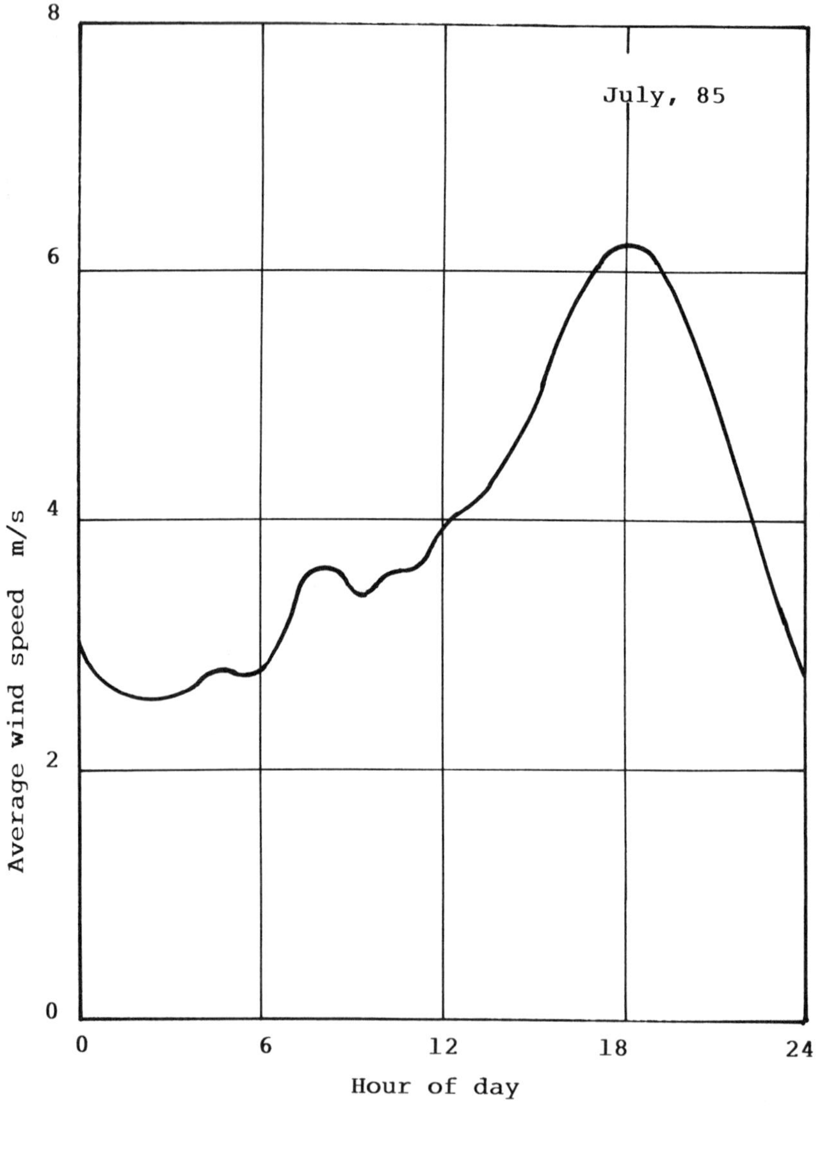

FIGURE 2: Typical variation of wind speed with time of day during the month of July, 1985 at DDC Sadat City site

the solar and wind energy data for July 1985, while Figures 3 and 4 show those for December 1985. Of particular interest here is the shift between the time of maximum available solar energy and the time of maximum available wind energy. The time shift ranges from three to seven hours which is not optimum but is beneficial, as will be discussed later, in reducing the size of the storage facility. Another point of concern is the low level of wind speed during most of the night and early morning, particularly for the month of December when the wind speed hardly reaches the level of 4 m/s.

OVERALL SYSTEM DESCRIPTION

The envisaged hybrid energy system is composed of the following components:

i- Photovoltaic panels which transform solar radiation into electrical energy of the direct current type.

ii- A wind turbine which converts wind energy into mechanical rotational energy, then into electrical energy through appropriate conventional electro-mechanical devices.

iii- Storage batteries which act as a buffer between the renewable energy sources and the load.

iv- Water pumps which represent the electrical load which must be served by the hybrid energy system.

v- A controller which assesses the extent of the accumulated energy from the various sources and successively activates or deactivates the water pumps.

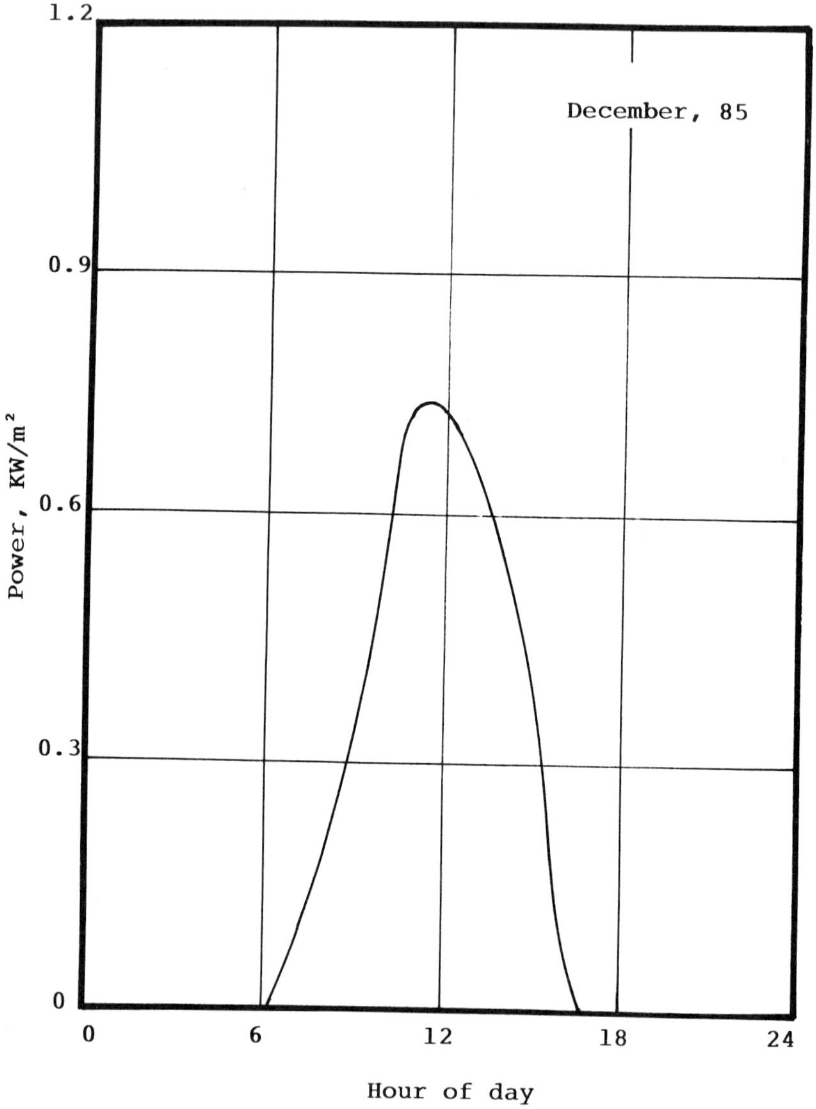

FIGURE 3: Typical variation of the solar intensity with time of day during the month of December, 1985 at DDC Sadat City site

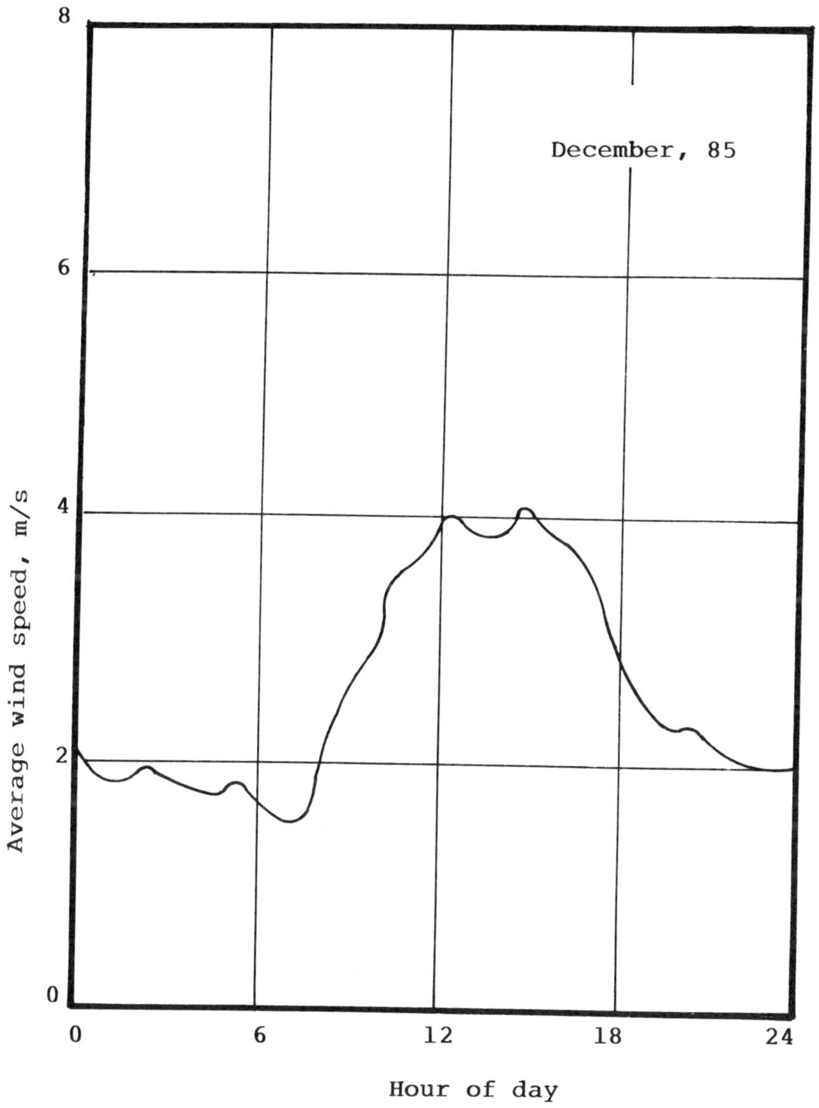

FIGURE 4: Typical variation of wind speed with time of day during the month of December, 1985 at DDC Sadat City site

CHARACTERISTICS OF THE SYSTEM COMPONENTS

PHOTOVOLTAIC MODULES

The typical characteristics of the modules selected for the system are shown in Figure 5. The electrical power output from the model is dependent on the amount of solar radiation falling on the panel and the cell temperature. The temperature effects, however, are significant only at high voltage. For a given voltage below 16 volts, the output current varies linearly with solar insolation. For an operating temperature of 25 °C this relationship is given by:

$$I = 3.25\ Ws \quad \text{(Amp.)}$$

Where Ws is the solar intensity in KW/m^2.

The maximum power output from the module for an incident solar radiation of 1000 $watt/m^2$ is typically 53 watts.

WIND TURBINE

The typical mechanical characteristics of the wind turbine characteristics are given in Figure 6. The turbine is 7 m in diameter with three blades. The typical power curve of the turbine is shown in Figure 7. In the range of wind speeds up to 10 m/s the electrical power output dependence on wind speed C closely is approximated by the relationship:

$$P = 140\,(C - 2.5)^2 \qquad C \geq 2.5\ m/s$$
$$P = 0 \qquad C < 2.5\ m/s$$

BATTERIES

The typical cell performance characteristics of the batteries are given in Table 1. The significant aspect to note is the effect of discharge current on battery efficiency. Analysis of the data shown

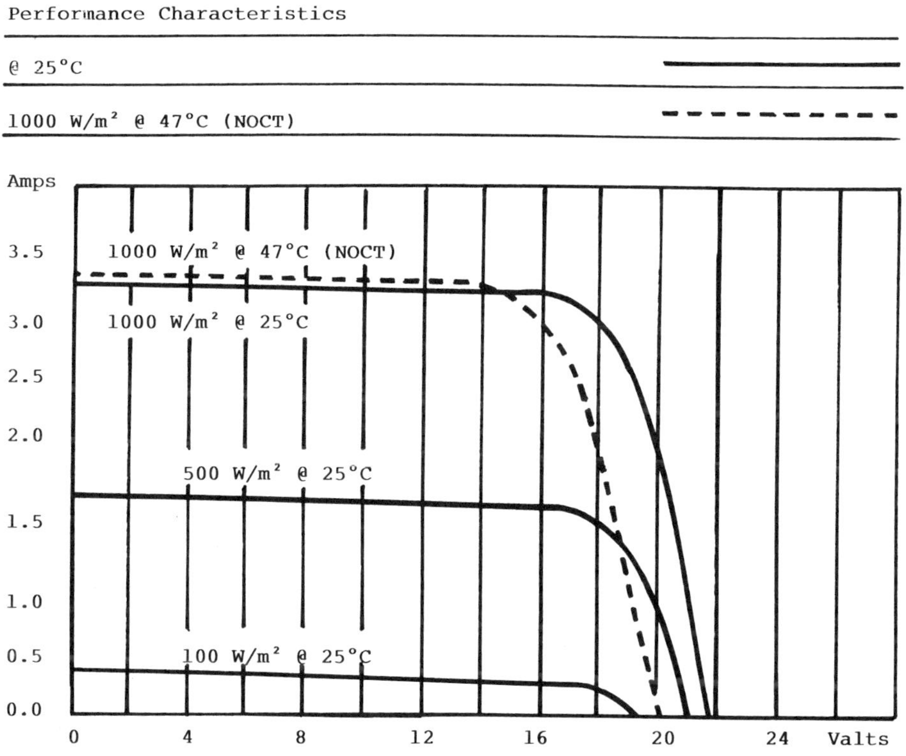

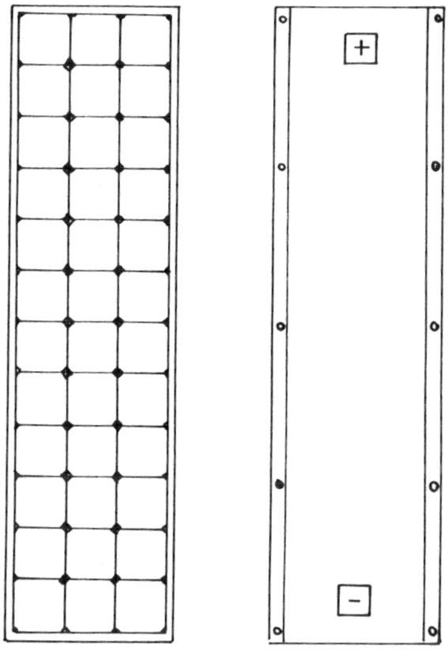

Power Specifications*	
Model	M55
Power (typical ± 10%)	53.0 Watts
Current (typical at load)	3.05 Amps
Voltage (typical at load)	17.4 Volts
Short Circuit Current (typical)	3.27 Amps
Open Circuit Voltage (typical)	21.8 Volts

Physcial Characteristics	
Length	50.9 in/1293 mm
Width	13 in/330 mm
Depth	1.4 in/36 mm
Weight	12.6 lb/5.7 kg

*Power specifications are at standard test conditions of: 1000 W/M², 25°C cell temperature and spectrum of 1.5 air mass.

IGURE 5: Characteristics of the photovoltaic modules modeled in the present study

BWC EXCEL SPECIFICATIONS

PERFORMANCE	
START-UP WIND SPEED	3.4 m/s (7.5 mph)
CUT-IN WIND SPEED	3.1 m/s (7.0 mph)
RATED WIND SPEED	12.1 m/s (27.0 mph)
CUT-OUT WIND SPEED	none
FURLING WIND SPEED	15.7 m/s (35.0 mph)
MAXIMUM DESIGN WIND SPEED	54 m/s (120 mph)
RATED POWER	10 kw
ROTOR SPEED	60-350 RPM
MECHANICAL	
TYPE	3 Blade Upwind
MOTOR DIAMETER	7.0 m (23.0 ft)
WEIGHT	463 KG (1020 lbs)
BLADE PITCH CONTROL	POWERFLEX®
OVERSPEED PROTECTION	AUTOFURL™
GEARBOX/BELTS	none
TEMPERATURE RANGE	-40 to 60°C (-40 to 140°F)
ELECTRICAL	
OUTPUT FORM:	
BWC EXCEL-S	240 VAC, 1Ø,60 Hz*
BWC EXCEL-R	Nominal 120 VDC
GENERATOR	Permanent Magnet Alternator
OUTPUT CONTROL SYSTEM:	
BWC EXCEL-S	POWERSYNC® Inverter
BWC EXCEL-R	VCS-10 Controller

*3 wire, with neutral - 220 VAC, 50 Hz option available

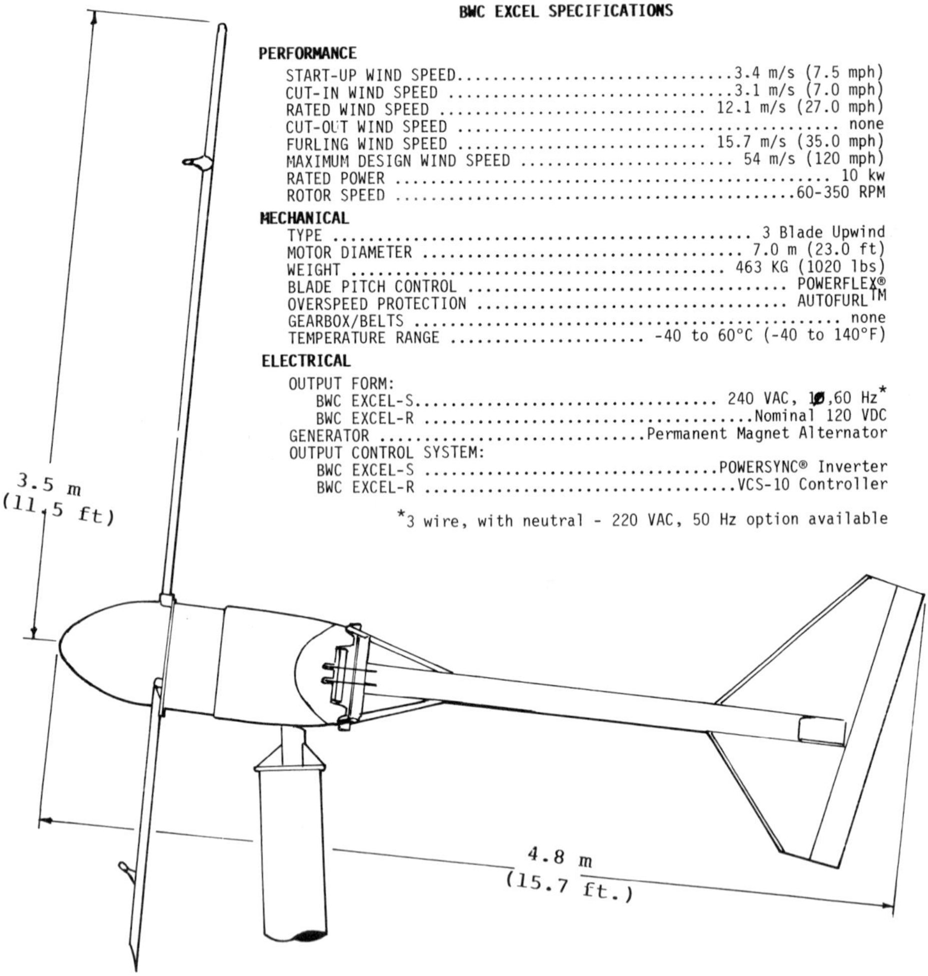

FIGURE 6: Mechanical characteristics of wind turbine modeled in the present study

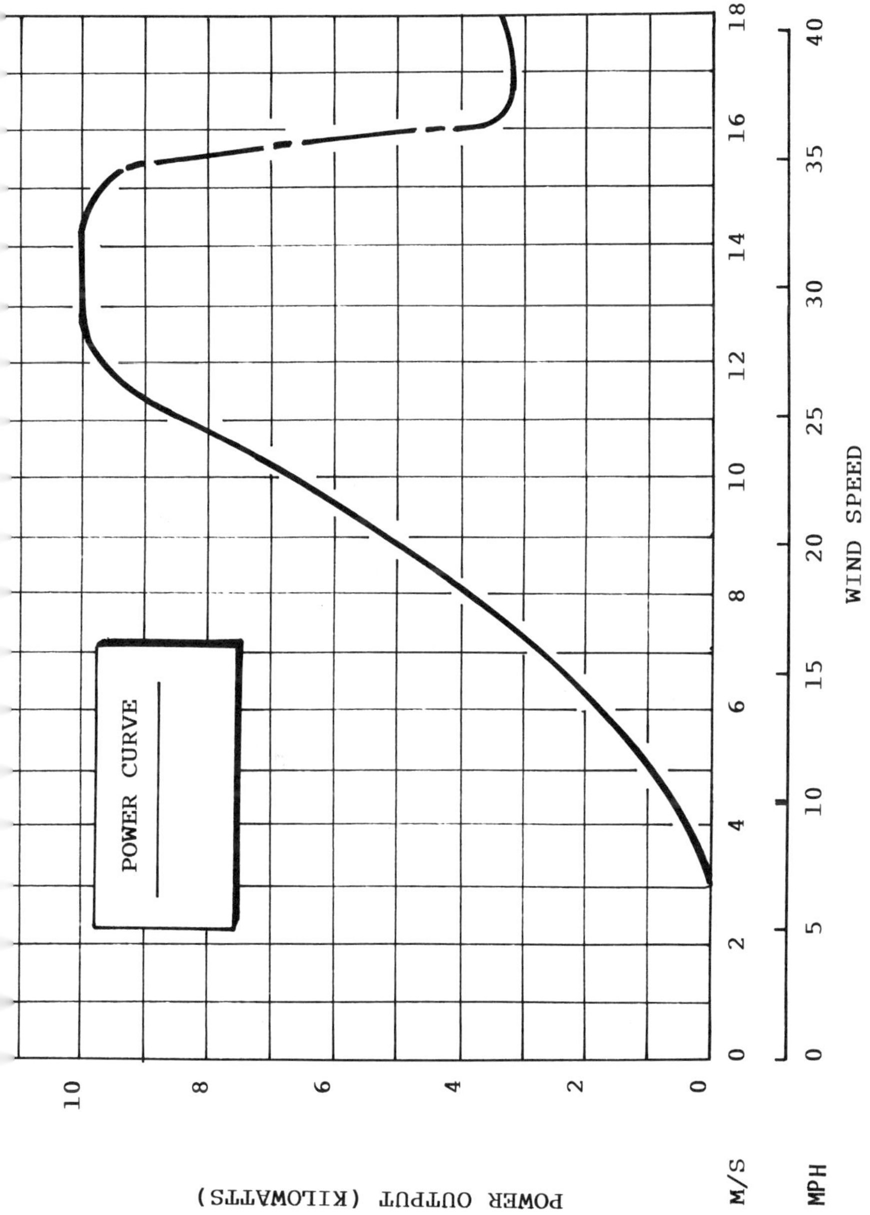

FIGURE 7: Variation of the electrical power output from the wind turbine with wind speed

AVERAGE CELL PERFORMANCE DATA

Discharge Rates in Ampere Hours and Amperes.

- 1.300 Specific Gravity Electrolyte
- 1.250 Specific Gravity Electrolyte Available For Annual Temperatures Greater Than 90°F (32°C)— Reduce Ratings by 10%.

			77°F (25°C)												32°F(0°C)
			TO 1.90 VOLTS PER CELL				TO 1.75 VOLTS PER CELL								TO1.75VPC
			3 DAY		2 DAY		1 DAY								1 DAY
CELL TYPE	VOLT. PER UNIT	24 HOUR A.H.CAP.	72 HOURS		48 HOURS		24 HOURS		16 HOURS		12 HOURS		8 HOURS		24 HOURS
			A.H.	AMPS	A.H.	AMPS	A.H.	AMPS	A.H.	AMPS	A.H.	AMPS	A.H.	AMPS	A.H.
E 95-5	2	180	192	2.6	180	3.7	180	7.5	173	10.8	163	13.6	153.6	19.2	153
E 95-7	2	270	288	4.0	270	5.6	270	11.2	259	16.2	245	20.4	230.4	28.8	230
E 95-9	2	360	383	5.3	360	7.5	360	15.0	346	21.6	326	27.2	307.2	38.4	306
E 95-11	2	450	478	6.6	450	9.3	450	18.7	432	27.0	408	34.0	384.0	48.0	382
E120-9	2	500	538	7.4	500	10.4	500	20.8	474	29.6	456	38.0	422.4	52.8	425
E120-11	2	625	673	9.3	625	13.0	625	26.0	592	37.0	570	47.5	528.0	66.0	531
E120-13	2	750	808	11.2	750	15.6	750	31.2	710	44.4	684	57.0	633.6	79.2	637
E120-15	2	875	942	13.0	875	18.2	875	36.4	829	51.8	798	66.5	739.2	92.4	743
E120-17	2	1000	1077	14.9	1000	20.8	1000	41.6	947	59.2	912	76.0	844.8	105.6	850
E120-19	2	1125	1212	16.8	1125	23.4	1125	46.8	1066	66.6	1026	85.5	950.4	118.8	956
E120-21	2	1250	1346	18.7	1250	26.0	1250	52.0	1184	74.0	1140	95.0	1056.0	132.0	1062
E120-23	2	1375	1481	20.5	1375	28.6	1375	57.3	1302	81.4	1254	104.5	1161.6	145.2	1168
E120-25	2	1500	1616	22.4	1500	31.2	1500	62.5	1420	88.8	1368	114.0	1267.2	158.4	1275
E120-27	2	1625	1750	24.3	1625	33.8	1625	67.7	1540	96.2	1482	123.5	1372.8	171.6	1381
E120-29	2	1750	1885	26.2	1750	36.4	1750	72.9	1658	103.6	1596	133.0	1478.9	184.8	1487

FINISHING CHARGE VOLTAGES: 2.65 VPC @ 77°F (25°C)

SELF DISCHARGE RATES: 2.0% PER MONTH AT 77°F (25°C)
12.0% PER MONTH AT 132°F (55°C)

MAXIMUM CONTINUOUS DISCHARGE RATES: E 95-5 to 11, E120-9 to 15 } 150 AMPERES
E120-19 to 29 250 AMPERES

TABLE 1

in Table 1 shows that the internal cell resistance is approximately 0.03 Ohm for the range of operating conditions expected in the present study.

Another important aspect to consider is the effect of depth of discharge on the life of the battery. As shown in Figure 8, the deeper the daily depth of discharge, the shorter the life of the battery. Previous experience with banks of batteries shows that the battery life is much shorter than what is normally indicated by the manufacturer.

WATER PUMPS

The performance characteristics of the DC pumps selected for the present investigation are shown in Figure 9. For the site location used in the present study, the total head required to pump water from the well to the tank is 50 m. The operating DC voltage of the system was selected to be 59 V which resulted in a water flow rate of one liter/second and a power consumption 1.2 kw.

SYSTEM PERFORMANCE SIMULATION

The characteristics of the various system components were modeled and incorporated in a computer program to simulate the performance of the overall system. The results of the simulation calculations are given below.

Electrical Output from Solar-Wind Hybrid System

Figure 10 shows the variation of the current output from a single panel of four modules in series by time of day for typical days in July and December, at an operating voltage of 59 V. Similarly, the electrical output from the wind turbine is shown in Figure 11 for the same two typical days.

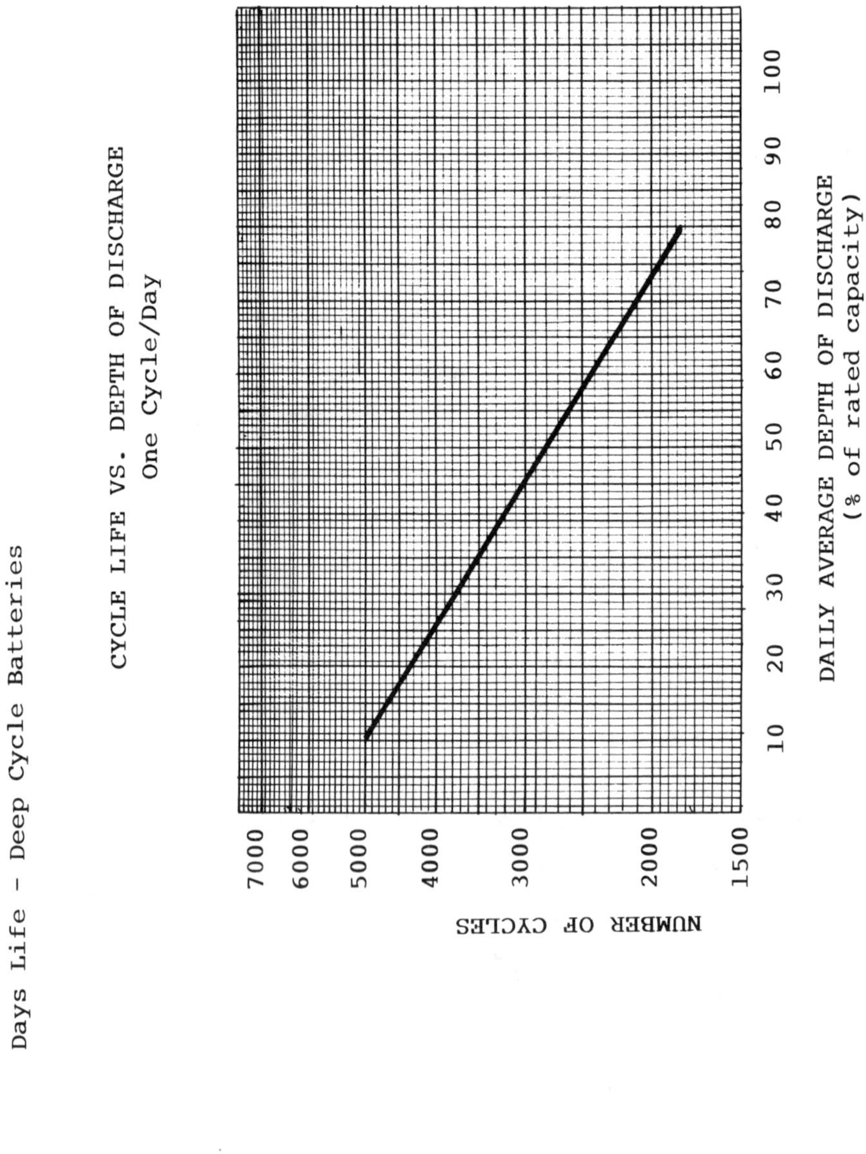

FIGURE 8: Effect of depth of charge on battery life

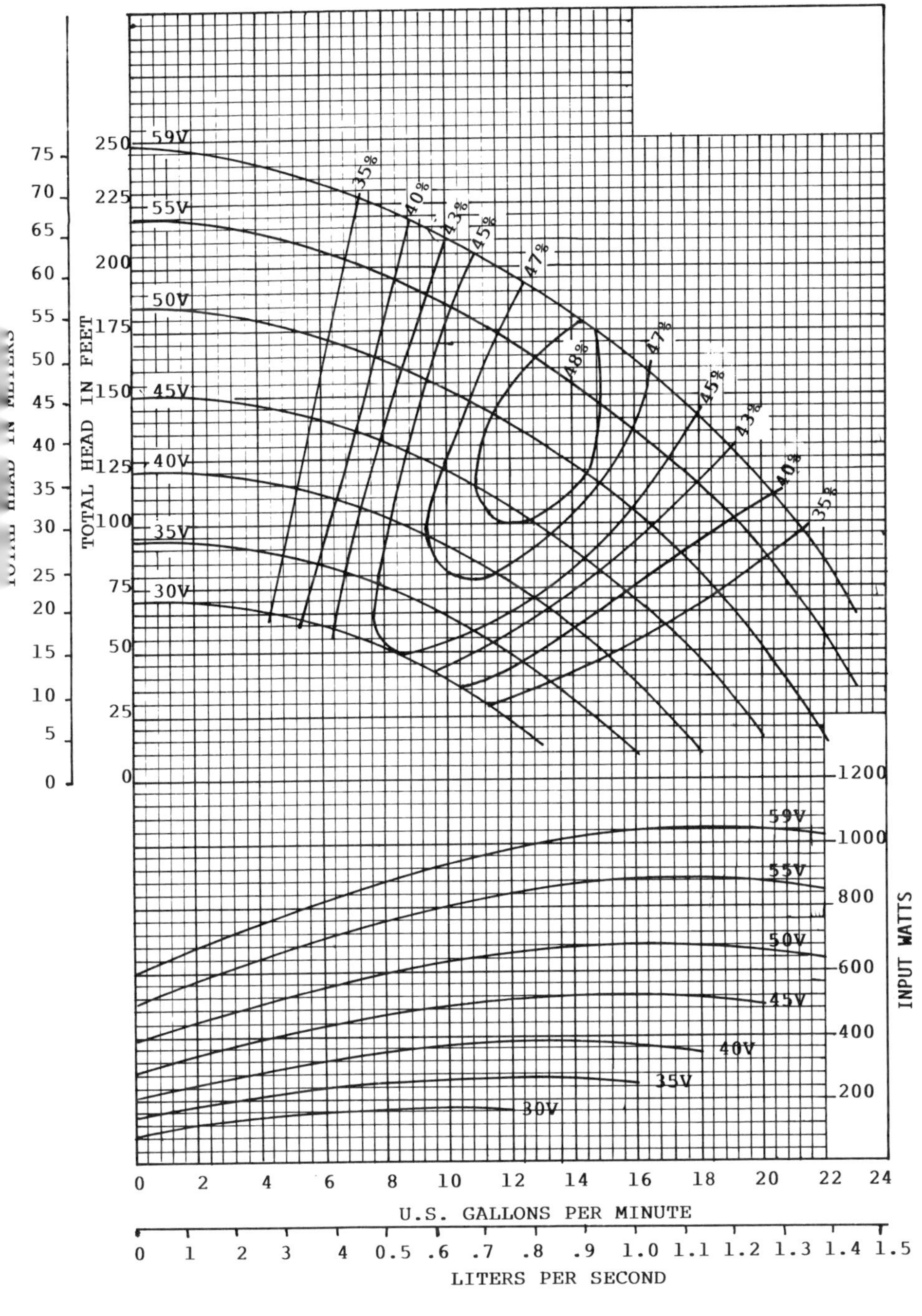

FIGURE 9: Characteristics of the water pump modeled in the present study

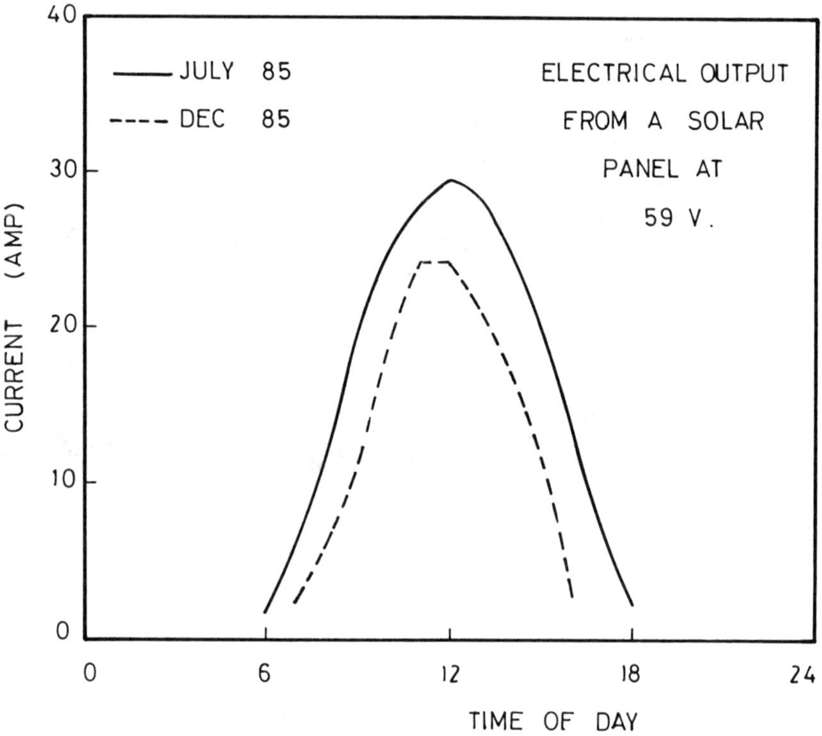

FIGURE 10: Variation of the current ouptut from a panel of four modules in parallel operation at 59 V with time of day

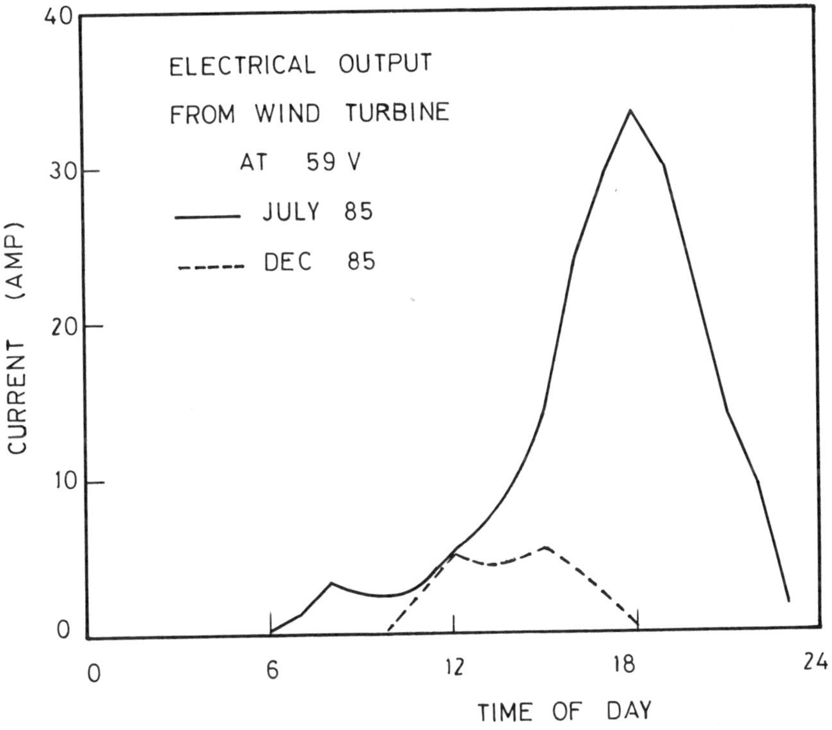

FIGURE 11: Variation of the current output from the wind turbine operating at 59 V with time of day

The total electrical output from the hybrid system will depend on the number of panels used. Figure 12 shows the current output of the hybrid system with 25 panels and one wind turbine. The use of the solar and wind energy sources clearly extends the on-line power availability to well after sunset.

Pump Operation Strategy

The total number of pumps which can be incorporated in the system is dependent on the maximum available electrical output from the hybrid system. For example, if 25 solar panels were used in the hybrid system without electrical storage batteries, the maximum number of pumps which can be simultaneously operated in July and December are four and three, respectively.

To minimize the dependence of the system on the batteries, it was decided to operate the pumps in a sequential fashion. The sequence control is dependent on the state of charge of the batteries. A minimum depth of discharge is selected, below which the batteries can not be further discharged. For example, as shown in Figure 13, at 10 A.M. during the typical July day, the available on-line electrical output from the hybrid system is about 66 Amperes at 59 V. This is more than the power needed to operate three pumps, but less than that required for four pumps. At that instant, the state of charge of the batteries is checked. If the charge was above the minimum level, the fourth pump would be operated and the extra power needed would be taken from the battery bank. If not, only three pumps are allowed to operate and excess power is used to charge the batteries. The situation continues for a period of 12 minutes after which the energy balance is checked again under the same logic control.

Figures 14 and 15 show the pump operation sequence for the two typical days of July and December, as measured by the rate of water delivery. The number of solar panels used in the

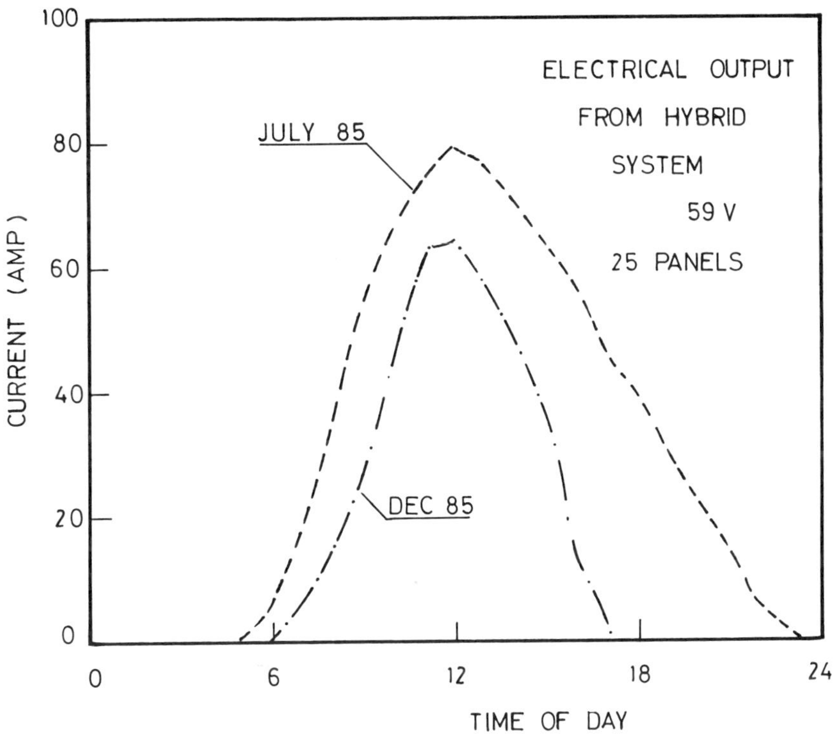

FIGURE 12: Variation of the total current output from the hybrid system at 59 V with time of day. Number of panels = 25

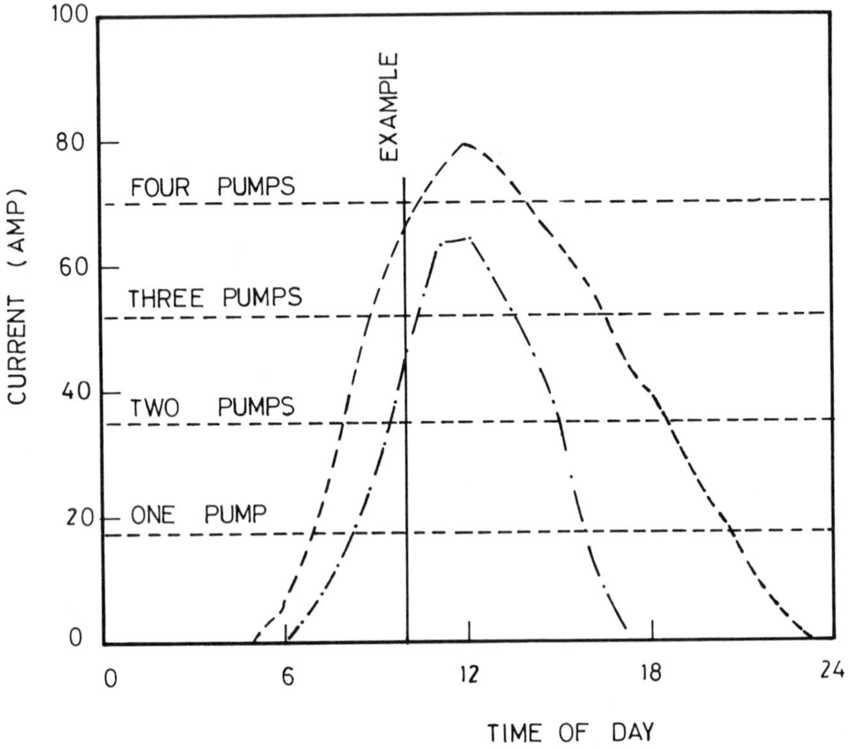

FIGURE 13: Example showing the control strategy for pump operation

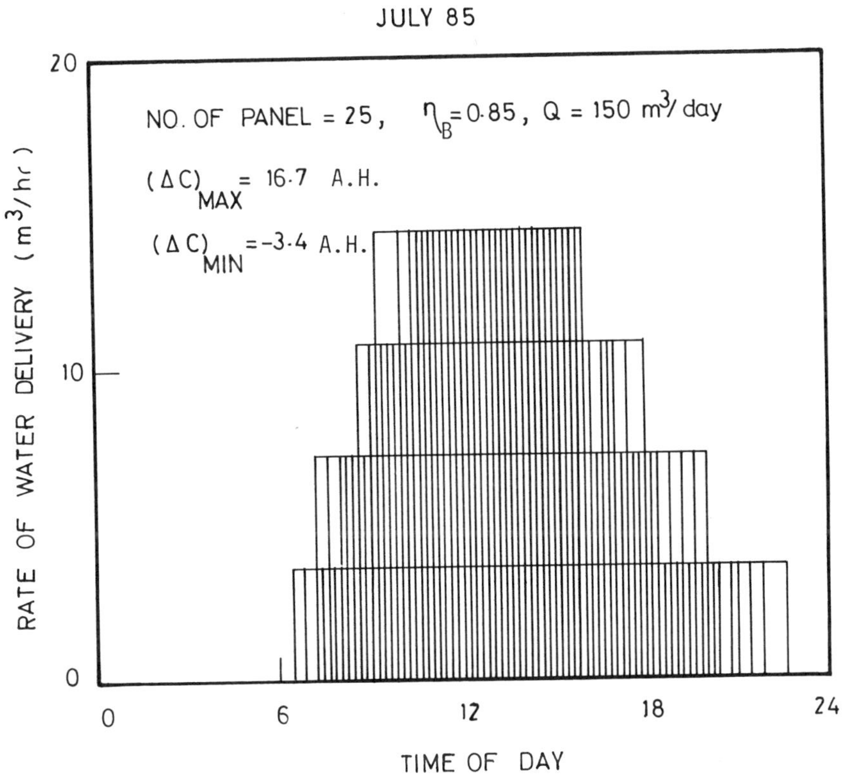

FIGURE 14: Pump operation sequence for a typical day in July, 1985

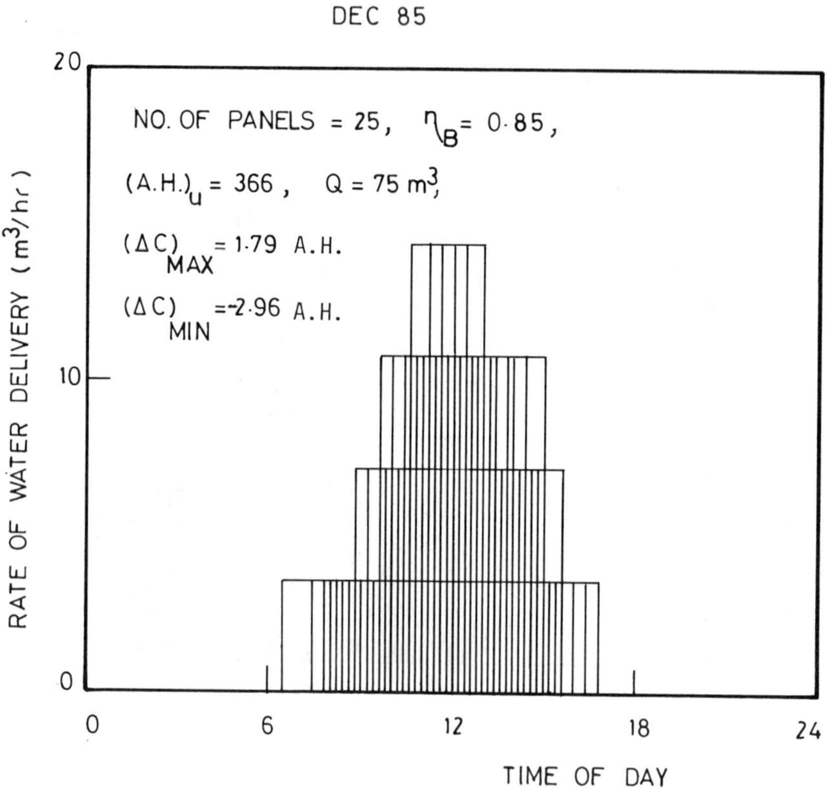

FIGURE 15: Pump operation sequence for a typical day in December, 1985

simulated hybrid system was 25. The battery efficiency was assumed constant at 85%. For July, the amount of water delivered was 150 m^3/day. The maximum battery overcharge was 16.7 A.H., and discharge was 3.4 A.H. For December, the water delivery was reduced to 75 m^3/day.

Effect of Number of Panels on Water Delivery

Figure 16 shows that if the number of pumps was fixed at 4, which can be due to limitation of well size, the water delivery increases linearly with number of panels. Depending on the amount of water needed per day, this diagram can be used to determine the number of solar panels which must be used. It should be noted, however, that the maximum amount of battery overcharge also increases very rapidly with number of panels used, as shown in Figure 17.

Effect of Battery Efficiency

Since the efficiency of the batteries decreases over time, it was decided to check its effect on water delivery. Figure 18 shows that the delivered water quantity decreases almost linearly with the decrease in battery efficiency. If no batteries were used whatsoever, the July and December water delivery would be reduced to 124 and 58 m^3/day respectively.

CONCLUSIONS

1- Simulation of a hybrid energy system to pump underground water in remote areas was found to be very useful in determining the effects of various system variables on water delivery.

2- For a given well size, physical dimensions of pumps, and temporal variation of renewable energy source, the appropriate number of photovoltaic panels should be selected such

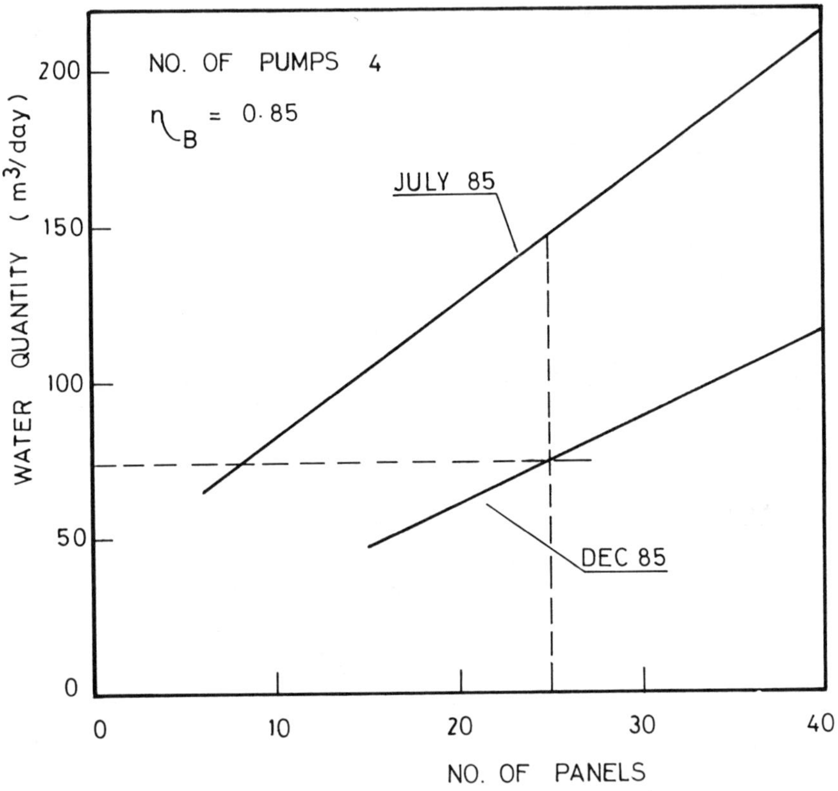

FIGURE 16: Effect of number of panels on water delivery for a given number of pumps

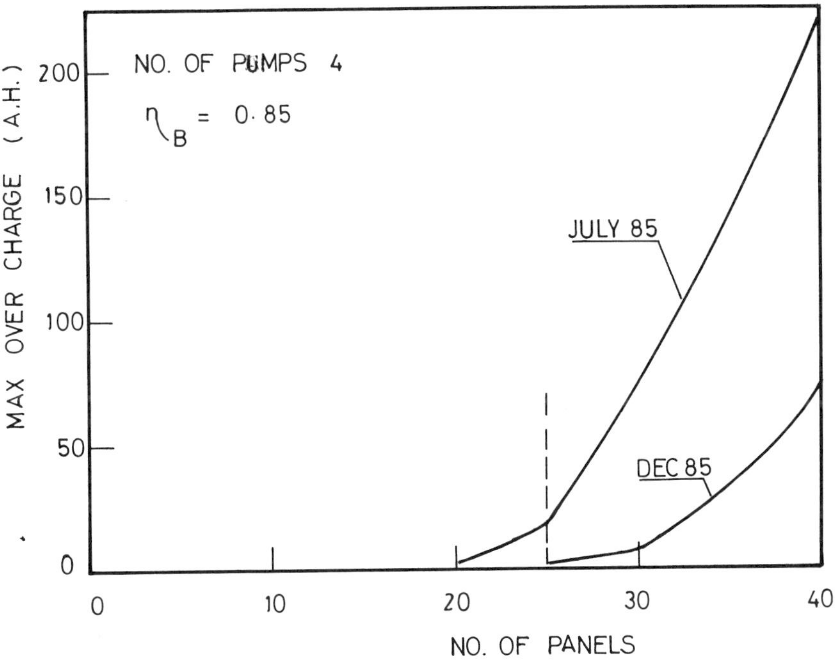

FIGURE 17: Effect of number of panels on battery overcharge for a given number of pumps

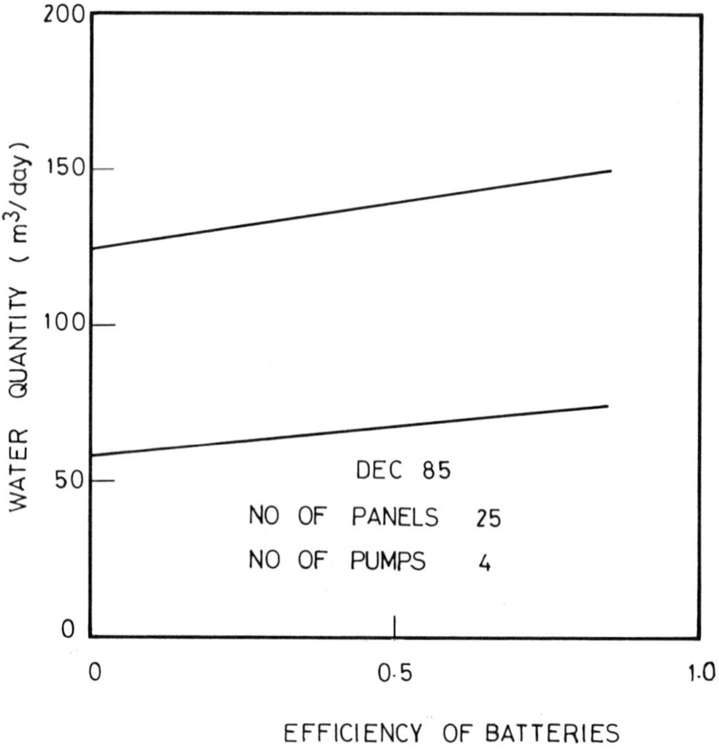
200
150
100
50
0
WATER QUANTITY (m3/day)
DEC 85
NO OF PANELS 25
NO OF PUMPS 4
0
0.5
1.0
EFFICIENCY OF BATTERIES

that the maximum electrical power available is slightly more than the power required by the maximum number of pumps to be used.

3- A limited-size battery bank can be used to maximize the utilization of available renewable energy sources. The use of batteries however, should be balanced by problems of maintenance and upkeep. In the present system, whenever power is available it is used to pump water. Limited excess power resulting from the step load demand of the pumps is either wasted if no battery banks are used or stored in limited size battery bank.

REFERENCE

1- Mobarak, A., Safwat, H., and Abdel Hamid, A., "Availability of Wind Energy in a Typical Desert Location in Egypt" presented at the Second International Conference, Desert Development, Jan. 25-31, 1987.

EGYPTIAN-ITALIAN RENEWABLE ENERGY SETTLEMENT (EIRES)

A. ZUFFETTI
AGIP S.p.A., Milano, Italy

Abstract The Italian and Egyptian governments are implementing a large development project (LDP) involving construction and operation of an agricultural settlement in Egypt based on the exploitation of renewable energy sources (EIRES: Egyptian-Italian Renewable Energy Settlement). This project is financed by the Department for Development in cooperation with the Italian Ministry of Foreign Affairs. This cooperation aims particularly at carrying out the recovery of desert areas which allow the production of foods by the utilization of local resources.

The Italian AGIP SpA and the Egyptian General Petroleum Company (G.P.C.) are two companies which lead the project. AGIP is providing the basic design, detailed engineering, technology, procurement of main equipment and project supervision. GPC provides detailed engineering, procurement of materials, site construction and project management. Aeritalia, AGIP-GIZA, Life, Pragma and SES have been appointed to carry out different tasks within the project. The implementation of the project has started. The total cost is estimated at 15 million US dollars.

The main aims of the project are:

- To introduce modern agricultural and animal breeding techniques and to ensure comfortable conditions for a permanent human settlement in a desert or semi-desert area.

- To produce human food.

- To provide a high level of self-sufficiency.

INTRODUCTION

The Egyptian-Italian Renewable Energy Settlement (EIRES) project consists of an integrated experimental agricultural and cattle breeding farm, based on the use of energy from renewable sources. Located in the Western Desert of Egypt, this project is financed by the Department for Development Cooperation of the Italian Ministry of Foreign Affairs. It aims to recover desert areas with emphasis on utilizing local resources for food production.

The Italian AGIP S.p.A. and the Egyptian General Petroleum Company and the two companies leading the project. On the Italian side, PRAGMA, AERITALIA, SES and LIFE have been appointed to carry out different tasks within the project. AGIP is providing the basic design, detailed engineering, technology, procurement of main equipment and project supervision. On the Egyptian side are the General Petroleum Company, a company of the Egyptian Oil Ministry, and the Military Technology Institute. The G.P.C. is providing detailed engineering, procurement of materials in Egypt, site construction and project management.

OBJECTIVES

The EIRES Project, located in the remote and hostile environment of East Oweinat, aims to produce food despite the environment's negative characteristics and, once the settlement is fully operational, to sell its products at a competitive price in the market. While the above is a hard task to achieve, the immediate objectives are:

a. to introduce in the desert, or semi-desert areas, modern agricultural and animal-breeding techniques and ensure comfortable conditions for a permanent human settlement.

b. to make use of renewable energy sources such as

photovoltaics, wind and biogas in an integrated system.

c. to produce food under a system that is largely self-sufficient.

Once the settlement is a full-scale operation, its expected foods will be meat, chicken, fresh milk, eggs and vegetables grown in greenhouses.

FARMS

GENERAL

The economic base of the settlement is the farm. Cropping, animal husbandry and management must be compatible with energy resources. Experts have determined the needs of proposed vegetable species and the needs of cattle breeds able to acclimatize to this area. These parameters, taking into account the seasons, the life cycle and the adopted methods, were introduced in mathematical models and worked out by means of computers to define optimized solutions. From these, taking into consideration some constraints (climate, temperature, wind, evapotranspiration, pedologic characteristics, and available energy), two models were selected: one with maximum agricultural optimization and one with absolute energy autonomy.

Thereafter, the crops of the various fields were determined. On these fields, subject to sowing rotation and spacing, the main crops will be fodder plants such as alfalfa, sorghum and barley.

For selection of a cattle breed, cattle acclimatized to the New Valley will be considered. They will be transferred and bred with particular care so as to assure "industrial" yields. Cattle are unprecedented in this area.

AGRICULTURE

Modern developments in agronomy allow plants to overcome unfavorable environmental conditions caused by:

- total absence of rain,

- hot and dry winds that are almost constant,

- a sandy soil configuration almost totally lacking in nourishing elements.

It is possible to compensate for lack of rain by providing a suitable irrigation system and pumping from the underground water table. The presence of rows of trees for windbreaks limits the negative action of the wind and, by adding organic manure and vegetable refuse to fertilizers, essential nourishing elements are introduced into the sandy soil.

The farm is devoted mainly to breeding cattle for production of milk and meat. A goal for the future is to cultivate and supply the necessary products for balanced feeding of the animals.

Among the various cultivars possible, those which show great adaptability to unfavorable environmental conditions and which produce acceptable yields in environmental conditions similar to those of the East Oweinat region were selected. The cultivars selected are alfalfa, berseem, sudangrass, sorghum and barley.

The total agricultural surface, of which 67.2 hectares can be cultivated, will be subdivided into 96 plots and joined two by two forming 48 monoculture production units. The 48 units are surrounded, on all four sides, by roads and windbreaks that are arranged crosswise as to the north-south direction to provide a greater number of windbreak bands in the direction of the prevalent winds.

Before beginning agricultural activities, deep ground cultivation will be carried out to break up the underlying compacted soil, allowing for uniform water drainage. About 40 or 50 tons of animal manure will be distributed to fertilize the soil before the plowing.

CATTLE BREEDING

Animals will be bred for both milk and meat production. Their manure will be used in an anaerobic digestion plant to produce biogas and fertilizers. Local cows and Dutch bulls will be crossbred to produce strong cattle that can tolerate the arid climate. The herd will consist of 161 head of cattle: 70 cows, 1 bull, 60 covering animals, and 30 animals for fattening. The farm's expected total yearly milk production will be about 2000 hl. Meat production will equal 30 animals per year, each with an average weight of 440 kg.

BIOGAS PLANT

The biogas production plant was designed to utilize the amount of animal waste that will be produced when the farm is working at full capacity. An anaerobic digester, of 99 cubic meter capacity was built from sheet steel, painted inside and insulated outside to keep the internal temperature at an optimum temperature of 35 - 36°C and to minimize heat leakage.

GREENHOUSE

Vegetables for the farm personnel will be grown in the greenhouse throughout the year. In addition, experiments with particular types of vegetables will be conducted there.

IRRIGATION PLANT

The irrigation plant was designed to meet agronomical needs and allow easy management of cultivation rotations and irrigation turns on the various plots as well as to satisfy water demand during the months of maximum utilization. The need has been calculated on the basis of evaporation data obtained for the East Oweinat site.

The water is lifted from two wells. A main well has a 200 cm/h water flow rate nominal capacity and a standby well has 100 cm/h. The lifted well water equals 1,200,000 cm/year. The lifted water is stored in a water basin with 35,000 cm capacity and then pumped to the users by the irrigation pumps. The basin has been designed to store water for seven days.

Field irrigation is done through a feeding system with pipes having a 400 - 200 mm diameter and a rain distribution system of 38 sprinklers for each of the 48 cultivation units.

WINDBREAKS

Wind disturbs agricultural cultivation. It is necessary to use efficacious instruments to avoid damage that may occur directly (damage to fruit trees and cultivations) and indirectly (scoring due to sand and excessive evapotranspiration).

Windbreak rows of quick-growing trees, such as Casuarina (Cassowary) and Tamarisk are provided on and within the farm perimeter. The total number of trees required for the whole windbreak band is about 21,500. The windbreak and orchard irrigation plant is a "drip" type with semiportable PVC drip pipes able to supply a maximum of 2 l/h per tree.

ENERGY

Energy production normally comes from a wind and photovoltaic hybrid system. Integration with a diesel unit has been considered. The choice of the photovoltaic unit is based on insolation, temperature and site location. The choice of the wind unit is based on observations of the wind duration and speeds at the site. These data were introduced in a computerized mathematical model and daily energy values were obtained and used to design the system and its electrical components. The estimate of energy consumption was based on an appropriate evaluation of the farm needs and other uses.

The comparison between generation and consumption allowed both specification of users' sizing and definition of the design criteria of energy storage, in batteries as well as the water basin. The design energy production is 628,000 kWh/y. The expected energy production is 1,063,000 kWh/y. The nominal voltages are 380 Volts in alternate and 300 Volts in direct current. Also, the 12 Kwe biogas cogeneration experimental system is connected to the plant.

Energy from the PV-wind-biogas-battery is collected on a common 300 V DC main bus and then distributed to inverter panels. The inverters feed a single big user or a motor control center that, in turn, feeds the single small user. The inverter panels can receive energy from both DC sources and AC sources by diesel unit.

Two diesel units are used for (1) battery charging, (2) energy integration, and (3) emergency situations. Units can be synchronized to work in parallel.

Four battery units (15 elements each) for a total of 1800 kwh are intended, not merely as energy storage but also as an electrical flywheel to smooth the power peaks (generation/load) and absorb the transients. The battery also constitutes the voltage reference point (300 V) for the load regulation/management. In the case of a full blackout from generation, the battery storage could be sufficient for one or to two days depending on whether it is necessary to irrigate and pump water from the well or not.

A master computer unit monitors or regulates the load and the energy production making use of the microprocessors' secondary unit. Daily energy produced under normal conditions is greater than that needed for loads. Energy surplus is shared between the water basin filling and battery charging, while a temporary deficit can be covered by the diesel unit and by resorting to batteries. The total yield of the system is about 43%.

POWER STATION

The power station consists of a package modular unit preassembled and wired with the following equipment:

- A DC panel collects all energy produced, i.e., PV-wind-biogas. The batteries with their balancing function and the inverter feeding the users.

- Seven inverter panels. Four are pump-dedicated (well pumps and storage basin pumps) and three charge the several small loads; one further panel is an installed standby.

- A diesel system. Diesel engines are located in a room separated from the power station and built with provisions to avoid heat and vibration transmission to the other part of the building and its equipment.

- A control room containing all computers, an office, sanitary services, and a corridor complete the station.

- The battery room. A shed for protection and an insulated modular unit helps to maintain average temperatures and lengthen the life and efficiency of the accumulators.

PHOTOVOLTAIC FIELD

The total power of the photovoltaic field will be 200 kwp (AM 1.5; t = 25°C), utilizing monocrystalline and polycrystalline modules. The latter has a slope which changes according to an east-west daily tracking system; the remaining strings (69) are mounted with a 15° fixed slope.

WIND SYSTEM

Four maximum power machines with 31,2 KVA each and one 280 KVA are provided. They yield 7 kw each and 128 kw, respectively, at design wind speed of 7 m/s.

Totally, the installed gross power is, therefore, 313 kw and the delivered one, at design conditions is 156 kw. Fundamental considerations are wind speeds and their statistical incidence during the year, as well as air density, types of turbulence and sand transport possibilities in the atmosphere.

At East Oweinat, it is necessary to ensure that enough energy will be produced under the conditions referred to and that energy production will be assured at lower wind speeds as well as at higher ones.

COOLING AND BIOCLIMATIC TECHNIQUES

The principles of passive solar architecture has been applied to all buildings. Cooling is based on comfort needs for personnel's lodging, working, or resting quarters. A special case may be cattle in sheds or running equipment.

It is important to achieve mass cooling and where necessary, heat exhaust systems as close as possible to the source (electrical equipment). It was decided to resort to water evaporative cooling in air currents of the "residential building". For other cases, air/water heat-exchangers will be used. Well water is drawn at 21°C.

UTILITIES

The infrastructure and utilities of a farm consist of the drinking water distribution plant with relevant pumping stations and surge tanks, the cooling water pumping station, the road system, the fuel and gas cylinder storehouse, the irrigation water lifting and distribution plant, the cooling store, the workshop, the stores and social buildings (residences, common rooms).

WATER LIFTING AND DISTRIBUTION

The system consists of a water-storage basin and a water-lifting station. The basin is the compacted earth type with its bottom and walls coated with waterproof membrane. The water pumping from the table to the basin is carried out by two wells. From the basin the water is pumped to the irrigation network for field, orchard, windbreaks and cooling/climatization of buildings.

STATUS OF THE PROJECT

The basic design of the project was completed at the end of 1984. Detailed engineering is almost completed and procurement of equipment is complete. Late deliveries are expected by the second quarter of 1987. Erection activities began in July 1985 and will continue until the end of 1987. Water basin and pump houses are nearing completion. Water lifting and storage are expected to be complete in May 1987.

SUMMARY OF TECHNICAL DATA

THE SITE

Location:	East Oweinat, southwestern Egyptian desert. Latitude 22° 24'N. Distance from the nearest populated area: 400 km.
Soil:	Sterile Nubian sands
Altitude:	250 m a.s.l., flat area
Temperature:	From -2°C to 45°C
Insolation:	SKWh.sqm day - summer; 5-6 KWh/spm day - winter
Wind:	Frequent, medium-low speed; some sand- storms; average wind speed 7 m/sec.
Humidity:	25-40% average summer/ winter
Evaporation:	10-25 mm/day average winter/summer

THE FARM

Cultivated Fields:	67 ha
Orchard:	1.5 ha
Garden:	0.5 ha
Greenhouse:	100 sqm
Cows:	70
Bull:	1
Covering animals:	60

Fattening animals:	30
Milk production/year:	2000 hl
Meat production/year:	13 tons
Cold store:	70 cm
Population:	70 persons
Casuarina trees:	21,500

THE WATER

Sources:	From two wells at a level below 30 mt, having 300 cm/h total design capacity
Water storage:	35,000 cm
Lifted water:	1,200,000 cm/year

THE ENERGY

Expected production:	1,063,000 kWh/yr.
Design production:	628,000 kWh/yr.
P.V.:	200 KWp equal to 366,100 kWh/yr.
Wind:	320 kW equal to 261,900 kWh/yr.
Biogas:	12 KWe on experimental basis
Battery:	1,800 kWh for 10 h, Pb type
Voltages:	300 V d.c.; 380/220 V 3F+N - 50 Hz
Diesel unit:	Two, 80 KW each
Energy consumption:	500,000 kWh/yr.

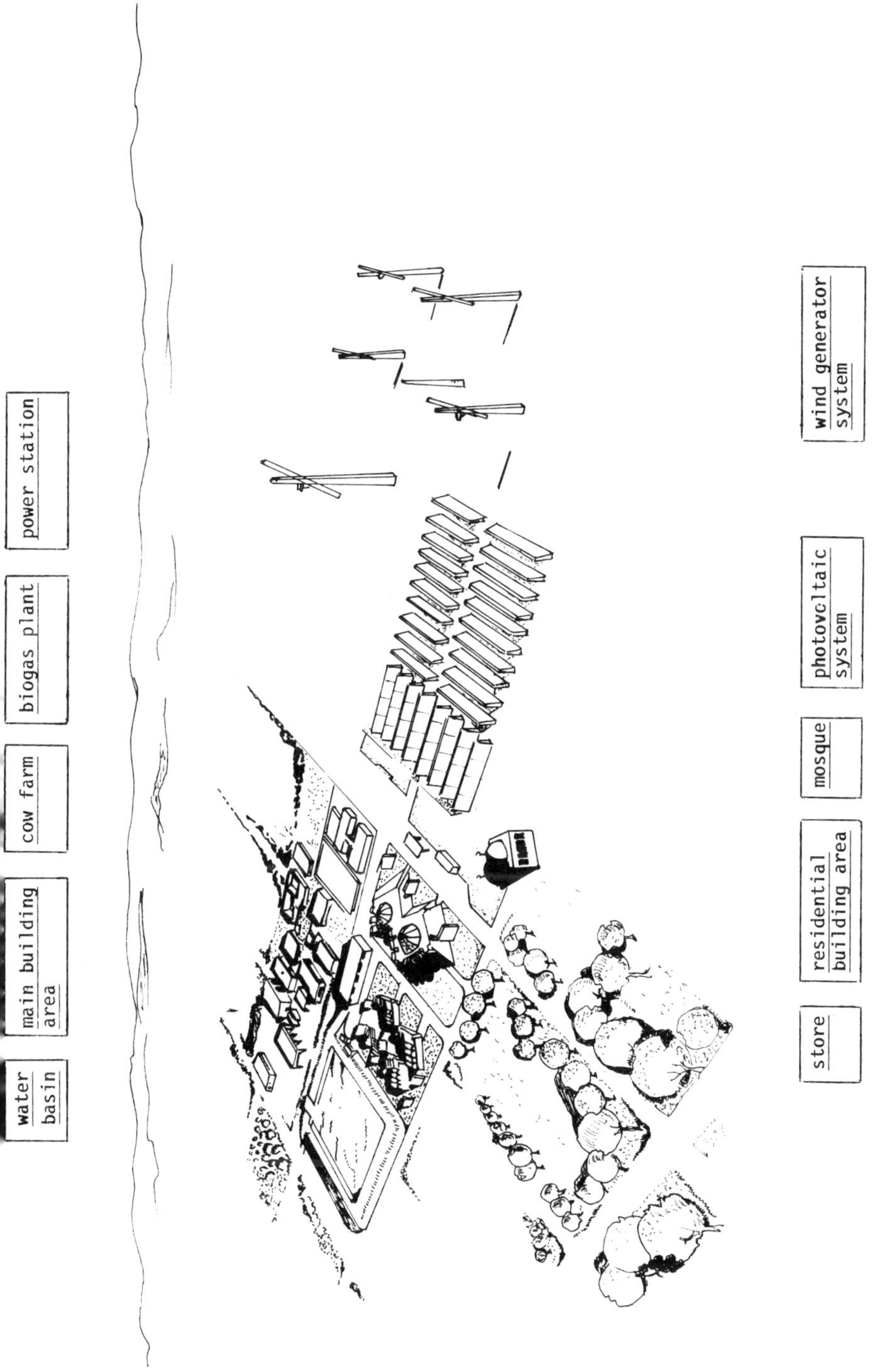

General view, East Oweinat

DEMONSTRATION AND FIELD TESTING OF PHOTOVOLTAIC TECHNOLOGIES IN SOME APPLICATIONS HAVING POTENTIAL FOR USE IN EGYPT

REDA BOTROS
Egyptian Electricity Authority, Cairo

PETER BORGO
Meridian Corporation, Falls Church, Virginia

Abstract Pros and cons in selecting applications of photovoltaic energy for remote villages in Egypt are discussed. Village lighting, water pumping, and desalination are considered.

INTRODUCTION

The Egyptian Electricity Authority, US AID/Cairo, and a group of U.S. consultants form a team responsible for conducting field test demonstration projects of eleven renewable energy applications in Egypt. These include the use of photovoltaic, wind and solar thermal systems for water pumping, ice making, desalination, industrial process heat, small village power, and grid-connected electricity. The specific objectives of the four-year program are two-fold: (1) to demonstrate the viability of renewable energy technologies in Egypt, and (2) to establish the infrastructure necessary to ensure that renewable technologies, which have proven successful, are available for widespread use in the country.

The village power applications result from a desire to provide electricity for the basic domestic needs of people living in small, remote villages that are not connected to the Egyptian national grid. These villages are generally comprised of less than 100 residents, approximately 10 to 15 households. There are hundreds of these villages throughout Egypt; many are in coastal, mountainous or other topographically-enhanced regions where the wind resource may be sufficient for the application of wind electric systems and all are in areas where the solar resource is sufficient for the use of photovoltaic power systems.

There are three photovoltaic (PV) field tests as follows:

1. a PV/diesel hybrid system for ice production of six tons per day for fish preservation at Wadi El-Raiyan, El-Fayoum, which is a remote area 140 km southwest of Cairo,

2. PV power for electricity production for pumping, lighting and domestic use at Baharia Oasis in El-Ezz village,

3. a PV desalination plant backed by diesel at El-Kasr village near Mersa Matrouh Governorate which is 600 km northwest of Cairo beside the Mediterranean Sea.

LIGHTING AND WATER PUMPING AT BAHARIA OASIS

El-Ezz village in Baharia Oasis was selected for the demonstration of photovoltaics for village lighting and water pumping. The government had no plans to supply this area with electrical services either by grid or remote diesel generator.

At most places within the Baharia Oasis, water is obtained from artesian wells that carry groundwater to the surface under natural pressure. The water table depth in this area is less than five meters from the surface.

The average water requirement per feddan is estimated to be 25 m^3/day during the summer. The minimum farming area for agricultural self-sufficiency for a household is designated as approximately three feddans.

PV PUMPING SYSTEMS

Shallow pumping is the best application of PV systems for such areas because:

1. The system is simple. It is composed of a PV array directly coupled to a permanent magnet DC motor that drives a centrifugal pump as shown in Figure 1.

2. There is no need for energy storage. One may store water instead of energy. Thus battery banks and a battery-charged controller can be eliminated from the system.

3. It is low cost and reliable.

4. The I-V characteristics of the DC motor centrifugal pump group are well - matched with those of the PV array characteristics so that the pumping system extracts nearly the maximum power from the PV array at any radiation level. The pump speed simply increases with the solar radiation resulting in higher discharge around noon. The behavior of the pumping system is indicated in Figure 2.

5. Water demand is naturally matched with the level of solar radiation, i.e. more water is needed during summer when solar radiation is maximum.

6. The reasons previously mentioned make the PV pumping system the most preferable of PV applications, especially when the water depth is not great and a centrifugal pump can be used.

7. Many PV pumping systems are performing successfully throughout the world. The East Owienat project in the southwestern part of Egypt is one example.

These positive aspects of PV systems led the working team to install a PV pumping field test in

FIGURE 1.

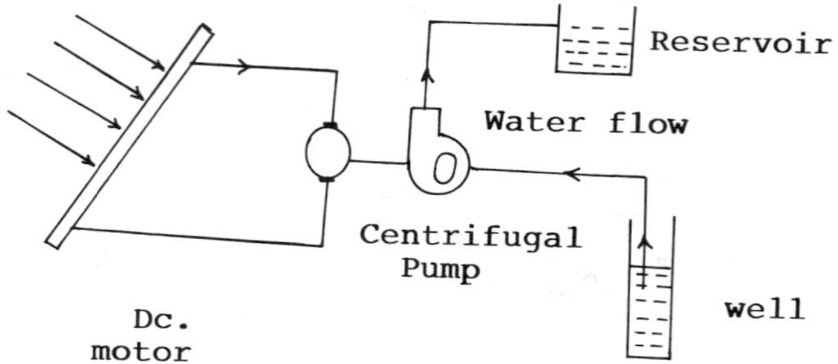

Schematic representation of directly coupled to P.V.

FIGURE 2.

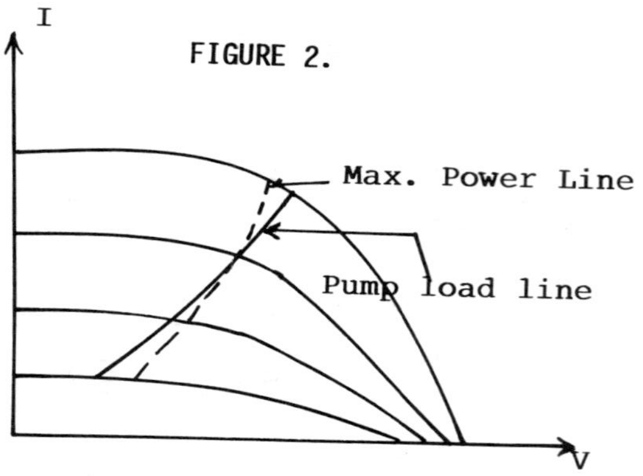

Maximum power line and pump load line at various radiation level

Baharia Oasis 380 km southwest of Cairo. A PV pumping system planned for Baharia Oasis will serve both land irrigation and domestic use. A nominal head of 5 meters depth will be used for system sizing. Since the average water requirement per feddan is about 25 m^3/day and the land area is 60 feddans, the overall water demand is 1500 m^3/day from 5 meters depth. Multiple modular systems are planned to cover the water demand of 1500 m^3/day.

The system includes the following subsystems:

1. A primary water - pumping system with:

   a. A PV power source (about 9 KWp) to provide sufficient water to flood irrigate a 60 feddan area every other day plus a total of 10 cubic meters per day for domestic use.

   b. A water source with a minimum of two new wells.

   c. Two inter - connected water storage components to provide:

      i. Two hours of water supply for irrigating early in the day, prior to the PV system's becoming operational with sufficient piping and valving to transport and control water flow into the existing irrigation system.

      ii. Up to a one-day supply of water for domestic use with piping to allow a central location of a single, manually - operated valve for extracting domestic water.

2. An option of two portable PV water pumping units. These produce about 130 Wp capable of providing sufficient water to flood irrigate one feddan per day from shallow wells.

PV LIGHTING SYSTEM

1. Twenty PV power systems are included. Each is about 260 Wp and is capable of providing an individual household with sufficient power to meet the energy demand of 1000 watt-hours/day. Each system will include two 20-watt and two 40-watt florescent lights with fixtures; four 220 vac electrical outlets; batteries; inverter; and sufficient wiring and hardware to allow for a light and outlet in each of four rooms. The contractor will be required to install the systems, assure that each is operational, and demonstrate to the user that the systems can be used for appliances such as a television, refrigerator, small power tools, sewing machine and battery charger.

2. Ten individually PV - powered street lights of about 72 Wp each are to be located where required in and near the village of El-Ezz.

The electrification of houses and streets may be achieved by systems which are either centralized or decentralized. Using a centralized system, all PV panels, battery storage, and battery charger controllers are installed in a central station from which electrical power is distributed to households and streets. This technique decreases installation costs and facility maintenance. A significant power loss results, however, if DC power at low voltage is transmitted.

Another centralized option would be to convert DC power generated at the central station to AC power through a DC/AC inverter and then transmit the power as shown in Figure 3. Since an AC system would have an overall 10% efficiency penalty, the array size would need to be increased by 10%.

In a decentralized system of electrification, PV panels would be placed on each of the twenty houses and on each street lamp. In this option, the lengths of cables are minimized but the installation costs increase. Both centralized and decentralized systems are being studied for their

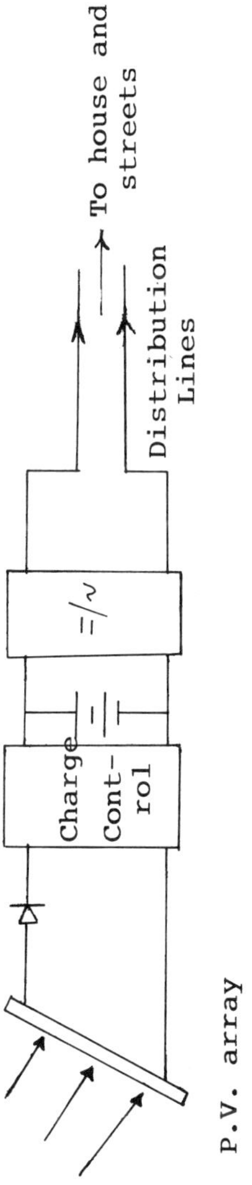

FIGURE 3: Centralized P.V. Plant

technical reliability and cost.

PHOTOVOLTAIC DESALINATION PLANT

Fresh water is a basic need for development in Egypt, particularly in remote areas. The Government of Egypt is mobilizing efforts to develop coastal and desert areas by providing them with fresh water and necessary energy. To achieve this goal, the project group of EEA/MC consultants conducts studies before preparing system specifications for reliable systems.

Egypt's northern coastal area has some old Roman wells with brackish water. They contain a stratified layer of potable water during the rainy season and high-salinity water during the dry season. These wells are open channel wells rather than conventional deep wells (Figure 4). Their construction permits contamination by particulate matter. Such matter can be measured by the salt density index (SDI) in order to specify necessary pretreatment to make the water potable. Final equipment specifications can be prepared accordingly.

In these remote areas, the lack of an adequate water distribution network implies that smaller plants distributed close to demand centers would be advantageous.

The objectives of this project at El-Kasr are to:

1. Demonstrate, in a remote Egyptian environment, the viability of a PV-powered stand-alone desalination system.

2. Demonstrate a capability to desalt brackish water with variable salinity, changing it to potable water with salinity of no more than 500 ppm total dissolved solids.

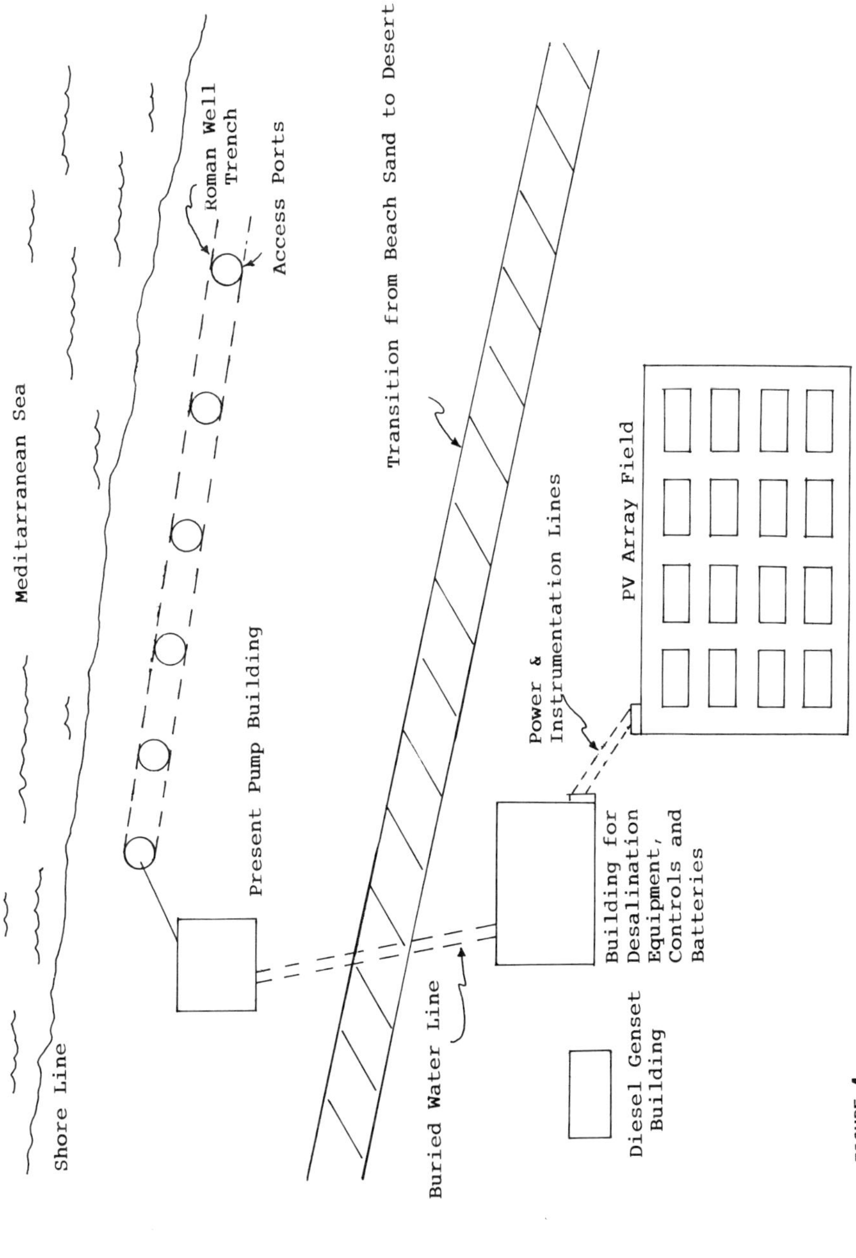
Shore Line
Meditarranean Sea
Roman Well
Trench
Access Ports
Present Pump Building
Transition from Beach Sand to Desert
Buried Water Line
Power &
Instrumentation Lines
PV Array Field
Building for
Desalination
Equipment,
Controls and
Batteries
Diesel Genset
Building

FIGURE 4.

3. Demonstrate that such a system can be maintained in a remote location.

A plant capacity of 10 to 20 m^3/day can satisfy the potable water needs of 500 people. A reliable plant of this size easy to operate and maintain should have wide application throughout Egypt.

The availability of PV power more or less complements the need for fresh water in the northern region. The periods of lower insolation occur during the rainy season matching water needs very well. In summer when water needs are greater, insolation is higher.

DESALINATION TECHNIQUES

There are three techniques frequently used for water desalination:

1. The multi-flash system (MFS), is a thermal technique by which water is desalinated by a distillation process. The water is simply evaporated leaving the salt in the basin, and the condensed vapor is collected as fresh water. In this technique, the specific energy (energy required to desalinate one cubic meter of water) is constant regardless of the water salinity. The heat energy required in MFS is about 11 Kwh per cubic meter.

2. In reverse osmosis (RO), the saline water is subjected to high pressure so that the fresh water passes through a selective membrane leaving the dissolved solids on the other side of the selective membrane.

   The pressure required is approximately linearly proportional to water salinity. Thus for low salinity (such as brackish water), the energy demand is less than that required for the distillation technique. Another important difference between RO and

MFS is that the energy in MFS is in the form of heat, while for RO it is an electrical energy. This electric energy drives the High Pressure Pump (HPP).

3. Electro - dialysis (ED) depends on the existence of a selective ion-permeable membrane placed in the salt solution in such a manner that a current flowing in the solution must pass through the membrane. The passage of current through the system will cause the formation of a more concentrated layer on one side of the membrane and a diluted layer on the opposite side. The specific energy of this technique is smaller than that of MFS and RO for very low water salinities (up to 2000 ppm). For higher salinities (up to 15000 ppm), RO has the least specific energy and for much higher salinities MFS is the optimal choice (See Figure 5).

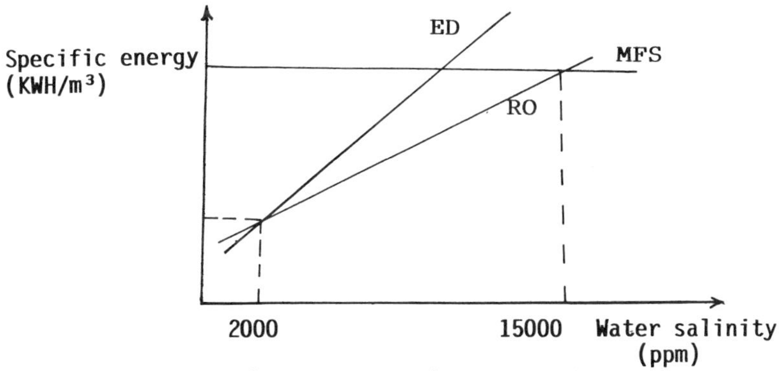

FIGURE 5: Schematic representation of specific energy of different techniques

Hence, the choice of the optimal technique depends on the water salinity. In the site under consideration, the water salinity varies from one season to another. When rain falls, the underground water salinity decreases to 1200 ppm during winter, while in the summer dry season, the water salinity rises to 12000 ppm. The annual average water salinity is considered to be about 7000 ppm; thus the optimum desalination technique will be by RO according to facts mentioned previously.

Since the RO system requires electric power, and the system will be installed in a remote area, the PV system is the suitable solution for generating electricity. Accordingly, the decision was made to install a PV-powered RO desalination plant near Mersa Matrouh on the Mediterranean coast.

REVERSE OSMOSIS SYSTEM

Saline water conversion systems are generally divided into two classes: (1) seawater conversion (high salinity systems) and (2) brackish water conversion (low salinity systems). Seawater RO systems have to be operated at high pressures (80 - 120 bars), while brackish water systems are usually operated at a pressure of about 40 bars. Most brackish water can be treated in a single RO process to obtain potable water. For seawater, a single-stage RO process would produce water of salinity of the order of 1000 - 1500 ppm which is not suitable for human consumption. To overcome this problem, the water produced in the first stage has to be fed into a second RO process, similar to the single stage used for brackish water. Therefore, brackish water requires a single-stage RO system while seawater generally needs a multistage system.

A schematic representation of a single-stage RO system is shown in Figure 6. Water is introduced into the system via a water suction pump (WSP). Between the feed source and the high pressure pump (HPP) there is a pretreatment unit. The objectives of the pretreatment unit are twofold: (1) controlling the pH of the raw water by acid injection, and (2) disinfecting the raw water to prevent growth of organisms within the system. Injection of acid is desirable to maintain a pH of between 3 and 5 in the raw water, thus increasing the working lifetime of the membrane. Periodic additions of chlorine prevent growth of organisms.

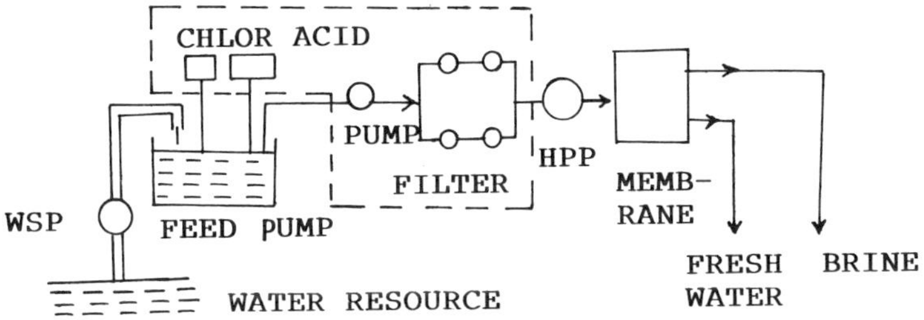

FIGURE 6: A single - stage RO system

A filter assembly unit is used to eliminate fine particles or emulsions of heavy greases or waxes which otherwise may cause serious problems, as such materials tend to stick to the membranes.

After filtration the water is pumped to the HPP where it enters the membrane separator unit. At the outlet of the membrane unit there are two water streams, the fresh water stream for human use and the brine stream.

DESIGN PARAMETERS

The following parameters will be considered in establishing the design of a desalination plant:

1. A capacity of 10 to 20m^3/day is desired.

2. The product water quality will be 500 ppm or less

3. The design will minimize energy consumption while maintaining simplicity of operation

4. The system will be capable of treating water with 1000 ppm to 12000 ppm.

5. The plant should operate during sunlight hours to minimize battery storage.

6. A diesel genset will supply back-up power

7. The system will use only proven, commercially-available products to improve reliability and maintainability.

CONCLUSION

1. The electro-dialysis (ED) desalination technique has the following advantages:

   a. The ED is a direct current process which is well suited for solar applications because the system self-corrects for increases or decreases in available energy, thus it uses the PV power more efficiently.

   b. The ED system will not require the addition of chemicals because the reverse polarity technique will minimize scaling potential.

2. In spite of the above mentioned advantages of the ED desalination technique, the RO units are generally preferable over the ED at the salinity levels of El Kasr village for the following reasons:

   a. The power requirement for daylight operation of the RO is less than that needed for ED operation for high salinity levels.

   b. The RO system will be designed on the basis of daylight operation, to permit for a larger size and to increase product water.

c. Larger RO desalination units are more reliable and efficient and are more available at industrial standards.

The principal conclusion is that the RO system is preferred over the ED system at higher salinity levels.

AVAILABILITY OF SOLAR ENERGY WITH DIFFERENT TRACKING MODES IN THE WESTERN OASES OF EGYPT

M.A. MOSALLAM SHALTOUT
Nat. Res. Inst. of Astronomy and Geophysics
Helwan, Cairo, Egypt

Abstract The five main oases of the Western Desert of Egypt (Siwa, Farafra, Bahariya, Dakhla, and Kharga) have 450,000 feddans of cultivable area. Underneath each oasis is a large store of deep underground water. These oases are remote areas where the use of photovoltaic pumping of underground water may prove to be economically feasible. It is estimated that in 1991 the cost of pumping a cubic meter of water at a pumping rate of 40,000 m^3/yr. from a depth of 30 m by photovoltaic systems would be 63% of the cost of pumping by diesel engines, or 69% of the cost of pumping by a solar thermal system. Solar energy is clean and solar systems would occupy only 0.1% of the cultivable land[1] in the area.

This study estimates (1) the different components of solar radiation based on meteorological data from each oasis, and (2) the solar energy input in kwh/m^2/day for five modes of photovoltaic systems for water pumping.

The diffuse fraction of solar energy is ≤ 0.2 of the total solar energy for all oases throughout the year, except in Siwa and Bahariya in April due to the effect of the dust storms (El-Khamasin).

INTRODUCTION

The Western Desert in Egypt covers an area of about 600,000 sq. km (about 60% of the total area of Egypt). It extends to the Nile valley in the east and longitude 25° E in the west, and to the Mediterranean Sea in the north and latitude 22° N in the south (see Figure 1). Its population is approximately 500,000 (less than 1% of the total population of Egypt). It contains five main oases: Siwa, Bahariya, Farafra, Dakhla, and Kharga (see Table 1 and Figure 1). The oases of the Western Desert are characterized by low and sporadic annual rainfall, low values of relative humidity, high rates of evaporation, high degrees of solar radiation and strong summer winds.

There are 450,000 feddans (hectare = 2.38 feddan) of cultivable area in the five main oases (see Table 1).[1] Each oasis is located above a large storage of deep underground water. It is of utmost importance to our national growth, balanced regional development and attainment of a reasonably good standard of living for the people, that we should develop the arid zones and make them as productive as possible. These objectives can be achieved if the resources available are exploited using scientific knowledge and technological developments.

TABLE 1: The western desert oases of Egypt

Oasis	Location			Cultivable area (feddans)
	Latitude	Longitude	Elevation (m)	
Siwa	29° 12'	25° 29'	- 15.0	25,000
Bahariya	28 20	28 54	128.0	50,000
Farafra	27 03	27 58	90.0	140,000
Dakhla	25 29	29 00	106.2	100,000
Kharga	25 27	30 32	77.8	135,000

Detailed knowledge about the solar energy input to any oasis is required in a large variety of applications, including the fields of agriculture, crop drying, direct solar heating, photovoltaic electricity, bio-conversion, hydrology and others. The application of photovoltaic electrical power to water pumping has received increasing attention recently because of the expected cost reduction in photovoltaic solar panels. In recent

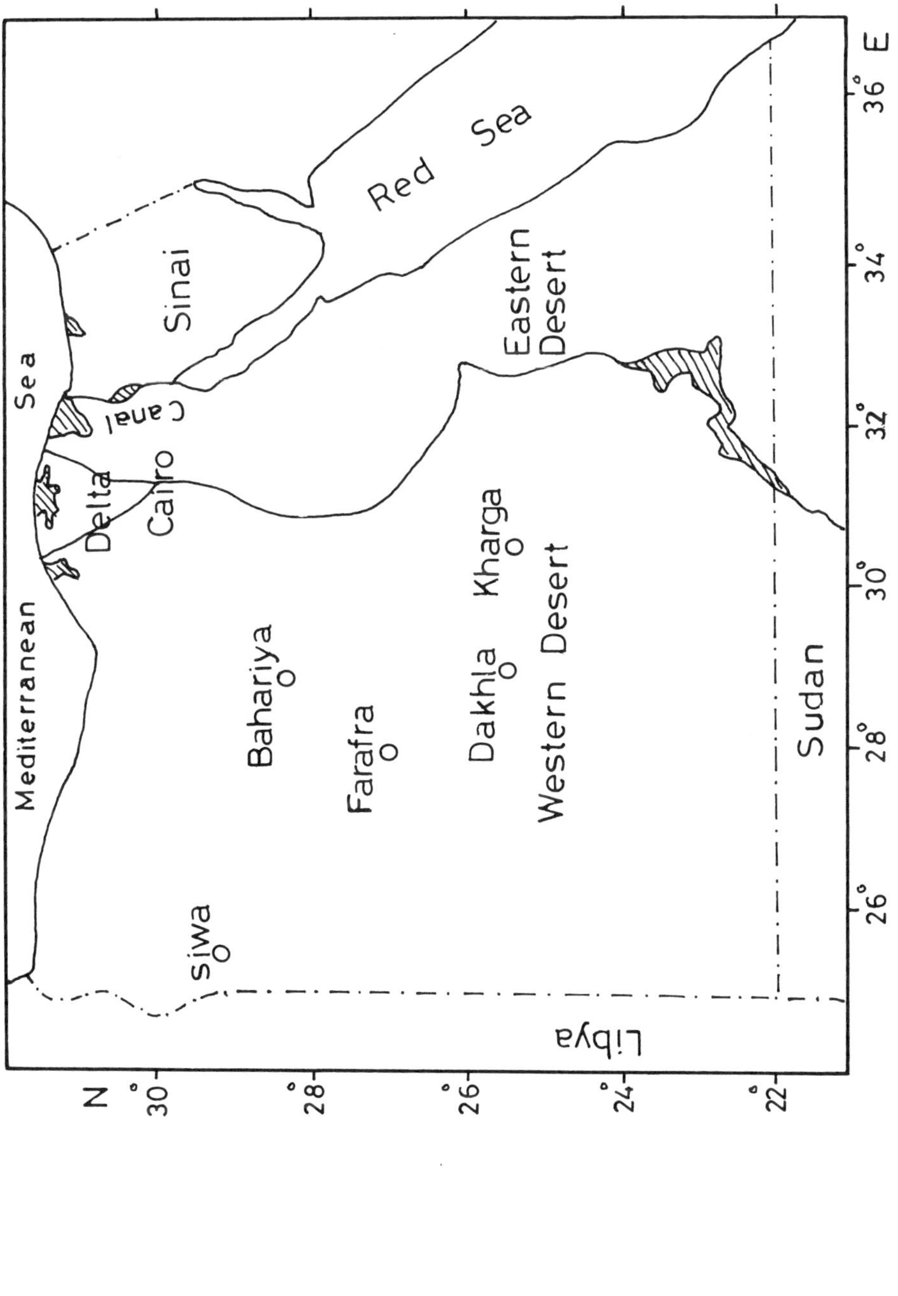

FIGURE 1: Location of the western oases in Egypt.

years, the prices of photovoltaic generators have decreased drastically.

The objective of this study is to estimate (1) the different components of solar radiation from the meteorological data for each oasis, and (2) the solar energy input in kwh/m^2/day for five modes of flat-plate photovoltaic tracking of water pumping systems.

DATA

Monthly mean values of the cloudiness and the relative humidity for the five main oases of the Western Desert of Egypt are obtained from "Climatological Normals for the Arab Republic of Egypt up to 1975" of the Egyptian Meteorological Authority, Ministry of Civil Aviation, Cairo. The values are given in Table 2.

TABLE 2: The monthly mean values of cloudiness (CC) and relative humidity (R) for the western desert oases

Oasis	Siwa		Bahariya		Farafra		Dakhla		Kharga	
Month	CC	R	CC	R	CC	R	CC	R	CC	R
Jan.	1.6	53	1.7	48	1.5	46	1.0	41	1.2	43
Feb.	1.4	45	1.6	43	1.5	39	1.0	34	1.0	36
Mar.	1.3	38	1.4	34	1.3	30	0.8	26	1.2	29
Apr.	1.6	33	1.7	29	1.5	24	1.1	20	1.2	22
May	1.2	30	1.2	28	1.0	23	0.8	19	0.8	21
June	0.3	30	0.3	31	0.2	23	0.2	20	0.2	21
July	0.1	34	0.3	35	0.3	26	0.2	23	0.2	24
Aug.	0.3	37	0.2	36	0.3	28	0.1	24	0.2	26
Sept.	0.4	41	0.4	42	0.4	33	0.1	29	0.3	30
Oct.	0.9	45	0.9	44	0.6	37	0.4	32	0.5	32
Nov.	1.6	52	1.5	51	1.3	45	0.7	40	0.8	42
Dec.	1.9	56	1.3	50	1.8	47	1.2	42	1.1	46
Annual	1.0	41	1.0	39	1.0	33	0.6	29	0.7	31

CF = CC/8

METHODOLOGY

There are two methods for estimating the different components of solar radiation (global, diffuse, and direct) from meteorological data. Both are adopted by the author for any site and any month in Egypt.

Method I: Estimate the different components of solar radiation from the monthly mean value of cloudiness and relative humidity of the site.[2]

Method II: Estimate the different components of solar radiation from the monthly mean value of sunshine duration at the site.[3]

The monthly value of global radiation and sunshine duration are available for Kharga Oasis only, while the monthly value of cloudiness and relative humidity are available for all oases. For that reason, we selected Method I for our computations and analysis.

In this method, the instance values of direct and diffuse components of solar radiation were calculated and then used to compute the mean monthly global radiation.

DIRECT COMPONENT

According to Paltridge and Proctor,[4] a single empirical curve relating direct radiation to the solar zenith angle in clear skies is:

$$I\ (\Theta) = C\ (1 - \exp\left[- 0.075\ (\tfrac{\pi}{2} - \Theta)\ \right])\ldots(1)$$

(I and C are in W m^{-2} and C is taken to be equal to 950.66 W m^{-2} for Australia).

The calculation of direct beam energy over a day is simply a matter of calculating as a function of time, integrating I (Θ) over a day, and reducing the integral value by the observed fractional cloud cover CF, since the presence of cloud in the solar beam normally reduces 1 to zero. Mathematically, the appropriate formula used in calculating the mean daily input of direct energy I_h on a horizontal plane is:

$$\bar{I}_h = (1 - CF) \sum_{T\,SR}^{T\,SS} I\,(\Theta) \cos \Theta \quad \ldots\ldots\ldots(2)$$

Where TSR and TSS are the times of sunrise and sunset, and

$$\cos \Theta = \sin \delta \sin \Phi + \cos \delta \cos \Phi \cos H \quad \ldots(3)$$

Where: Θ = the solar zenith angle, δ = the solar declination for the time of the year and H = the hour angle.

In the present work, the values of H were calculated by using the following relationship.

$$H = \frac{360}{24}(12-t) \quad \ldots\ldots\ldots\ldots\ldots\ldots\ldots\ldots\ldots(4)$$

Where t = the number of hours before and after solar noon.

Furthermore, equations (3) and (4) enable the sunrise and sunset hours to be calculated as Θ is equal to $\pi/2$ at sunrise and sunset.

DIFFUSE COMPONENT

In clear skies, D is assumed to be primarily a function of the solar zenith angle Θ:

$$D\,(\Theta, CF) = A_1 + A_2 \left(\frac{\pi}{2} - \Theta\right) \quad \ldots\ldots\ldots(5)$$

In order to introduce the effect of clouds on the diffuse radiation, a relationship of the form

$$D\,(\Theta, CF) = A_1 + A_2 \left(\frac{\pi}{2} - \Theta\right) + A_3\, CF (6)$$

was used where CF equal to zero indicates a clear sky and CF equal to 1 indicates the sky is fully covered with clouds. A_1, A_2 and A_3 were proposed to be equal to 2.535, 3.478 and 200.85 respectively for the United States by Kreider and Kreith,[5] where D is in W m^{-2}.

In order to calculate the mean daily value of the diffuse solar radiation D, the following relationship was used:

$$\bar{D} = \sum_{TSR}^{TSS} D\ (\Theta\ , CF) \quad \text{.........................(7)}$$

GLOBAL RADIATION

To compute the values of global solar radiation on a horizontal surface, the instantaneous value of G (Θ) is first calculated

$$G\ (\Theta) = (1 - CF)\ I\ (\Theta)\ \cos\ (\Theta) + D\ (\Theta) \quad \text{......(8)}$$

The mean daily global solar radiation on a horizontal surface G is obtained by a simple summation.

$$\bar{G} = \sum_{TSR}^{TSS} G\ (\Theta) \quad \text{...........................(9)}$$

According to Shaltout[2], the constant C in equation (1), and the constants A_1, A_2 and A_3 in equation (6) must be subject to regional and monthly variations in Egypt, and are given by the following equations:

$$C = 950.66\ [1.14 - 0.383\ (R/100 - 0.15)]\ Y_i \quad (10)$$
$$A_1 = 2.535\ [0.46 + 0.667\ (R/100 - 0.15)]\ Z_i \quad (11)$$
$$A_2 = 3.478\ [0.46 + 0.667\ (R/100 - .15)]\ Z_i \quad (12)$$
$$A_3 = 200.85\ [0.46 + 0.667\ (R/100 - .15)]\ Z_i \quad (13)$$

Where R is the annual mean of relative humidity at the site, and Y_i and Z_i are factors for the monthly variations, and can be obtained from Table 3, where i represents months numbered from 1 to 12.

A computer program was designed in order to calculate the values of the mean monthly direct, diffuse and global solar radiation on a horizontal surface.

The time step for the summations in equations (2), (7) and (19) was 15 minutes. The solar zenith angle calculations and integrations of equations (2), (7) and (9) were performed for the 15th day of each month.

TABLE 3: The factors Y_i and Z_i

i	Month	Y_i	Z_i
1	Jan.	1.1108	0.9970
2	Feb.	1.0978	1.0462
3	Mar.	1.0524	1.0450
4	Apr.	1.0058	1.1562
5	May	0.9556	1.0496
6	June	0.8994	0.9622
7	July	0.8988	0.9732
8	Aug.	0.9084	0.9832
9	Sept.	0.9388	0.9776
10	Oct.	0.9788	0.9586
11	Nov.	1.0492	0.9326
12	Dec.	1.1084	0.9166

The cloud factor CF was, however, a true monthly average obtained by using the meteorological data from past years. Therefore, the values of direct, diffuse and global solar radiation on a horizontal surface are those of monthly means.

ENERGY FLUX DENSITY

The instantaneous flux density F (Θ) received by flat plates can be calculated by a formula, from Paltridge and Proctor:[4]

$$F(\Theta) = I(\Theta)\cos B + D(\Theta)\cos^2\frac{\varepsilon}{2} + G(\Theta).A.$$

$$(1-\cos^2 \frac{\varepsilon}{2}) \quad \text{.......................(14)}$$

where B is the angle between the beam and the normal to the plate, and ε is the angle of inclination of the plane, and A is the surface albedo. In the present work, A has been assumed to be 0.2 and the surface to be a non-specular reflector.

The incidence angle B can be calculated for the various tracking modes according to [5,6] as follows:

1. Stationary (Horizontal) plate
   $\cos B = \sin \Phi \sin \delta + \cos \Phi \cos \delta \cos H$ ...(15)

2. South -facing flat plate but tilted at the latitude angle Φ
   $\cos B = \cos \delta \cos H$. ....(16)

3. Fully tracking plate
   $B = O$ ....(17)

4. N-S polar axis and E-W tracking plate
   $\cos B = \cos \delta$ ....(18)

5. E-W horizontal axis and N-S tracking plate
   $\sin B = \cos \delta \sin H$ ....(19)

The inclination of the plate ε can be instantaneously estimated by the following formula,

$$\varepsilon = \Theta - B \quad \text{..................................(20)}$$

The mean monthly energy flux density on each tracking plate F is obtained by a simple summation.

$$\bar{F} = \sum_{TSR}^{TSS} F\,(O) \quad \text{............................(21)}$$

The instantaneous values of energy taken every 15 minutes for Θ, δ, H calculated by the previous program are used to calculate the incidence and angles B and ε respectively in equations (15 to

20). Also, the resultant I (Θ), D (Θ) and G (Θ) are used to calculate the energy flux density in equation (14). The integrations of equation (14) for the various tracking modes were performed on the 15th of each month and the resultant values are those of monthly means.

RESULTS AND DISCUSSION

Using the available data from the five oases in Table 2, and applying our methodology, we obtained the following results for each oasis:

I. Solar Radiation Components:

- I the monthly mean value of the direct solar radiation of normal incidence.
- I_h the monthly mean value of the direct solar radiation on a horizontal surface.
- D the monthly mean value of the diffuse solar radiation.

II. Solar Energy Flux for Different Tracking Modes:

- Mode 1: Stationary (Horizontal) plate. We must mention here that the solar energy flux on the horizontal surface is the global solar radiation G, and

  $G = I_h + D$ ....(22)

- Mode 2: South - facing flat plate but tilted at the latitude Φ.
- Mode 3: Fully tracking plate. We must also mention here that the solar energy flux on a fully tracking plate is the sum of I and D.
- Mode 4: N - S polar axis and E - W tracking plate.

- Mode 5: E - W horizontal axis and N - S tracking plate.

The resultant values are given in kwh/m^2 day for each oasis in Tables 5 to 9.

Table (4) shows a testing of our methodology in the Western Desert of Egypt. For mode 1 flux "global radiation," the difference between the observed and estimated radiation does not increase more than 4.5% and is equal to 1.5% for the annual average. Also, for mode 2 flux, the maximum difference is 11.6% in January, while it is -0.3% in May and equal to 4.1% for the annual average. In our opinion, these differences in percentages are acceptable and the method of analysis is reliable.

From Table 5 through 9, we note the following:

1) The direct and global solar radiations are very high for all oases. In fact, the Western Desert is in the richest zone of solar energy input in the world with sunshine duration ≥ 3600 hour/year.

2) The diffuse fraction (D/G ratio) of solar energy is ≤ 0.2 of the total solar energy for all oases throughout the year, except for Siwa and Bahariya in April, due to the effect of dust storms (El-Khamasin).

3) The solar energy flux for the different tracking modes is high for all oases. It reaches 11.5 kwh/m^2/day in the full tracking mode in summer.

For these reasons, solar energy in the Western Desert of Egypt is a strong and renewable source for the power supply for all desert development projects.

Flat-plate PV power systems have attributes which make their use advantageous for a variety of electrical energy needs in the Third World.[7]

TABLE 4: Testing the methodology by solar radiation data of Kharga oasis

Month	Horizontal mode			South facing mode by tilting angle Φ		
	ob.*	EST.	Diff. percent	/1/**	EST.	Diff. percent
Jan.	4.67	4.77	2.1%	5.81	6.57	11.6%
Feb.	5.70	5.97	4.5	6.74	7.45	9.5
Mar.	6.69	6.85	2.3	7.21	7.64	5.6
Apr.	7.59	7.82	2.9	7.42	7.75	4.3
May	8.11	8.28	2.1	7.54	7.52	- 0.3
June	8.57	8.43	-1.7	7.43	7.29	- 1.9
July	8.31	8.35	0.5	7.44	7.36	- 1.1
Aug.	7.94	7.98	0.5	7.51	7.59	1.1
Sept.	7.18	7.20	0.3	7.47	7.66	2.5
Oct.	5.97	6.08	1.8	7.04	7.28	3.3
Nov.	5.01	5.02	0.2	6.19	6.74	8.2
Dec.	4.34	4.45	2.5	5.78	6.34	8.8
Annual	6.67	6.77	1.5%	6.97	7.27	4.1%

* The observed global solar radiation on horizontal surface are obtained from the Egyptian Meteorological Authority, for the period (1976-1984) for Kharga.

** The solar energy flux in kwh/m^2/day are given in /1/ for different tilting angles 10, 20, 30, ..., 90 for Kharga, for comparison the author made interpolation for Φ = 25° 27' of Kharga for the given values of 20° and 30°.

TABLE 5: Solar energy input to Siwa oases in Kwh/m^2/day

Month	S.R. Components			S.E. Flux for different tracking modes				
	I	I_h	D	1	2	3	4	5
Jan.	6.91	3.24	0.85	4.09	5.95	7.73	7.27	6.56
Feb.	7.86	4.25	1.03	5.28	6.87	8.87	8.68	7.15
Mar.	8.49	5.20	1.23	6.43	7.36	9.70	9.69	7.37
Apr.	8.49	5.68	1.69	7.37	7.38	10.12	10.01	7.64
May	9.11	6.31	1.62	7.93	7.20	10.70	10.22	8.16
June	10.00	6.98	1.36	8.34	7.16	11.38	10.54	8.69
July	10.13	7.06	1.31	8.37	7.34	11.47	10.74	8.67
Aug.	9.51	6.49	1.25	7.74	7.41	10.78	10.48	7.96
Sept.	8.96	5.74	1.07	6.81	7.40	10.06	10.04	7.41
Oct.	7.87	4.50	0.94	5.44	6.75	8.82	8.73	6.88
Nov.	6.78	3.34	0.84	4.18	5.87	7.60	7.26	6.35
Dec.	6.35	2.85	0.78	3.63	5.45	7.10	6.59	6.11
Annual	8.37	5.14	1.17	6.31	6.84	9.53	9.19	7.41

TABLE 6: Solar energy input to Baharia oases in Kwh/m^2/day

Month	S.R. Components			S.E. Flux for different tracking modes				
	I	I_h	D	1	2	3	4	5
Jan	6.93	3.31	0.86	4.17	5.96	7.76	7.30	6.58
Feb.	7.74	4.24	1.06	5.30	6.79	8.77	8.58	7.07
Mar.	8.45	5.22	1.23	6.45	7.33	9.66	9.66	7.34
Apr.	8.42	5.66	1.69	7.35	7.33	10.04	9.93	7.59
May	9.15	6.36	1.59	7.95	7.21	10.71	10.23	8.17
June	10.03	7.02	1.33	8.35	7.18	11.38	10.54	8.70
July	9.91	6.92	1.32	8.24	7.24	11.26	10.55	8.54
Aug.	9.68	6.63	1.21	7.84	7.49	10.92	10.61	8.06
Sept.	9.04	5.83	1.06	6.89	7.44	10.12	10.11	7.45
Oct.	7.97	4.61	0.93	5.54	6.81	8.91	8.82	6.94
Nov.	7.01	3.51	0.82	4.33	6.01	7.82	7.46	6.50
Dec.	7.11	3.26	0.69	3.95	5.91	7.81	7.22	6.66
Annual	8.45	5.21	1.15	6.36	6.89	9.60	9.25	7.47

TABLE 7: Solar energy input to Farafra oases in Kwh/m^2/day

Month	S.R. Components			S.E. Flux for different tracking modes				
	I	I_h	D	1	2	3	4	5
Jan.	7.42	3.63	0.79	4.42	6.24	8.22	7.71	6.91
Feb.	8.11	4.52	0.99	5.51	6.98	9.10	8.89	7.27
Mar.	8.80	5.50	1.15	6.65	7.50	9.95	9.94	7.51
Apr.	8.86	6.00	1.55	7.55	7.51	10.37	10.25	7.77
May	9.59	6.68	1.45	8.13	7.38	11.03	10.51	8.35
June	10.32	7.23	1.22	8.45	7.28	11.58	10.71	8.82
July	10.07	7.05	1.24	8.29	7.29	11.34	10.61	8.59
Aug.	9.74	6.71	1.16	7.87	7.50	10.93	10.61	8.06
Sept.	9.25	6.02	1.01	7.03	7.54	10.29	10.27	7.54
Oct.	8.55	5.02	0.84	5.86	7.13	9.43	9.32	7.27
Nov.	7.48	3.83	0.76	4.59	6.28	8.26	7.86	6.81
Dec.	6.86	3.22	0.74	3.96	5.74	7.59	7.03	6.47
Annual	8.75	5.45	1.08	6.53	7.03	9.84	9.48	7.61

TABLE 8: Solar energy input to Dakhla oases in Kwh/m^2/day

Month	S.R. Components			S.E. Flux for different tracking modes				
	I	I_h	D	1	2	3	4	5
Jan.	8.27	4.15	0.71	4.86	6.74	9.02	8.43	7.50
Feb.	8.96	5.09	0.89	5.98	7.48	9.88	9.64	7.80
Mar.	9.63	6.10	1.03	7.13	7.98	10.70	10.69	7.98
Apr.	9.53	6.50	1.41	7.91	7.84	10.93	10.97	8.11
May	9.94	6.96	1.34	8.30	7.53	11.30	10.75	8.52
June	10.39	7.30	1.17	8.47	7.31	11.60	10.71	8.84
July	10.28	7.22	1.16	8.38	7.38	11.49	10.73	8.69
Aug.	10.10	7.00	1.08	8.08	7.68	11.23	10.89	8.35
Sept.	9.77	6.43	0.92	7.35	7.82	10.75	10.73	7.83
Oct.	8.98	5.36	0.79	6.15	7.38	9.82	9.71	7.53
Nov.	8.40	4.41	0.66	5.07	6.83	9.13	8.66	7.44
Dec.	8.94	4.33	0.51	4.84	7.00	9.55	8.76	7.89
Annual	9.43	5.91	0.97	6.88	7.42	10.45	10.04	8.04

TABLE 9: Solar energy input to Kharga oases in Kwh/m^2/day

Month	S.R. Components			S.E. Flux for different tracking modes				
	I	I_h	D	1	2	3	4	5
Jan.	7.98	4.01	0.76	4.77	6.57	8.76	8.20	7.30
Feb.	8.89	5.06	0.91	5.97	7.45	9.84	9.60	7.78
Mar.	9.03	5.73	1.13	6.86	7.64	10.17	10.16	7.64
Apr.	9.33	6.36	1.46	7.82	7.75	10.77	10.63	8.02
May	9.87	6.91	1.37	8.28	7.52	11.25	10.71	8.51
June	10.31	7.25	1.19	8.44	7.29	11.55	10.67	8.81
July	10.21	7.17	1.19	8.36	7.36	11.44	10.69	8.66
Aug.	9.91	6.86	1.12	7.98	7.59	10.06	10.74	8.15
Sept.	9.45	6.23	0.98	7.21	7.66	10.48	10.46	7.66
Oct.	8.80	5.26	0.82	6.08	7.28	9.67	9.56	7.42
Nov.	8.23	4.32	0.70	5.02	6.74	8.98	8.52	7.33
Dec.	7.85	3.80	0.65	4.45	6.34	8.54	7.87	7.19
Annual	9.16	5.75	1.02	6.77	7.27	10.21	9.82	7.87

Solar-energy operated water pumps have attracted considerable attention from engineers, solar technologists, manufacturers, and other prospective users. There are two basic design approaches for photovoltaic water-pumping systems. One is to use a backup battery with the pumping system, and the other is to power the pump directly from the photovoltaic system without a battery. In the second approach, the pump can be powered either by directly coupling the PV array and motor or by using a peak power tracker to interface the motor and the PV array.[8]

As we mentioned before, there are 450,000 feddans of cultivable area in the five oases of the Western Desert of Egypt. If we assume that each feddan needs 5000 m^3/year of water,[1] then we must secure 2230 million m^3/year of water for all of the cultivable area. Assuming that there will be wells of underground water of 40 m depth with each well serving 8 feddans, i.e. 40,000 m^3/year water supply, then the electricity supply needed is 395 million Kwh/year for the entire cultivable area, (taking into consideration Western Desert conditions where the depth of water is more than 100 m in many sites). The generation capacity for the project is 180 MW/[1] If all of this power is supplied by PV units, then a total area of 0.1% of cultivated land is needed for a PV system. The economics of solar water pumping in Egypt,[9] and the cost of pumping 1 m^3 of water from 30 m depth are compared in Table 10.

TABLE 10: Cost of pumping 1 m^3 of water at a pumping rate of 40,000 m^3/year and pumping depth of 30 m

Technology	1981 Prices P.T. $\left(\frac{L.E.}{100}\right)$	1991 Prices P.T. $\left(\frac{L.E.}{100}\right)$
Diesel	3.8	6.0
PV	18.7	3.8
Solar thermal	9.8	5.5

We must take into consideration the fact that the oases are remote areas, and transporting fuel and maintenance are two problems. The potential of the various irrigation devices must be assessed at both market and shadow prices. Shadow prices take into account factors such as cost of generation (import of fuel), etc.[10] At the same time, flat-plate PV systems have desirable features: modularity, and no moving parts, low maintenance requirements and potential long life.

Therefore, we conclude from the scientific, technological and economic points of view that PV water pumping systems are the best solution for the development of the oases, and for agriculture in the 450,000 feddans in the Western Desert.

CONCLUSION

From our previous results and discussion we conclude the following:

1. The Western Desert of Egypt is a very rich region in solar energy input. The duration of sunshine in this zone is ≥ 3600 h/year. The solar energy flux on a fully tracking mode reaches to 11.5 $Kwh/m^2/day$ in the summer.

2. PV systems with flat plates are the best solution for supplying power for underground water pumping. From the scientific, technological, and economic points of view such plates would best solve the problem of 180 MW needed for the hydrology of 450,000 feddans of the cultivable area in the five oases.

3. Tables 5 to 9 are reliable for any plan or project of solar energy utilization in the development of the Western Desert of Egypt.

REFERENCES

1. A.T. El-Kheshin, S.E. Habib and M.F. El-Fouly. 1985. Photovoltaic potential in Egypt, economic and market data. Private print.

2. M.A. Mosallam Shaltout. 1985. Estimation of the different components of the solar radiation over Egypt from the meteorological data. Solar and Wind Technology Vol. 1, No. 3: 175-185.

3. M.A. Mosallam Shaltout. March 23-26, 1986. Atmospheric transmission and diffuse fraction of Solar Energy over Egypt. International Symposium on Applications of Solar and Renewable Energy, Cairo, Egypt.

4. G.W. Paltridge and D. Proctor. 1976. Solar Energy Vol. 18: 235.

5. J.F. Kreider and F. Kreith. 1975. Heating and cooling, engineering practical design and economics. McGraw-Hill, New York.

6. C. Wyman, J. Castle and F. Kreith. 1980. Solar Energy Vol. 24: 517.

7. L. Rosenblum. 1983. Status of flat-plate photovoltaic systems for applications in developing countries. Solar Energy Vol. 31: 381.

8. Y.R. Hsiao and B.A. Blevins. 1984. Direct coupling of photovoltaic power source to water pumping system. Solar Energy Vol. 32: 489.

9. T. El-Tablawi. 1982. Optimal strategy for solar energy in Egypt towards the year 2000. Part II, Internal Report.

10. V.V.N. Kishore, M.R. Gandhi, N. Pathak, S.D. Gomakle, K.S. Rao, P. Jaboyedoff, W. Lehman and Ch. Marquis. 1986. Development of a solar (thermal) water pump proto-type: an indo-Swiss experience. Solar Energy Vol. 36: 257.

A SOLAR POWERED NFT SYSTEM FOR DESERT DEVELOPMENT

F. ASSABGHY, I. EL-BAGOURI, S.A. SEIF, AND K. EL-KHESHEN
Desert Development Center, American University in Cairo, Egypt

Abstract The paper describes the installation and operation of a stand-alone solar powered Nutrient Film Technique (NFT) system for vegetable production. It is a field experiment situated in the western desert of Egypt at the Sadat City site of the Desert Development Center (DDC) of the American University in Cairo. Assessment of the system from the technical and economic points of view is undertaken to explore the potential of the technique as an alternate approach to small scale high-return desert farming systems. Hydroponic cultivation of plants in desert environments presents attractive features such as inherent efficiency in the use of water and nutrients -- both scarce commodities in arid lands. Furthermore, such intensive agricultural practices under protective structures usually insure outstanding plant growth and productivity. Land use is minimized with possible savings in the scale of the infrastructure of desert development. In NFT systems the nutrient solution is continuously flowing down plastic gullies by gravity. Recirculation pumps are necessary to maintain the flow. The required energy is an important input to the system and the feasibility of supplying this energy from photovoltaic panels is investigated. Another important input is the nutrient solution itself and development of an appropriate solution from locally available chemicals is attempted and described. The results of the field experiment are presented.

INTRODUCTION

The imperatives for developing the deserts of Egypt

are only too evident. There remains the overriding consideration of a rapidly expanding population crowded on a narrow strip of fertile land, that does not exceed 4% of the national territory. Strategies for desert development are continuously being refined as new technologies come on line. The relatively recent past has seen the traditional flood irrigation methods supplanted by more water efficient systems such as drip, sprinkler, sideroll and centerpivot. While these techniques have resulted in considerable savings in water, they are however, intensive in their energy requirements. Indeed, the water-energy aspects of desert agriculture are in reality inseparable and together they constitute a major constraint in desert development. The so- called closed hydroponic systems are potentially the most efficient water-energy cropping systems that can be devised. For this reason a growing interest exists in such techniques for growing crops in the desert. The Nutrient Film Technique (NFT) is classified as a closed hydroponic system.

THE NFT CONCEPT

The concept of NFT is simple[1,2]; a nutrient solution is circulated in a shallow stream over the roots of growing plants. The solution which contains all of the nutrients necessary for plant growth is regularly monitored and fresh nutrients are added when depletion occurs. This solution is allowed to flow down channels (or gullies) on an inclined plane and is collected at the exit of the gullies in a catchment tank. The solution is then pumped up from the catchment tank to a header tank which feeds the input to each channel. This is a closed system. The amount of water and nutrients are limited to essentially what is needed for plant growth with very little loss.

The roots in the NFT provide little anchorage and a protective structure, such as a plastic tunnel, is necessary to shield the plants from the wind - always an important aspect of desert agriculture. In addition, these protective structures allow relative independence from seasonality.

The main inputs to an NFT system are therefore: WATER, NUTRIENTS and ENERGY. In addition there are the inputs normally associated with horticultural practices such as seedlings, pesticides, etc.

The energy required is necessary to maintain the recirculation pumps in operation. The static heads involved are not large and do not exceed a few meters. Solar powered water pumps are a realistic possibility in desert sites somewhat removed from the electric grid. Indeed, even at today's prices, small-scale solar water pumps powered with photovoltaic (PV) cells may be competitive with conventional diesel-fueled or gasoline-fueled pumps. Solar water pumps offer a number of attractive advantages, namely, reliability, low maintenance and the control flexibility inherent in electrical systems[3].

OBJECTIVES

The objectives of the work were as follows:

a) To set up a field trial of the NFT cropping system.

b) To demonstrate the possibility of growing a variety of crops.

c) To evaluate the economic viability of NFT as a farming system in the context of the Egyptian Desert Development effort.

d) To investigate ways by which the technology can be simplified in order to enable user acceptance and to facilitate replication.

e) To investigate the possibility of using solar water pumps in the operation of the systems.

THE PHYSICAL PLANT

THE PROTECTIVE STRUCTURE

The design of a plastic greenhouse for warm climates described by Virhammar[4] was adopted. The essential features of the tunnel were as follows: 36 meters long, 6 meters wide, 3 meters high. The structure of the tunnel was based on standard 3/4 inch water pipes. All construction steps (with the exception of those requiring welding) were carried out on site. The tunnel was covered with UV stabilized polyethylene plastic film 200 microns thick. The energy needed to power the unit was supplied by a PV system (described below), so that the unit could be described as a stand-alone solar powered NFT system, which in our work is referred to as NFT I.

Within the framework of this project, a second structure was also erected. This was a commercially-available locally-manufactured tunnel of sturdier construction. The piping for the arches was 1½ inches in OD and the longitudinal pipes were of 1 inch OD. The tunnel dimensions were 8.5 meters X 36 meters. This unit is referred to as NFT II. The power for this unit could optionally be derived from the sun although not as a stand- alone configuration but as one of the loads of a bigger solar powered system. Alternately, the unit could be powered from utility or standby generated power.

The NFT I unit (Figure 1), was operational by January, 1985 and the NFT II unit by July, 1985.

PV SYSTEM AND ASSOCIATED CIRCUITRY

The PV System* consists of six parallel branches; each is a series combination of four PV modules (Figure 2). Each module produces 2.2A at 15V for peak solar insolation (taken to be 1 kw/m²). The PV system's rated power is 792 Wp.

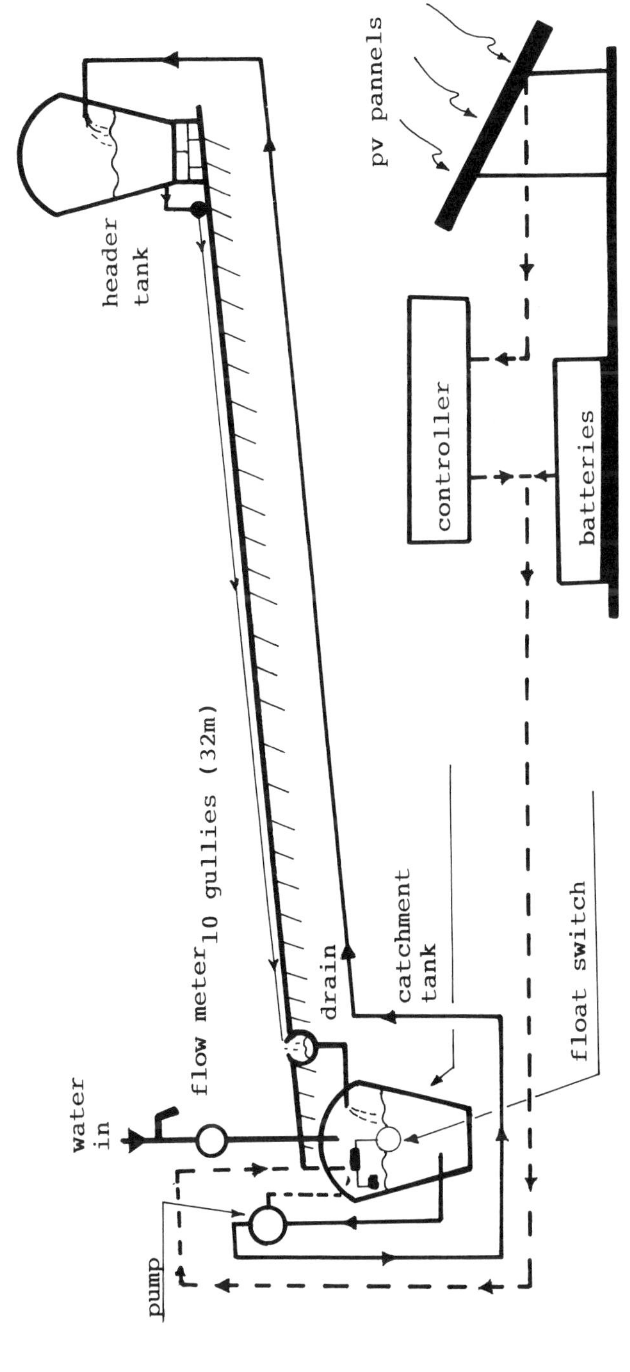

FIGURE 1

NFT

I

The battery bank consists of 4 Chloride automotive 12V batteries connected in series to produce a nominal system voltage of 48V.

The motor is a 2 Hp DC Honeywell motor [model BA 5332]. This drives a Jacuzzi pump.*

THE MONITORING SYSTEM

The monitoring system is based on the availability on the site of a computer/data logger facility. The parameters monitored are:

- The array current Ia; as a voltage across a one milliohm resistor in series with the array.
- The battery current Ib; also as a voltage across a one milliohm resistor in series with the array.
- The battery voltage Vs.

The voltage are inputted to the sensing channels of the data logger via telephone cables. The computer/data logger is programmed to read all of its channels every minute and to store the hourly average on magnetic tape. Analog monitoring is also possible by means of a dual channel pen recorder.

THE CONTROL SYSTEM

- To operate the pump when the header tank is fully discharged into the catchment tank.
- To stop the pump when the header tank is recharged.

---

* The PV modules, pump, and motor, were part of a donation to the Desert Development Center of AUC by ARCO SOLAR INC; Subsidiary of Atlantic Richfield Corporation, through the World Bank/UNDP.

- To protect the batteries from overcharge.
- To protect the batteries from overdischarge.

Relay R2 will control the operation of the motor/pump set. The switch R2 is normally open (no) and will close (to run the motor) when its coil is activated. The coil of R2 is in series with the float switch and the switch S2; both of which must be closed for the motor to operate.

- The float switch is normally open and will close when the header tank is emptied.
- The float switch is returned to the (no) state when the tank is again charged [this is adjusted to be 1/2 full to allow for some water to be taken up by the plants without system disruption due to inability of the pump to recharge the header tank to capacity.]
- Switch S2 controlled by the computer, is normally closed (nc) and will switch to open in the event of over-discharge of the batteries as indicated by a system voltage low of Vsl. The switch will return to its closed state when the batteries are recharged as indicated by a voltage rise to a value slightly above Vsl (to introduce some hysteresis), namely Vsly.
- Relay R1 (nc) controls the array current Ia. In the event of overcharge, as indicated by a system voltage high of Vsl, the coil of R1 is activated by S1 going from (no) to closed by computer control and R1 switches to open. R1 returns to its (NC) state when Vs falls below Vsl to a value of Vsly (again to introduce some hysteresis).

The switch labelled "computer control" in Figure 2 will override the computer control of R1 and R2 in the "OFF" position.

The voltage levels, Vsl, Vsly, Vsh and Vshy, are easily adjusted by the software for optimum system performance. Typical values are as follows:

FIGURE 2

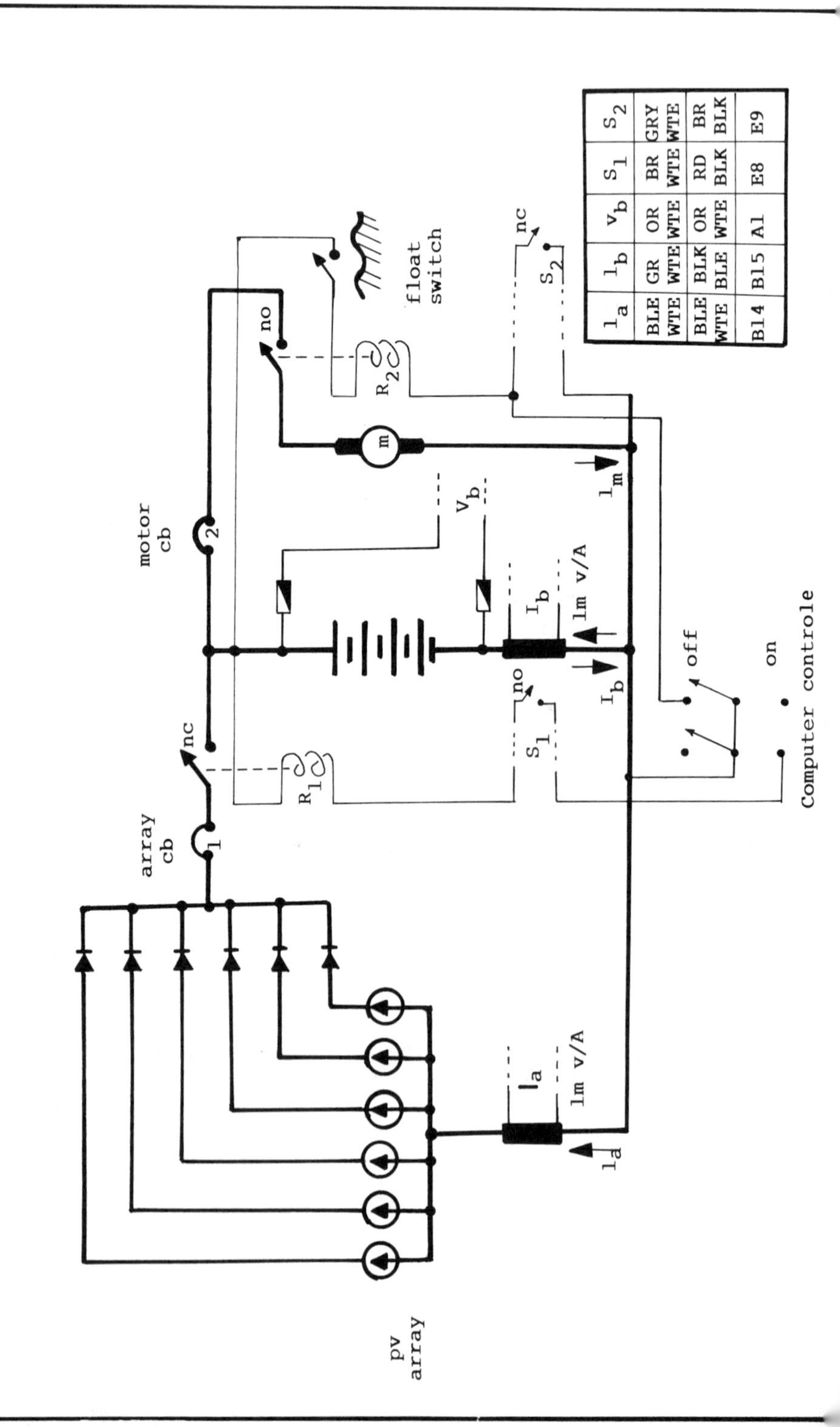

I_a	I_b	V_b	S_1	S_2
BLE WTE	GR WTE	OR WTE	BR WTE	GRY WTE
BLE WTE	BLK BLE	OR WTE	RD BLK	BR BLK
B14	B15	A1	E8	E9

$Vsl = 48V$ $Vsly = 49V$

$Vsh = 60V$ $Vshy = 54V$

Typical chart recorder traces are shown in Figure (3). One channel is monitoring the battery current Ib (polarity indicates whether batteries are charging or discharging). The other channel indicated the system voltage Vs.

The system allows the possibility of seasonal adjustment of the water flow down the gullies to match the availability of the solar resources.

THE NUTRIENT SOLUTION

The major input in NFT systems is the Nutrient Solution.

The adopted nutrient solution was basically that developed by Wye college. However, since no ready micronutrient mixes were available in Egypt (e.g., Librel BMX chelate mix), this was substituted by the quantities of micronutrients recommended in the ADAS solution, as reported by S.W. Burrage, 1982. The adopted solution was as follows (Table 1) (stock solution to be diluted 1:100):

ANALYSIS OF SADAT CITY WATER RESOURCE

Water at the Sadat City site is available from an underground well. The depth of water is around 40 meters. Water salinity is quite low (around 315 ppm of total soluble salts). The following is the average composition of the underground water being used for preparation of NFT stock solutions as well as for supplying growing plants with water through dilution of the stock solution and compensation of water losses.

Table 2 shows that water quality is good and no major problems should arise from the use of such water for NFT experiments.

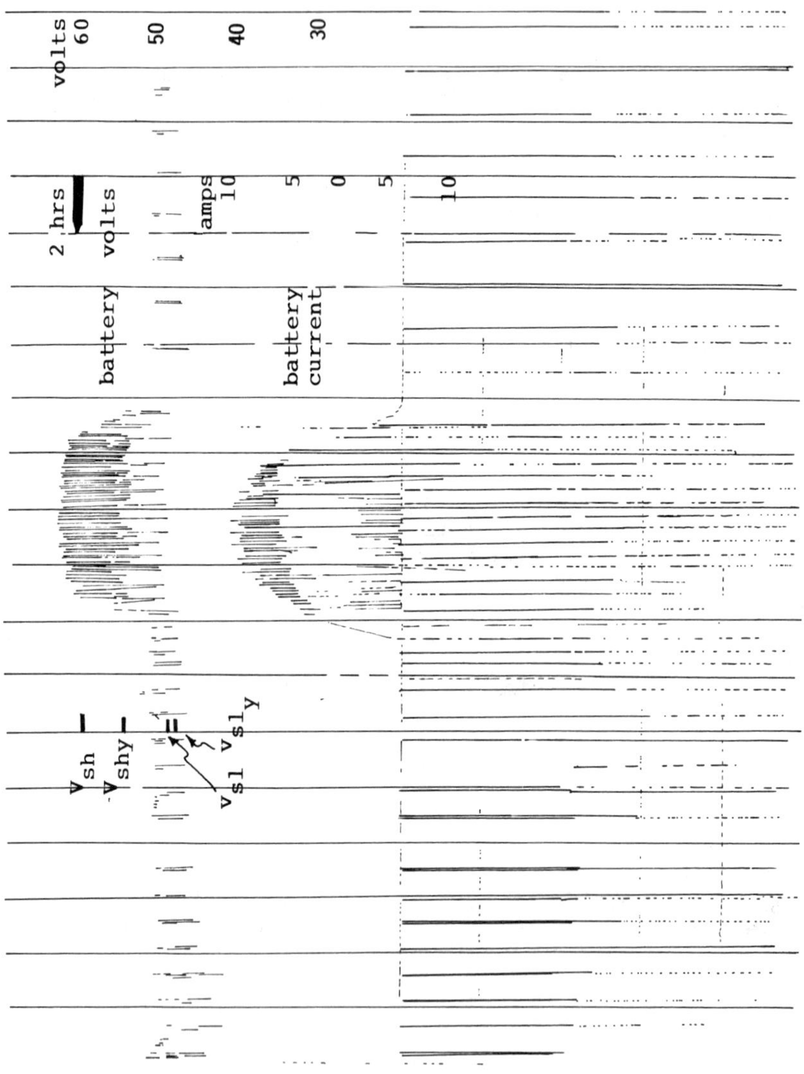
volts
60
50
40
30
2 hrs
battery volts
amps
10
5
0
5
10
battery current
V_{sh}
V_{shy}
v_{sl}
v_{sly}

FIGURE 3

TABLE 1

COMPOUND	FORMULA	Kg./200L
Solution A:		
Potassium nitrate	$K\ NO_3$	16.59
Magnesium nitrate	$Mg\ (NO_3)_2 .6\ H_2O$	6.54
Potassium dihydrogen ortho phosphate	KH_2PO_4	4.14
Potassium sulfate	K_2SO_4	7.33
Iron chelate	Fe EDTA	0.80
Manganese sulfate	$Mn\ SO_4 .4H_2O$	0.101
Copper sulfate	$Cu\ SO_4 .7H_2O$	0.019
Zinc sulfate	$Zn\ SO_4 .7H_2O$	0.022
Ammonium molybdate	$(NH_4)_6\ Mo_7O_{24} .4H_2O$	0.018
Boric acid	H_3BO_3	0.083
Solution B:		
Calcium nitrate	$Ca\ (NO_3)_2$	3% solution

TABLE 2. Average values of major characteristics and composition of water resource at Sadat City site

pH	EC mmhos/cm	Cations				Anions			
		Ca^{++}	Mg^{++}	K^{+}	Na^{+}	$CO_3^{=}$	HCO_3^{-}	CL^{-}	$SO_4^{=}$
		<-------------------- M.Eq./Liter -------------------->							
7.78	0.49	0.90	0.70	0.05	2.42	-	3.02	0.95	0.10

COMMERCIAL GRADE CHEMICALS

The nutrient solution constitutes an important input to the NFT system. Any decrease in the cost of the chemical constituents making up the solution is expected to reflect favorably on the economics of the technique. At the start of the project, AR grade chemicals were used in the preparation of the solution. Plant growth was vigorous. At a later stage, attempts were made to replace some of the more expensive AR grade chemicals by less expensive commercial or fertilizer grade chemicals. The chemicals which were gradually replaced were as follows:

K_2SO_4 AR grade was replaced by the commercial fertilizer, K_2SO_4 $Ca(NO_3)_2$. K_2HPO_4 AR grade was replaced by the commercial fertilizer, triple phosphate.

The previous substitutions were highly successful with the exception of triple phosphate which was met with mixed successes depending on the source of the triple phosphate used. Therefore, K_2HPO_4 of chemically pure CP grade was used instead of AR grade. This replacement was successful and led to reduction in cost of one of the most expensive components of the nutrient solution.

SOLUTION MONITORING AND NUTRIENT UPTAKE

Solution monitoring for temperature, electrical conductivity (EC) and pH was carried out using a single handled instrument developed by S.W. Burrage of Wye College. Adjustment of the pH was carried out by applying appropriate amounts of 10% nitric acid. The (EC) is an indication of the total dissolved salts. For more comprehensive monitoring of the nutrient solution, it is now necessary, from time to time, to run a complete chemical analysis. Table 3 shows such an analysis.

Samples for analysis were taken at 24 hour intervals. Two samples were taken at each interval, one immediately before and one shortly after addition of the supplementary solution. The results in Table 3 are therefore an indication of the range of nutrient variations over a 24 hour period as well as nutrient consumption.

HORTICULTURE

Horticultural practices for plants grown in NFT do not differ significantly from the standard techniques for growing crops in greenhouses or plastic tunnels. In most cases, F_1 hybrid seeds were used for the main crops, namely tomatoes, cucumbers, green pepper and cantaloupe. Different approaches for seedling production were tested. In general, plant performance was somewhat uneven and ranged from "excellent, truly outstanding" to "poor". As a result of this, the average yields were well below the anticipated high yields reported for NFT crops. The economic evaluation presented in a subsequent section of this paper should be viewed with the above considerations in mind.

The reason for the uneven performance could be attributed to a number of factors of which the most likely are:

a) The experimentation with different nutrient solutions.

b) The rise in temperatures during the hot season and,

Each of these factors is currently being investigated in order to achieve optimum results.

INDEPENDENT GULLIES

Five independent gullies with independent closed circulation systems have been installed to facilitate experimentation with various nutrient solutions, principally with regards to starting materials that are available locally.

TABLE 3. Nutrient concentrations in NFT-I nutrient solution (in ppm) (14-16 March 1986)

Sample Date	Design	P	Ca^{++}	Mg^{++}	K^{+}	Fe	Mn	Zn	Cu	Na^{+}
14/3	Just before adding supplementary sol.	8.5	89	45	140	1.5	0.1	0.4	0.19	130
14/3	After adding ½ m^3 supplementary sol.	17	119	96	265	2.18	0.81	0.58	0.33	135
15/3	Before adding supplementary sol.	10	82	60	130	1.48	0.16	0.39	0.02	125
15/3	After adding ½ m^3 supplementary sol.	18.3	156	75	270	2.32	0.70	0.74	0.04	145
16/3	Before adding supplementary sol.	11.3	81	48	125	1.3	0.1	0.36	0.02	115
16/3	After adding ½ m^3 supplementary sol.	18.3	141	62	245	1.84	0.60	0.46	0.03	125

TEMPERATURE CONSIDERATIONS

The economy in the consumption of nutrients and, particularly, of water makes the NFT technique very attractive for use in areas where water is scarce, namely deserts. Most deserts, however, lie in areas of the world where high solar radiation and high temperatures present particularly difficult problems in the successful operation of NFT systems.

Initial attempts to alleviate the problems associated with operating the NFT system in hot climates comprised a) Shading by replacing the plastic film by a black netting and b) Using white on black plastic film for the gullies (white on the outside to reflect part of the solar radiation). These initial attempts failed to insure the establishment of a tomato crop. It became clear that more elaborate measures were necessary.

Temperature sensors were placed at various points in the closed circulation path of the nutrient solution as follows: (i) output of the tank, (ii) Input to the gullies, (iii) input to the tank. In addition, a sensor was placed inside the shadehouse to measure the ambient temperature. The sensors were linked to the data logger/computer facility available at the site via telephone cables. The computer was programmed to take a reading every hour on the hour and print the temperature at the points selected, together with the incident global solar insolation in kw/m^2.

The temperature monitoring system started operation just before noon on June 15. A glance at the data in Figure 4, clearly indicates that the closed plastic channels behave as solar collectors. The nutrient solution temperature reached a peak of 50°C on a day when the maximum ambient temperature was 42°C.

In order to limit the rise in solution temperature, the plastic gullies were insulated by means of 1" thick polystyrene foam (styrofoam). This also allowed taking advantage of the large

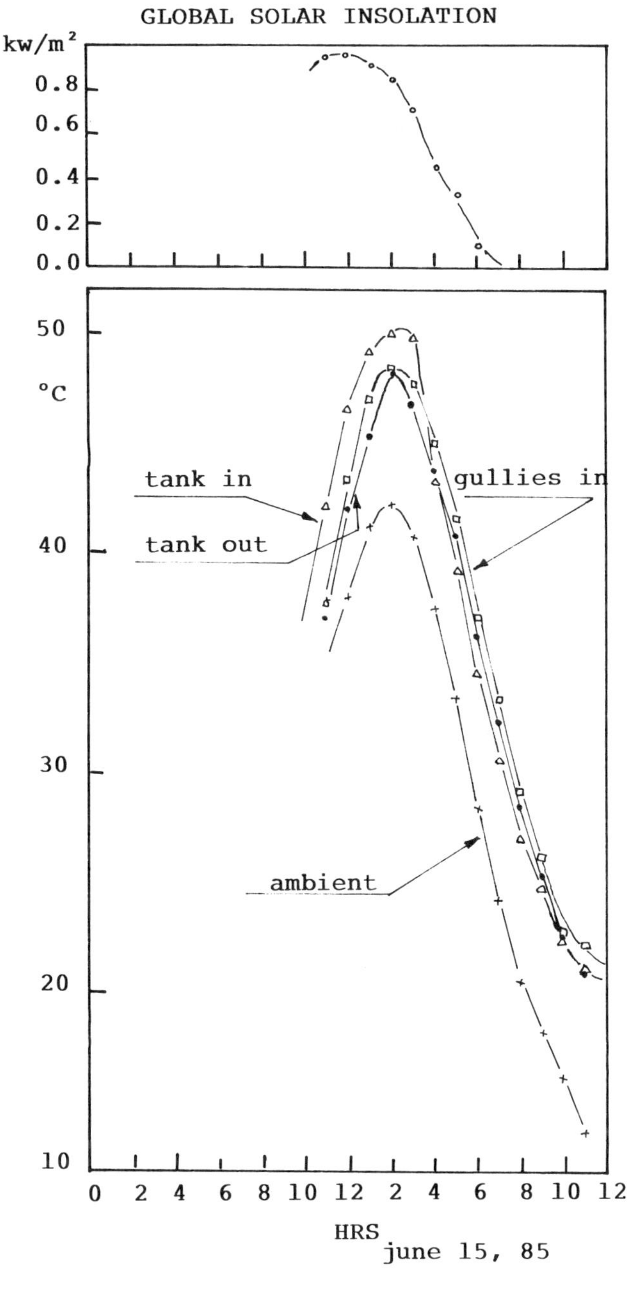
GLOBAL SOLAR INSOLATION
kw/m²
0.8
0.6
0.4
0.2
0.0
50
°C
tank in
gullies in
40
tank out
30
ambient
20
10
0 2 4 6 8 10 12 2 4 6 8 10 12
HRS
june 15, 85

FIGURE 4

diurnal variation in temperature in desert climates during the night. Circulation was allowed during the day. (The piping and water tank were, of course, also insulated). By operating the system in this way, an improvement was obtained and the solution temperature remained below the ambient for the hottest part of the day. (See Figure 5). However, the nutrient solution temperature still reached a high of about 33°C on a day of maximum ambient temperature of 36°C.

In order to bring the temperature down still further, a cooling spray was installed. The intake of the spray was the drainage tank and the spray was allowed to fall back into the tank. The spray was operated only during the hottest part of the day. The resulting cooling effect is clearly demonstrated in Figure 6. Between 50 and 100 litres per day were estimated to be consumed by evaporation of this spray.

The results of these efforts in operating the NFT II System during the hot months were, unfortunately, inconclusive.

In the following summer, different approaches were attempted in order to alleviate the effects of the rise in temperature.

(1) The plastic film covering the tunnel was whitewashed.

(2) Different insulating and reflecting surfaces were applied over the exposed sides of the plastic gullies.

(3) Gravel-filled gullies were tested.

Typical sets of data for the thermal behavior of some of these gullies are presented in Figures 7 and Figure 8. The data show that these measures led to decreases in the solution temperatures to a greater or lesser extent.

The main result emanating from this phase of the work became evident when the gullies were

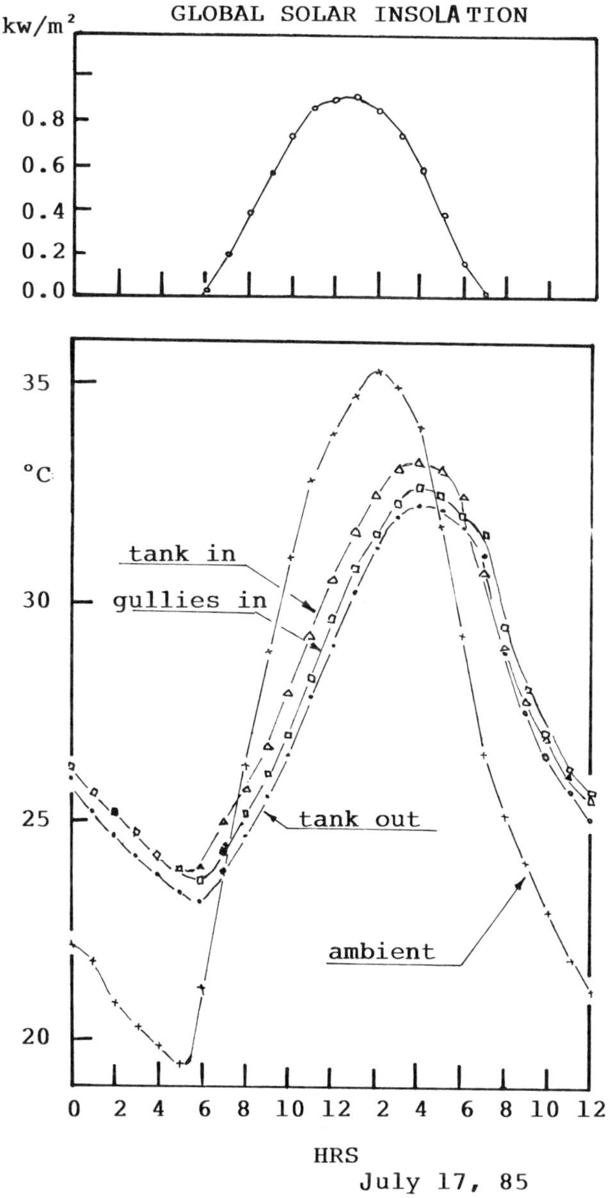

GLOBAL SOLAR INSOLATION
kw/m²
0.8
0.6
0.4
0.2
0.0
35
°C
30
25
20
tank in
gullies in
tank out
ambient
0 2 4 6 8 10 12 2 4 6 8 10 12
HRS
July 17, 85

FIGURE 5

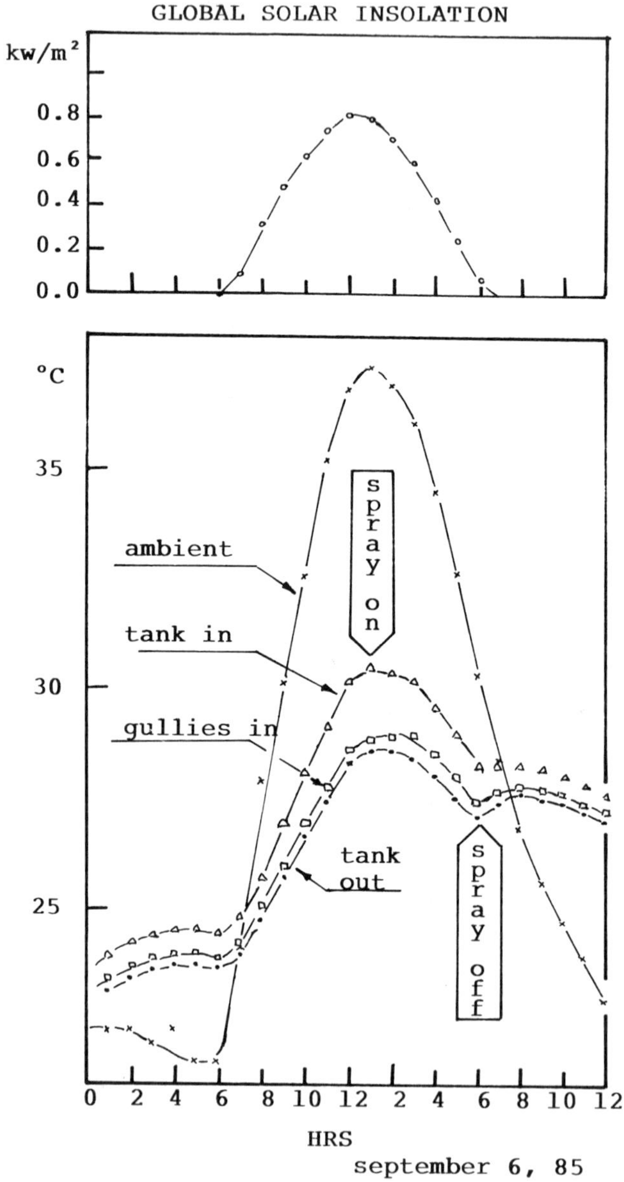
GLOBAL SOLAR INSOLATION
kw/m²
0.8
0.6
0.4
0.2
0.0
°C
35
30
25
ambient
tank in
gullies in
tank out
spray on
spray off
0 2 4 6 8 10 12 2 4 6 8 10 12
HRS
september 6, 85

FIGURE 6

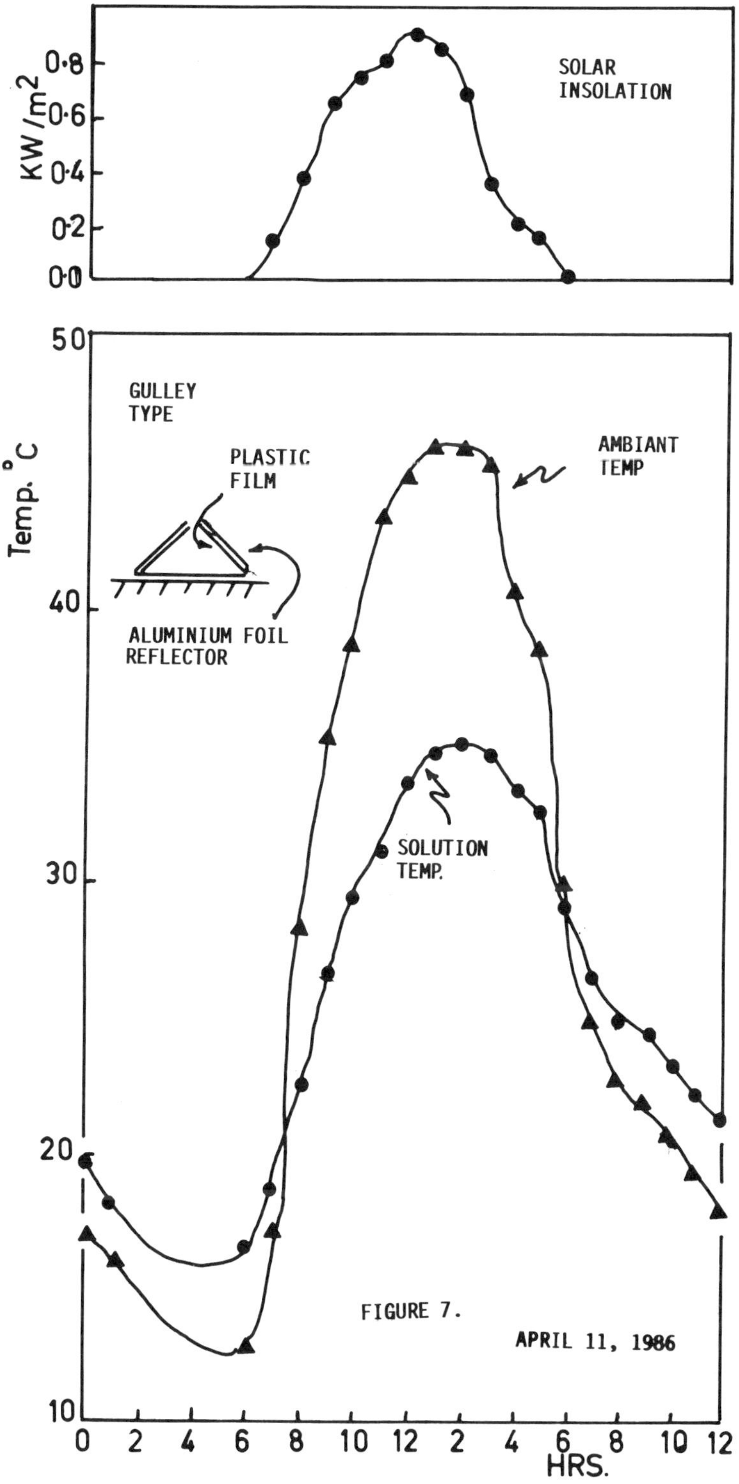
SOLAR INSOLATION
KW/m²
0·8
0·6
0·4
0·2
0·0
50
40
30
20
10
Temp. °C
GULLEY TYPE
PLASTIC FILM
ALUMINIUM FOIL REFLECTOR
AMBIANT TEMP
SOLUTION TEMP.
APRIL 11, 1986
0 2 4 6 8 10 12 2 4 6 8 10 12
HRS.

FIGURE 7.

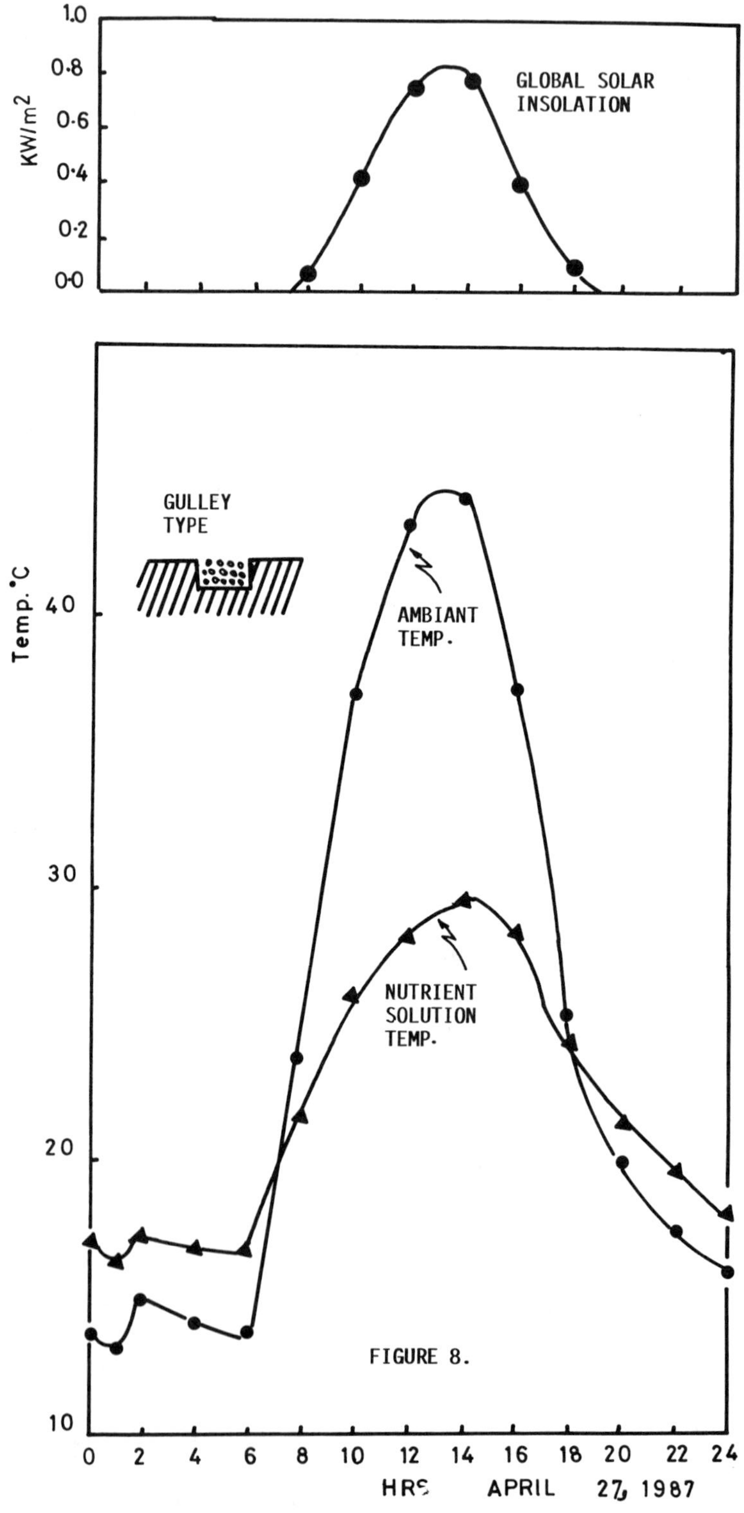
1.0
0·8
0·6
0·4
0·2
0·0
KW/m2
GLOBAL SOLAR INSOLATION
GULLEY TYPE
AMBIANT TEMP.
NUTRIENT SOLUTION TEMP.
Temp. °C
40
30
20
10
0 2 4 6 8 10 12 14 16 18 20 22 24
HRS APRIL 27, 1987

FIGURE 8.

planted. There was no possible comparison between plant performance in the gravel-filled gullies and the other gullies. The plants in gravel performed much better than the plants in the gullies not filled with gravel. Indeed, the plants in gravel showed no visible signs of stress, even after experiencing repeated heat-waves when the ambient air temperatures reached 45°C and above for several days in succession. This was interpreted as a result of the increased aeration of the solution in the open gravel structure.

ECONOMIC EVALUATION

The importance of careful economic evaluation of the technique cannot be overemphasized. Total costs which include operating and non-operating costs are fairly straightforward computations. The figure for the revenue is arrived at by extrapolating the gullies planted with a certain crop to total tunnel capacity. Bearing in mind that to date, average plant performance has fallen short of the expected potential for such systems, the computed net revenue does not, in our view, reflect the true potential of the technique. Table 4 shows the details of a crop budget for a cucumber crop. Of interest is the percentage composition of NFT costs (Table 5). Clearly, the major cost component is the nutrient solution which is about 50% of total operating costs. The increase in cropping output together with the decrease in cost of nutrient solution, should ensure the economic viability of the technique. Figures 9 and 10 are sensitivity analyses for a cucumber and a tomato crop showing increases in productivity of 3, 4 and 5 times together with a decrease in the cost of the nutrient solution to ½ and 1/3 actual cost. This sets the target for future work.

TABLE 4: Crop budget study for: Cucumbers

Activity : NFT
Date : 2/11/1986
Site : Sadat City
Date start/end: 7/9/86 - 1/11/86
Duration : 7 weeks

Greenhouse : II
Number of Gullies: 8
Dimensions (m) : 36 X 8
Volume (net) : 905 m^3

This budget represents only part of the crop since the enti[illegible] season extends over a period of 20 -24 weeks.

Item (per season)	Unit	No. of Units	Price or value/ unit L.E.	Total L.E.
A. REVENUE				
Yield (main)	Kgs	1265.6	0.60	759.3
Yield (secondary)				
TOTAL REVENUE				759.3
B. OPERATING COSTS				
Horticulture materials				199.9
Energy	Kwh	49.83	0.10	4.9
Water	m^3	38.18	0.02	0.7
Nutrient solution	m^3	24.50	18.60	455.7
Acid (for pH adjustment)	litres	14.64	0.60	8.7
Labor	man/day	24.22	5.00	121.1
Management	m^2/month	216	0.0210	7.9
Indirect costs (Depreciation)				95.3
TOTAL OPERATING COSTS				894.5
Returns above operating costs				(135
C. NON-OPERATING COSTS				
Overhead	Month	2	5.00	10.0
Rent	Month	2	1.80	3.5
Depreciation				52.1
TOTAL NON-OPERATING COSTS				65.7
D. NET REVENUE A-(B+C)				(201

TABLE 5: Percentage composition of NFT costs (Cucumbers)

ITEM	% OF COST
OPERATING COSTS	
Horticulture materials	22.36%
Energy	0.56%
Water	0.09%
Nutrient solution	50.94%
Acid (for pH adjustment)	0.98%
Labor	13.54%
Management	0.89%
Indirect costs (Depreciation)	10.65%
TOTAL OPERATING COSTS	100.00%
NON-OPERATING COSTS	
Overhead	15.22%
Rent	5.47%
Depreciation	79.32%
TOTAL NON-OPERATING COSTS	100.00%
Operating costs	93.16%
Non-operating costs	6.84%
TOTAL COSTS	100.00%

FIGURE 9

SENSITIVITY ANALYSIS OF CUCUMBER PRODUCTION

NFT

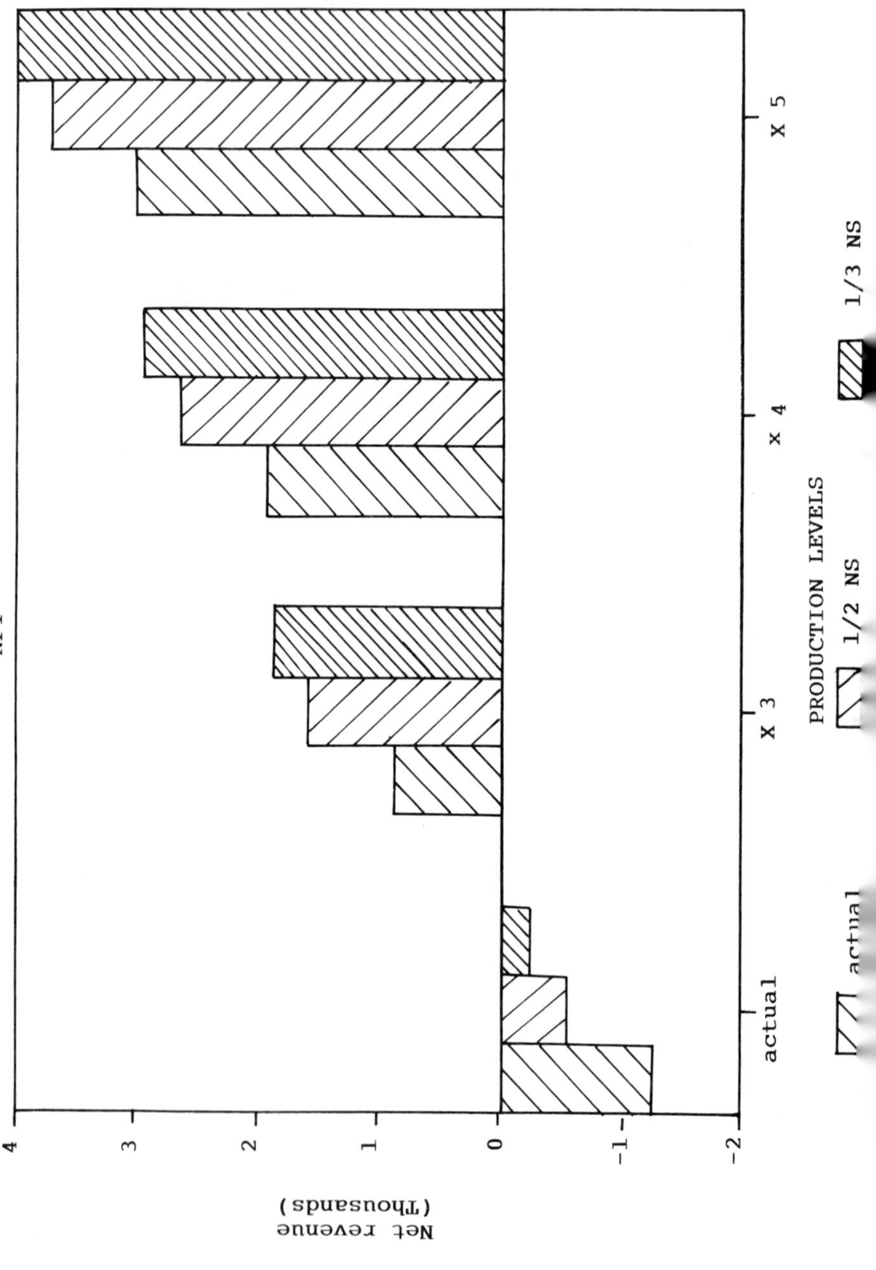

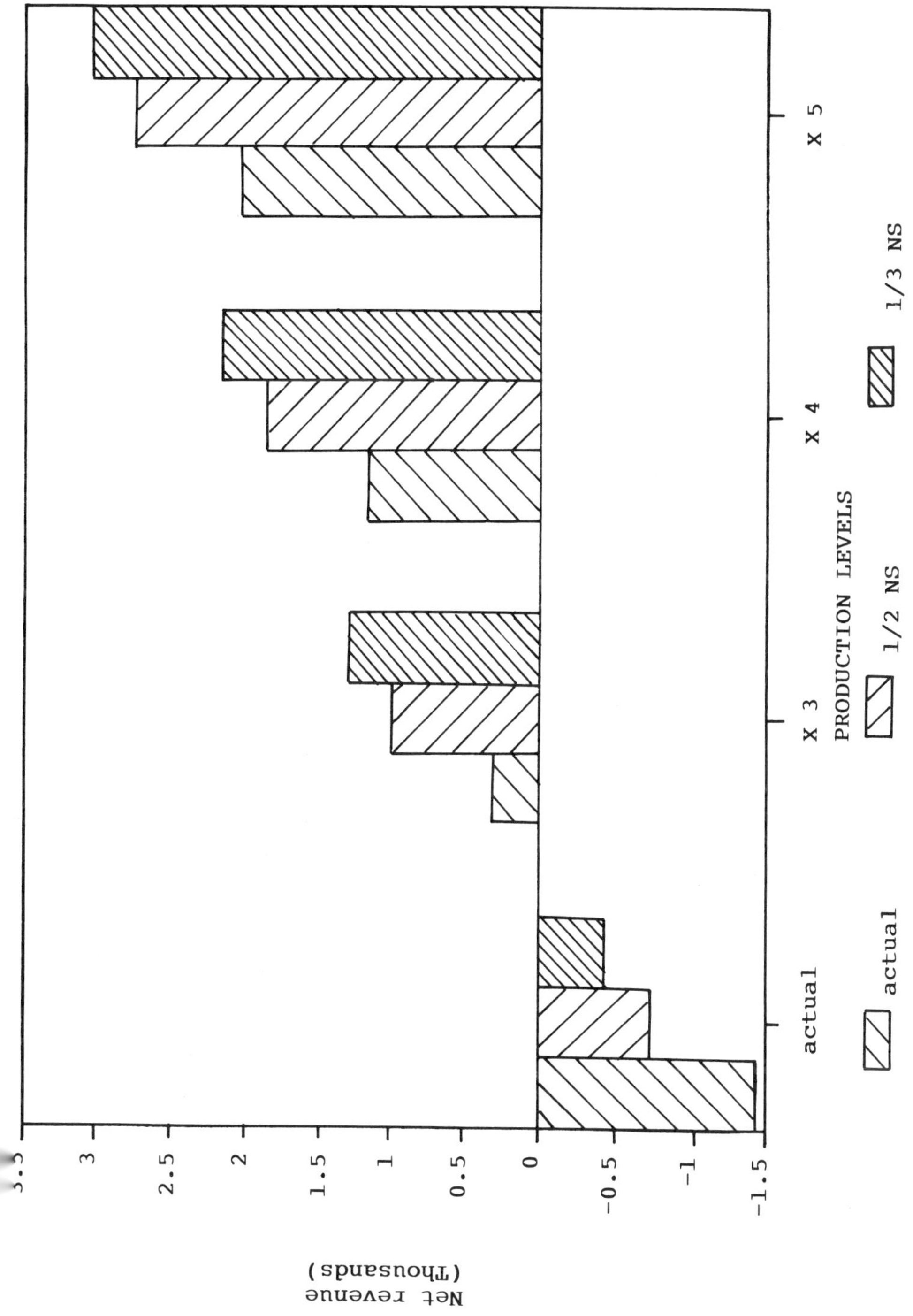
Net revenue (Thousands)
3.5
3
2.5
2
1.5
1
0.5
0
-0.5
-1
-1.5
actual
X 3
X 4
X 5
PRODUCTION LEVELS
actual
1/2 NS
1/3 NS

REFERENCES

1- The ABC of NFT - A. Cooper - Grower Books London, 1979.

2- Potential Technique For Small Scale High Return Desert Agriculture - Technical Considerations - s.W. Burrage, Wye College University of London. NFT Seminar, 1982.

3- Small-Scale Solar Powered Pumping Systems: The Technology, It's Economics and Advancement. Report by Sir William Halcrow and Partners in Association with Intermediate Technology Power Ltd. - UNDP Project GLO/801003 Executed by the World Bank - UK. June, 1983.

4- Plastic Greenhouses for Warm Climates - Kjell Virhammar FAO - Agricultural Services Bulletin - Rome, 1982.

ACKNOWLEDGEMENTS

This work was sponsored by financial support of the following agencies, FAO, UNDP, FORD FOUNDATION, and The U.S. AID through the FRCU of the EGYPTIAN SUPREME COUNCIL OF UNIVERSITIES. (MAXI LINKAGES PROJECT NO. 842083).

Also gratefully acknowledged is the continued interest and financial support of AUC trustee Mr. John Goelet.

The many valuable consultations with Dr. S.W. Burrage of Wye College, University of London have been of great help to the authors during the course of this work.

The following have contributed substantially to the field work Mr. Francio Chorin, Eng. Nasrat Goubrial, Eng. Mohamed Abbas, Eng. Hisham Nagati, and Mr. Peter Ritzer.

A PROMISING SOLAR THERMAL COLLECTOR FOR ARID AREAS

GHASSAN F. HAMAD and MAKAM C. GUPTA
MRI/Soleras, Kansas City, Missouri, U.S.A.

Abstract Availability of water for drinking and agriculture is one of the most vital needs in the development of arid areas. Solar energy is abundantly available in these areas; hence it is highly desirable to find a large-scale solar desalination process that is technically and economically feasible.

The solar desalination pilot plant at Yanbu, in the Kingdom of Saudi Arabia, employs an indirect heat transfer freeze technique for desalting seawater, and the solar power required to produce the refrigeration is provided by fresnel concept point focus dish collectors. The 200 m^3/day plant has been in operation for about two years and the resulting information is valuable in the search for a viable process in desert regions.

The two-year period of operation of the plant was planned to provide information of potential value relating to the design, operation, and maintenance of the collector system. Routine tests were intended for observing the effect of weather and extended operation on the controls, sensors, azimuth, and elevation drives. Special tests were performed to determine the system thermal losses, parasitic power requirement, reliability, and collector efficiency.

The paper compares the actual collector efficiency and thermal losses to those predicted in the design report and makes a first and second law analysis of the thermal cavity receiver. The paper also presents the behavior of the fluid used as the thermal transport media; the effect of the dusty, humid, seacoast environment on the collector; factors affecting tracking precision; and an outlook for the future.

INTRODUCTION

The SOLERAS Program is a unique effort in cooperative solar energy research, established in 1977 under the auspices of the United States-Saudi Arabian Joint Commission on Economic Cooperation. SOLERAS, one of more than 30 such projects under the direction of the Joint Commission, has initiated several major research projects: photovoltaic power system, solar-controlled environment agriculture, solar water desalination, solar thermal research and basic research in universities.

CBI'S INDIRECT FREEZE PROCESS

The Yanbu pilot plant in Saudi Arabia employs the CBI process, which is known as the indirect contact heat transfer, freeze desalination process. As the name implies, desalination is accomplished by the freezing or crystallization of water molecules from the brine into essentially pure water ice; the term, "indirect contact", refers to the fact that no mixing occurs between the process fluid or product water and the refrigerant, because all heat exchange occurs using heat transfer surfaces that separate the two fluids.[2] This process, under study and development by the Chicago Bridge and Iron Company since the late 1970s, resulted in their proprietary freezer design.

SOLAR THERMAL COLLECTOR

The cost of the solar equipment in a solar-powered desalination plant is about half the total cost of the plant. Minimum product water cost is achieved when energy consumption is minimized by selecting appropriate technologies on both the solar side and the desalting side. The major equipment on the solar side are the collectors. The Yanbu pilot plant design emphasized the critical importance of selecting an appropriate solar collector, which was not only at the forefront of concentrating solar collector technology, but which also had the best potential for an economical, mass producible design.

The paraboloidal concentrator, commonly known as the parabolic dish, is known to be the most effective solar collector. In addition to modularity, the dish, due to double axis tracking, enjoys the absence of any cosine losses, resulting in net power collection that is 30% higher than that of competing systems. Two-axis tracking, coupled with the ability to produce high solar concentration ratios, not only enables it to collect solar energy efficiently but makes it available for productive uses at temperatures higher than those of any other form of solar collector.[3]

The PKI point-focusing fresnel concentrating solar collector is comprised of three primary subsystems: the concentrator, receiver/fluid loop, and system controls.[2] The concentrator has identical curved, reflective columns utilizing a faceted fresnel design to support 864 small 0.3 m by 0.3 m (1 ft by 1 ft) second-surface, silvered glass mirrors. The net aperture area is 80.3 m^2 (864 ft^2) with a realized concentration ratio of 350 to 500:1. The columns are ganged together and rotated through their centers of gravity to provide elevation tracking. The concentrator is supported by a structure composed of steel tubing members and steel plate joints that distribute wind and gravity loads to the base supports. The base of the structure is an inverted track that rotates on wheels mounted on concrete piers. Azimuthal tracing is accomplished by rotating the entire structure from east to west throughout the day.

The receiver/fluid loop utilizes a monotube cavity mounted at the concentrator focal area in a stainless steel encased housing. Concentrated solar energy is transferred to a fluid circulated through the receiver.

A microprocessor-based package is used to automatically control the two-axis tracking and operation of the dish. Sensors are the basis for tracking and utilize concentrator-mounted shadow bands. A software program provides azimuthal tracking during cloudy periods so that collection of energy can begin immediately upon reappearance of the sun.

SYSTEM OVERVIEW

Figures 1 and 2 are the process flow diagrams of the solar energy power system and of the freeze desalination system, respectively. Included is a photograph giving a view of the pilot plant.

COLLECTOR EFFICIENCY: YANBU TEST RESULTS COMPARED TO ASHRAE STANDARDS AND KREPCHIN'S PREDICTIONS

The PKI collectors were tested for their performance as part of the desalination pilot plant test program. These tests do not conform to the ASHRAE Standard, since the plant was designed and set to operate at the specified design parameters. However, a comparison of the actual test results to the DSET laboratory report[7] is helpful in the evaluation of the collector overall performance. The DSET report relates to the results of tests conducted according to the modified ASHRAE Standard 93-77.

Krepchin[8] developed a computer program to simulate the steady-state performance of the desalination test facility, and his collector efficiency equations are based on the curves by PKI. Yanbu test results are also compared to Krepchin's predictions.

DSET LABORATORY TESTS

Results of the performance tests in accordance with

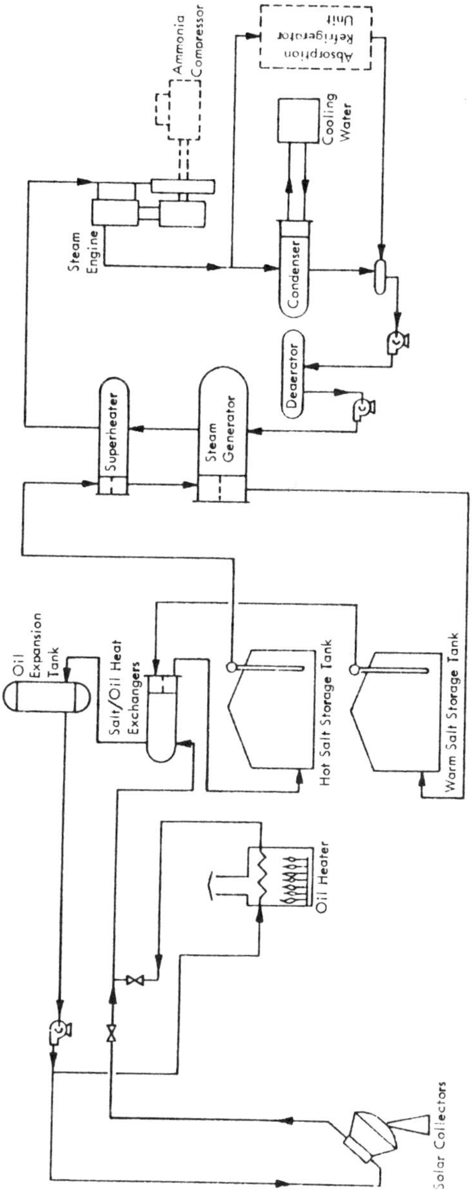

FIGURE 1: Process flow sheet (solar energy power system)

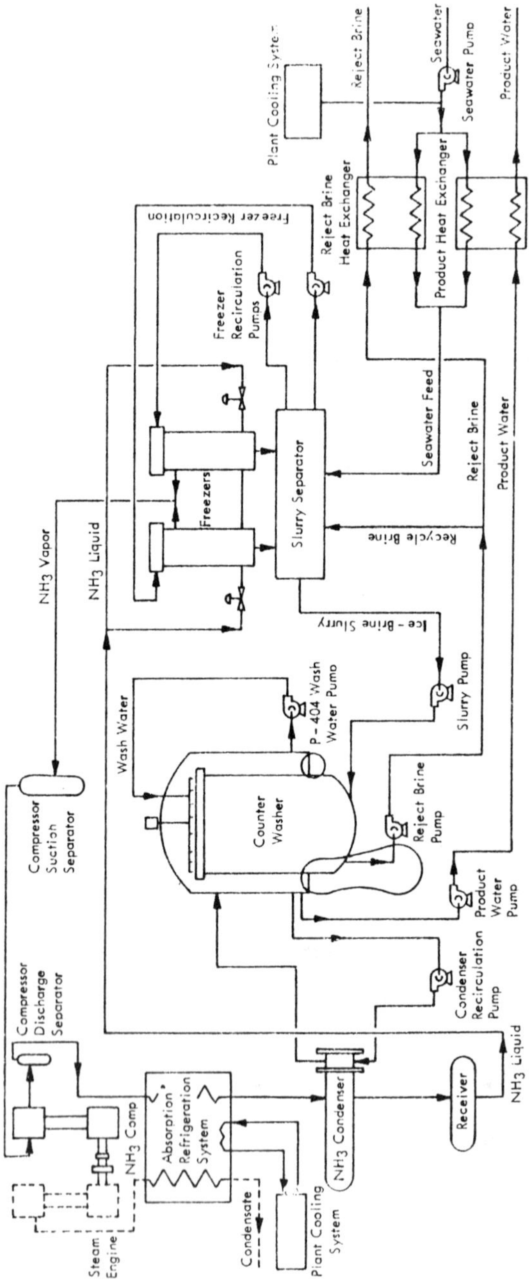

FIGURE 2: Process flow sheet (freeze desalination system)

A view of the solar-powered desalination pilot plant

the modified ASHRAE Standard 93-77 gave the following equations for the collector efficiency as a function of the inlet parameter.

$$\eta = 0.715 - 0.410\ IP \text{ (receiver uninsulated)}$$

$$\eta = 0.669 - 0.306\ IP \text{ (receiver insulated)}$$

KREPCHIN'S THEORETICAL ANALYSIS

Krepchin modeled the collector efficiency as a function of insulation only. For clean mirrors the efficiency equation is:

$$\eta = 0,6277 + 0.09699\ I + 0.35783\ I^2 - 0.2817\ I^3$$

For dirty mirrors the efficiency equation is:

$$\eta = 0.17602 + 0.99083\ I - o.44975\ I^2$$

$$\text{(I is in kW/m}^2\text{)}$$

YANBU TEST RESULTS

The performance tests at Yanbu give the following equation for the collector efficiency as a function of insolation:

$$\eta = -151.1357 + 838.7805 \times 10^{-3}\ I - 127.1034 \times 10^{-5}\ I^2 + 789.409 \times 10^{-9}\ I^3 - 156.71 \times 10^{-12}\ I^4$$

and the following equation as a function of the inlet parameter:

$$\eta = -169.1058 + 1923.2673\ (IP) - 5611.7044\ (IP)^2 + 5819.3918\ (IP)^3 - 1440.67193\ (IP)^4$$

COMPARISON

The efficiencies are based on the following parametric values:

Direct normal radiation = 850 W/m^2

Fluid mass flow rate = 0.27 kg/s

Fluid inlet temperature = 315°C

Ambient temperature = 30°C

Inlet parameter $= \frac{T_{in} - T_{amb}}{IDN} = 0.335°C\ m^2/W$

DSET Report	Krepchin's Prediction	Yanbu Tests
57.8% (receiver uninsulated)	79.5% (clean mirrors)	46%
56.6% (receiver insulated)	69.3% (dirty mirrors)	(receiver insulated and mirrors cleaned every three days)

It is to be noted that the efficiency equation derived from the Yanbu test data reflects its performance under actual conditions at the plant site over a period of about 18 months.

THERMODYNAMIC VIEW

The thermodynamic view provides the unique advantage of finding the feasibility of matching a source of energy with a task. This is particularly attractive in the utilization of solar energy. The first law of thermodynamics provides a quantitative measure of energy utilization, whereas the second law gives insight into the quality of energy.

The first law efficiency is the ratio of the desired energy output to the required energy input. When the magnitude of the energy output exceeds the input, as in the case of a heat pump, the efficiency (commonly referred to as the coefficient of performance) may greatly exceed 100%. Hence it would seem that the first law alone is of little benefit in assessing the performance of a system. On the other hand, the second law efficiency measures the effectiveness of energy utilization by using the concept of available energy, which can be defined as the maximum amount of work that can be produced from the energy of any system. The availability of a process transferring a quantity of heat q at temperature T is simply the product of the Carnot efficiency and the heat transferred.

In the following analysis, the first and second law efficiencies of the solar thermal cavity and of the PKI collector are determined to illustrate the usefulness of the availability concept.

FIRST LAW EFFICIENCY

CAVITY RECEIVER

The first law efficiency is defined as the ratio of the net energy absorbed by the receiver to the radiation energy passing through the cavity aperture. q_F is the radiation energy passing through the aperture, and q_R is the net energy absorbed by the receiver cavity.

$$q_F = I\ A\ \rho\ G = 39.74\ \text{kW}$$

q_R = energy absorbed by the receiver - heat loss

$$= I\ A\ \rho\ G\ \alpha_e - A_R\ [\varepsilon\ \sigma(T_R{}^4 - T_o{}^4) + h\ (T_R - T_o)] - A_e\ K_L\ (T_R - T_o)$$

The heat loss consisting of radiation, convection, and conduction terms is calculated to be 4.76 kW.

Hence,

$$q_R = 39.74 \times 0.974 - 4.760 = 33.95 \text{ kW}$$

First law efficiency = $\eta_F = q_R/q_F = 33.95 / 39.74 - 0.854$ or 85.4%

COLLECTOR SYSTEM

At a fluid flow rate of 0.27 kg/s, T_{in} = 303°C and T_{out} = 367°C, energy delivered by the collector is given by $mC_p \triangle T$, and equals 4.19 kW. Solar energy input to the collector at an insolation of 0.85 kW/m^2 is 68.25 kW. Hence the first law efficiency of the collector system is 58.9%. This does not account for the field losses that occur before the energy is delivered to the storage subsystem.

SECOND LAW EFFICIENCY

CAVITY RECEIVER

The second law efficiency is defined as the ratio of the availability gained by the working fluid to the available radiation energy entering the cavity.

The rate of availability gained by the working fluid is given by the product of the rate of heat transfer to the fluid and the Carnot efficiency between the temperature limits T_W and T_o.

$$b_W = q_R \left(1 - \frac{T_o}{T_W}\right)$$

$$= 33.95 \left(1 - \frac{303}{608}\right)$$

$$= 17.04 \text{ kW}$$

Available radiation energy entering the cavity is given by the following equation developed by Moynihan:[6]

$$b_F = \left[q_F \; 1 - \frac{4}{3} \frac{T_o}{T_F} + \frac{1}{3} \left(\frac{T_o}{T_F}\right)^4 \right]$$

$$= 39.74 \text{ X } 0.8878 = 35.28 \text{ kW}$$

Hence, the second law efficiency η_s = 17.04/35.28 = 0.483, or 48.3%.

COLLECTOR SYSTEM

The second law efficiency for the collector is given by the expression:[5]

$$\eta_s = \frac{\eta_F}{CE} - \frac{m\ c_p\ \ln \frac{T_2}{T_1}}{SE \quad CE}$$

where the first law efficiency is 58.9%, solar power input SE is 68.25 kW, and the Carnot efficiency CE between the temperature limits 3600 K and 303 K is 91.6%.

The second law efficiency of the collector is determined to be 32.1%.

TYPICAL DATA AND RECEIVER CHARACTERISTICS

I = direct normal radiation = 1000 W/m^2
A = projected area of the PKI concentrator = 80.3 m^2

p = reflectivity of the mirrored surface = 0.82
G = combined shadowing, blocking, and intercept factor = 0.71
$^{*}T_F$ = temperature of the focal region = 3600°K
T_O = temperature of the surroundings (dead state) = 30°C
T_R = average cavity wall temperature = 410°C
T_W = average working fluid temperature = 350°C
m = mass flow rate of the heat transfer fluid = 0.27 kg/s
T_{in} = inlet temperature of fluid = 303°C
T_{out} = exit temperature of fluid = 367°C
C_p = average heat capacity of fluid = 2.326 kJ/kg c
A_R = receiver aperture area = 0.255 m^2
A_E = effective conduction area = 1.354 m^2
L = insulation thickness = 0.1524 m
$^{**}h$ = convective heat transfer coefficient = 15.5 $W/m^2 °C$
k = thermal conductivity of insulation = 0.0694 W/m°C
$^{***}\alpha_e$ = effective absorptivity of cavity = 0.974
$^{***}\varepsilon_e$ = effective emissivity of cavity = 0.974

---

* Calculated by the experimentors[6]

** With a wind velocity of 10 km/hr the film heat transfer coefficient is determined from basic fluid flow and heat transfer relationships.[9]

*** $\alpha_e = \varepsilon_e = \dfrac{\alpha}{1 - (1 - 1/R)\ 1 - \alpha)}$

where R is the ratio of cavity inner surface area to the receiver aperture area.

CONCLUSION

One of the most important considerations in choosing the PKI collectors for the Yanbu plant was that the design had more promise for the practical application of solar technology. It was developed from experience on several earlier generations of equipment. The design uses many small interchangeable parts that are easily replaced if necessary. It incorporates rapid focus- defocus capability without excessive power consumption. The structural integrity of the collector is high and is capable of withstanding gusty winds.

During the actual period of operation, some difficulties were experienced. Tracking precision was adversely affected by overshoot of the hydraulic elevation drive, and diffuse tracking characteristics were affected by lack of shadowband control. Melting and failure of flux traps happened frequently. The collectors that were located very close to the sea suffered in mirror reflectance and edge corrosion. Improved edge protection is expected to reduce the problem.

For brevity, all the merits and demerits of the PKI collector are not discussed in this paper. However, it should be mentioned that the collector has demonstrated its capability to give effective performance in a humid, dusty, seacoast environment. The PKI collector, with some geometry redesign, may prove to be a very potential collector for application in solar desalination technology in desert regions.

REFERENCES

1. B.H. Khoshaim and J.S. Williamson. (eds.). March 1981. Proceedings of the Workshop on Solar Water Desalination, Denver, Colorado.

2. Technical Report by the Chicago Bridge & Iron Company, MRI Contract No. AF-1-9115-1, May 1982.

3. B.H. Khoshaim and J.S. Williamson. (eds.). April 1983. Proceedings of the Workshop on Solar Thermal Collectors, Lakewood, Colorado.

4. G.F. Hamad and M.C. Gupta. 1986. Energy considerations in a solar-powered desalination pilot plant, paper presented at the 9th World Energy Engineering Congress, Atlanta, Georgia, October 21-14.

5. M. Zarea and E. Mayer. 1986. Second law optimization procedure for concentrating collectors, paper presented at the International Congress on Renewable Energy Sources, Madrid, Spain, May 18-23.

6. P.I. Moynihan. 1986. Example of second law efficiency of solar thermal cavity receivers. Journal of Solar Energy Engineering, February Vol. 108/67.

7. L.A. Bard, T.J. Diel, W.J. Putman. January 1985. DSET Report No. 845011-4-F.

8. I.P. Krepchin. April 1984. Computer program PLANT for the performance analysis of solar power plants. Report No. MRI-8422.

9. G.F. Hamad. January 1987. PKI collector's performance. SOLERAS Technical Report.

A MODULAR POLYVALENT SOLAR DRYER FOR INDUSTRIAL PURPOSES

A. CHAMMA AND M. DAGUENET
L.T.E., University De Perpignan, France

J. RANDRIAMORASATA
E.E.S. Polytechnique, Antananarivo, Madagascar

M.H. GABR
Faculty of Engineering, Ain Shams University, Egypt.

Abstract An indirect solar dryer for polyvalent purposes was investigated theoretically and experimentally. Modular solar dryer prototypes were designed and constructed for this investigation. Each of these modules is composed of a flat plate air collector of a $2.3m^2$ surface area and a drying chamber of a $0.54m^3$ volume. The dryer is equipped with a centrifugal fan and a back-up heat source. The air expelled from the drying chamber can be partially or totally recycled, according to its state at exit. The whole system is a compact module which would permit the construction of a long drying tunnel by the assembly of more than one module. A numerical modelling of the system was done and its behavior was simulated by computer calculations. Prototypes were tested at Perpignan and at Antananarivo. An economical evaluation showed that the system is profitable for certain conditions.

INTRODUCTION

Drying is one of the common methods used for conservation of many agriculture products. A large portion of the world's supply of dried fruits and vegetables continues to be sun dried, using natural air currents under primitive conditions: being exposed to unexpected rains; wind-born dirt and dust; and infestation by insects, rodents and other animals. The quality of the resulting end-product is often seriously degraded, sometimes beyond edibility.

In the literature, many authors present different types of simple solar dryers,[1,2,3]. The cabinet dryer and the combination of drying chamber and solar air heater are two basic types of compact solar dryers which retain the advantages of the traditional method.

In the cabinet dryer, the necessary heat is supplied by directly exposing the crop to solar radiation. This process enhances the proper color 'ripening' of greenish fruits by allowing, during dehydration, the decomposition of residual chlorophyll in the tissue under direct solar radiation. The major drawbacks are the liability to overheat and cause crop damage, and the relatively slow drying rates that have so far been achieved due to poor rates of vapor removal.

In the combination of a drying chamber and air-heater, the crop is located on shelves in the drying chamber and heated by circulating air warmed during its flow through a low-pressure drop solar collector. Because the drying material is not subjected to direct sunshine, caramelization and heat damage do not occur. A further advantage is that the circulating air absorbs the water vapor from the crop, and thus accelerates drying.

Industrial drying, using large classical dryers, is working efficiently but with high energy consumption. They have not been adapted for use in arid areas in developing countries. The use of solar energy for the heating of air for drying has been found to be techno-economically viable[1,2], especially for arid zones with high insolation intensity.

The object of this study is to develop a viable solar crop dryer, scientifically conceived, polyvalent, and suitable for the industrial needs of a small collectivity. It proposed to use a modular-type solar dryer equipped with a back-up heat source and forced circulation of the drying air. The present work does not concern small and rudimentary dryers for familial needs.

The proposed design concept has the following advantages:

- it satisfies variable drying demands, qualitatively and quantitatively, using a series assembly of a sufficient number of modules. Required drying conditions, such as temperature, humidity and velocity of the drying air, could be fixed.

- it permits the realization of a multi-chamber drying complex with different drying conditions, for either continuous operation of intermittent loading and unloading, in each chamber.

- it allows standardization of the dryer elements making them identical for each installation. The initial cost of the dryer could be reduced by using mass-production procedures.

In this work, the design and construction of a solar-stimulated economically-viable crop dryer was investigated theoretically as well as experimentally. Such a dryer would be suitable for use with cash crops on a village scale. The drying process will be accomplished solely by the forced circulation of air, so there is a requirement for auxiliary power to drive a fan. Auxiliary power is needed also to provide additional heating. The experimental unit was of a modular design which enables changes in its operating mode to be accomplished easily and facilitates the integration of higher capacities. The predictions of the theoretical analysis were compared with the experimental results.

EXPERIMENTAL APPARATUS

A compact modular solar dryer was designed and constructed at the Laboratory of Thermodynamics and Energy, University of Perpignan, Perpignan, France. Perpignan is 420° north latitude. The modular dryer consists mainly of a flat plate solar air collector, and air-drying duct, and a drying chamber. The dryer is diagrammatically shown in Figure 1.

SOLAR COLLECTOR

The collector consists of the following elements assembled in a frame:

- a transparent tempered glass cover 4 mm thick;

- a black painted zinc absorber of 0.5 mm thick with absorption coefficient oc = 0.95 and emitting coefficient ε = 0.4;

- an isolating back plate constructed from an isolating panel, (steel sheet - polyurethane - steel sheet), 0.06 m thick.

The effective surface area is 2.3 X 1 m. The distance between the glass cover and the back plate is 0.05 m, and the absorber divides this distance into two equal parts each of 0.025 m.

HOT AIR DUCT

The hot air duct measures 1.2 m high, 1.0 m wide and 0.3 m deep. The duct is constructed from isolating panels that are 0.06 m thick. The duct is equipped with two centrifugal fans with a vertical shaft. Each fan has a maximum flow rate of 0.14 m^3/s. Thermostated electric heating resistances, with a total power of 4.5 kw, are mounted in the duct to permit the adjustment of the air temperature.

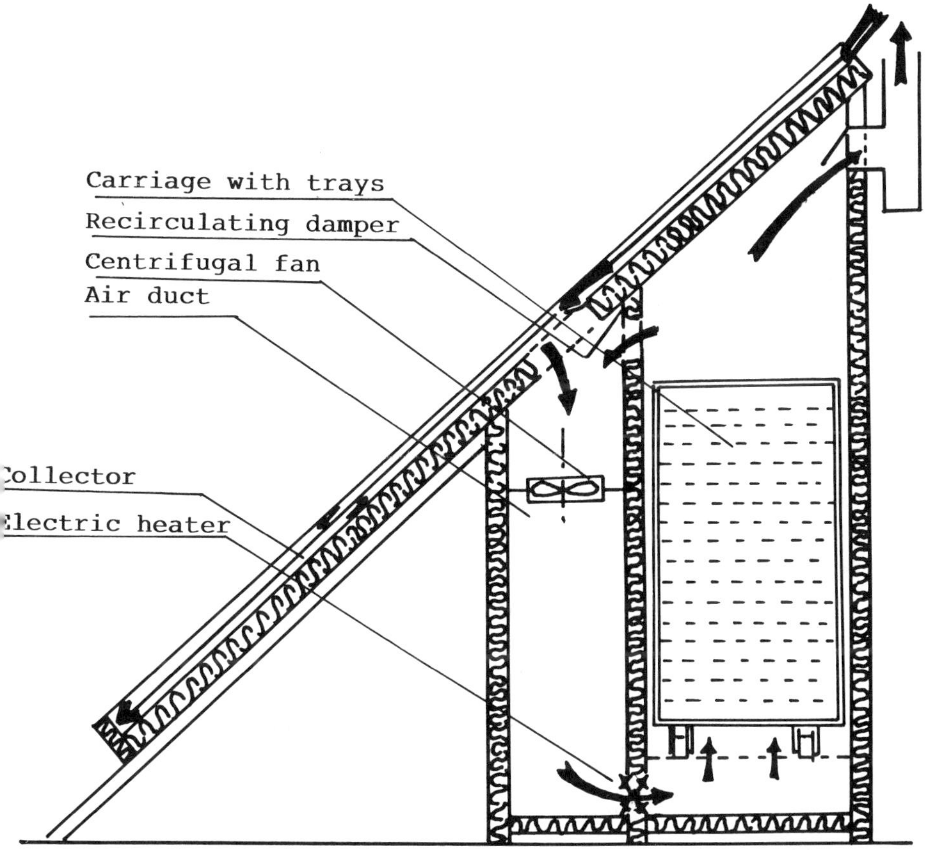

FIGURE 1: Schematic diagram of the modular dryer

DRYING CHAMBER

The drying chamber dimensions are 1.2 X 1.0 X 0.45 m. Its walls are constructed from the same panel as the hot air duct. It contains two moving carriages on which the product is to be dried; each has 16 horizontal trays. The total available exposing surface is 6.2 m^2. A perforated sheet is fixed as a base for the drying chamber and as a false floor. This false floor permits the carriage to be rolled on it and provides good air distribution.

INSTRUMENTATION

The measured parameters are temperature, humidity and velocity of the air at different locations in the system; the temperature and humidity of the product to be dried; the temperature of the walls; the meteorological data (insolation, temperature, wind speed); and the electric power consumption for the fans and the additional heat. All measured data are recorded using a Philips data recorder of type PM 9237A with 30 channels, connected to a central programmable data aquisition unit with 30 channels also.

EXPERIMENTAL ANALYSIS

OPERATING PRINCIPLE

The double pass solar collector consists of two distinct zones. In the top zone the flow is parallel while in the lower zone the flow is counter as shown in Figure 1. The air enters at the top of the collector, over and below the absorber, but in the lower zone all air passes first into the conduit formed by the glass cover and the absorber. Then it passes into the conduit formed by the absorber and the isolating back plate, and finally the air enters the air duct.

The centrifugal fan velocity is variable and the electric heater resistance is controlled by a thermostat. Thus, the air temperature at the inlet to the drying chamber could be independently adjusted between the ambient temperature and a maximum of 80°C. The air flow rate could be varied between 0 and 0.15 kg/s.

In order to conserve energy, two dampers are installed at the top of the drying chamber. These permit control of the quantity of recirculating air depending on the state of air at exit.

EXPERIMENTAL PROCEDURE

Many trials were conducted to study the influence of different parameters. Continuous drying (24h/24h) and discontinuous drying (10 h during a day) were investigated with and without air recirculation.

MODELING AND SIMULATION

AIR SOLAR COLLECTOR

In this study, the air flow in the collector is assumed to be one dimensional and the air temperature is uniform in the plane normal to the air flow. Thus, the problem depends on time t and space x. The collector is divided into layers of thickness Δx and the drying period in time intervals Δt.

The governing equations in each layer are written using the classical electric analogy method[2]; thermal energy balance was made at certain specified points and the Kirschoff laws were applied.

All of the heat transfer modes: conduction, convection and radiation, were considered using the following assumptions:

- conduction of the air is negligible in the flow direction;
- temperature gradients in the absorber are negligible;
- wind direction is parallel to the collector surface.

The different transfer coefficients are calculated using the available semi-empirical correlations[2].

In the collector simulation, meteorological data measured during the experiment conducted on the dryer prototype were used to facilitate the comparisons between the theoretical and the experimental results.

Different types of collectors were also simulated. It was found that the rise in temperature in the collector is much greater when using a semi-selective absorber as compared to the proposed air circuit.

DRYING CHAMBER

In the drying chamber calculations, the thermal radiation in the chamber and the conduction in the trays were neglected. The product to be dried was considered as a spherical shaped piece of apple with a uniform temperature and humidity.

The modeling procedure used for the collector, was also used for the drying chamber. In addition, mass transfer was considered using the mass transfer coefficient of a deduced correlation by Combes[2].

In the drying chamber simulation, the conditions relevant to experiments were applied to test the validity of the present model. Two cases were studied: the first was without air recirculation and the second was with 80% recirculating air. In all cases, the dryer was assumed to be filled with 12 kg of fresh apple exposed on 24 trays in two carriages. Tray No. 1 was at the bottom while the Tray No. 12 was at the top. The temperature at the chamber inlet was considered constant at 50°C.

EXPERIMENTAL AND THEORETICAL RESULTS

As an example, Figures 2 and 3 show the results of two different working conditions with and without air recirculation. The results are presented as the absolute humidity versus the drying time for apples. It was found that the recycling of air partially improves the homogenization of drying in the different trays and conserves energy. Good agreement is observed between the present model and the experimental results.

Air flow rate = 0.029 kg/s.
without recirculation

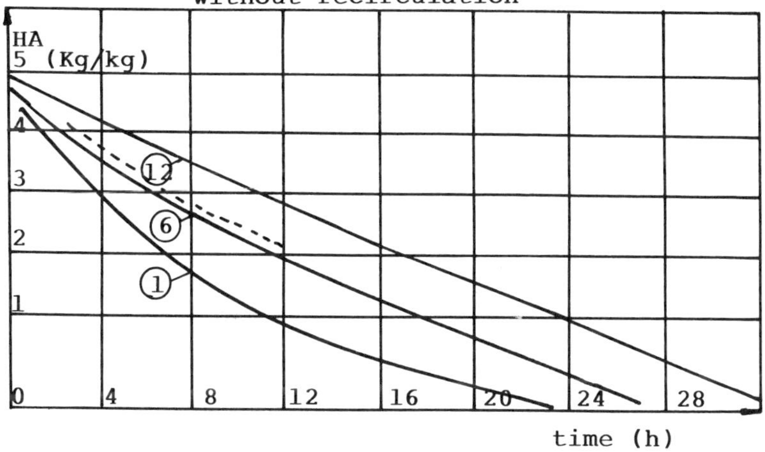

FIGURE 2: Variation of absolute humidity HA of apple (without recirculation)

Air flow rate = 0.075 kg/s, (0.06 kg/s recirculated).
with recirculation

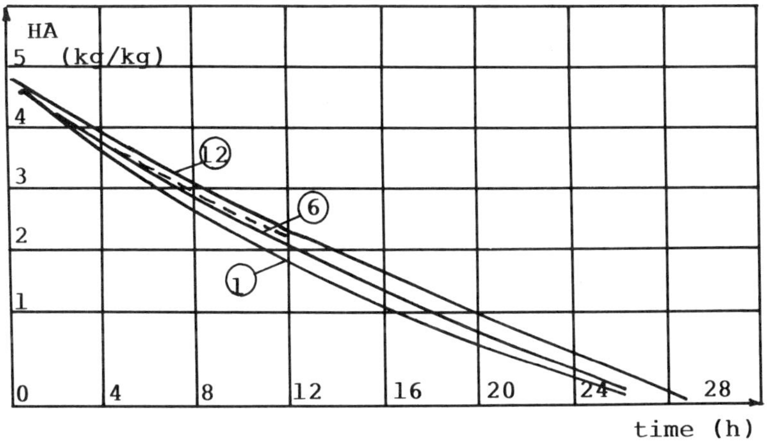

FIGURE 3: Variation of absolute humidity HA of apple. (with recirculation)

From the experimental results, it was found that the solar contribution ratio, defined as the ratio between the energy obtained from the collector and the total consumed energy, varied from 7 to 75%. As an example showing the distribution of the energy contribution during the drying cycle, Figure 4 shows the energy contribution versus time during apple-drying with 80% air recirculation. The solar contribution is 76%. The thermal efficiency of the dryer, defined as the ratio of the energy required to evaporate the water to the actually consumed energy, is from 46 to 49%.

drying with 80% recirculated air
Charge: 8.5 kg of fresh apple in slices

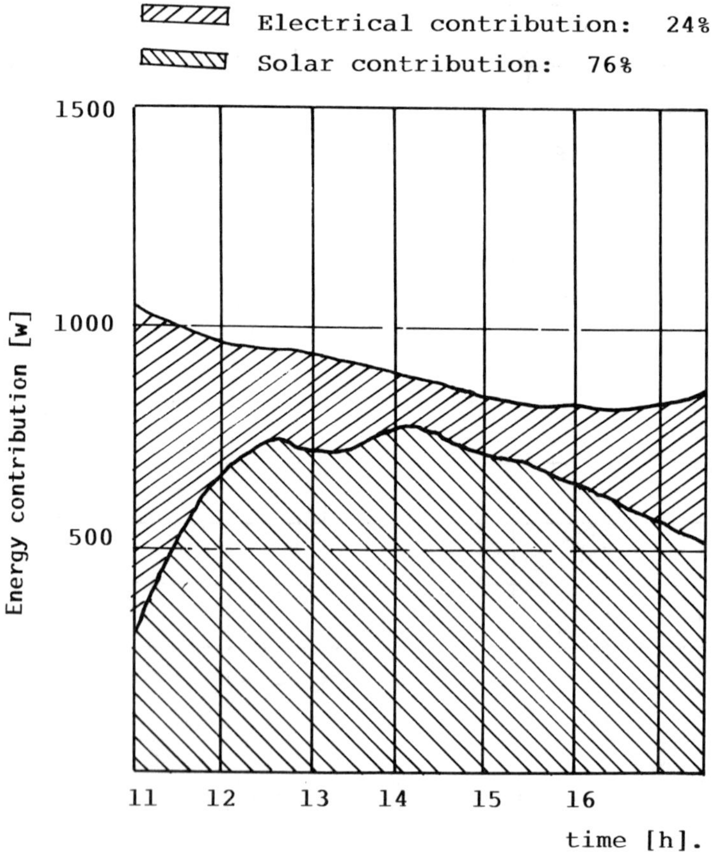

FIGURE 4 : Solar and elecrical energy contribution versus time.

CONCLUSIONS

The present modular dryer performs satisfactorily and is suitable for many drying needs. The model permits calculation of the performance of the drying installation. Good agreement was obtained between the theoretical and the experimental results.

REFERENCES

1. M. Daguenet, "Lee sechoirs solairs: theorie et pratique", Edition de 1 'UNESCO, (1986).

2. C. Combes, "Modelisation information d'un system de sechage solaire pour fruit et legumes", These de Docteur-ingenieur, Universite de perpignan, (1981).

3. J.Y. Quinette, "Etude experimental du sechage des denrees agro-alimentaire en vue de 1' optimisation solaire", These de docteur-Ingenieur Universite de Perpignan (1981).

NON-CONVENTIONAL ENERGY PLAN FOR DESERT DEVELOPMENT IN EGYPT

HUSSEIN K. ABDEL-AAL
Solar Energy Laboratory, National Research Center, Cairo, Egypt

Abstract In this paper, the solar-hydrogen approach is proposed for desert development in Egypt. Methods of converting solar energy along with the appropriate processes of splitting water for hydrogen production in Egypt are suggested. A master plan is then formulated for the integrated application of hydrogen both as a fuel and as a fertilizer ingredient for the development of arid land in Egypt. Such applications include (1) the use of hydrogen as a clean fuel for direct combustion or for electric power production, and (2) as a chemical feed stock in the fixation of atmospheric nitrogen to produce ammonia, which is basic for a number of down-stream fertilizer products.

INTRODUCTION

Egypt's resources of arable land, water, and energy are increasingly inadequate because of massive population growth. This has severely limited agricultural production, and created what is called the "food gap" between production and consumption, estimated to be around two billion dollars for the year 1980.[1] Figure 1 is a manifestation of this food gap.

Developing countries in general, in contrast to developed nations, need more fertilizers, and more water and energy to satisfy the demands of their increasing populations. A comparison between the two blocks worldwide shows that developed countries have 25% of the population, 45% of the arable land and produce 42% of the agricultural products, while developing countries make up the balance. On the other hand, developed countries utilize 70% of the fertilizers consumed, and 85% of the work done by tractors. This is illustrated in Figure 2.

In Egypt, projects of reclamation, desert development, and productive use of potentially arable land would all require tremendous quantities of capital and energy. In addition, high agricultural productivity depends upon the availability of energy which is generally supplied by fossil fuels.

The question is, how can we supply the energy needed for development in Egypt? The per capita consumption of energy exceeds 650 KWhr, and the total energy produced annually is more than 40 billion KWhr. One-fourth of this energy is generated by hydropower, while the balance is produced using natural gas and petroleum products. It is in the interest of our national economy to reduce - not to increase - our energy consumption. As a matter of fact, a target has been set to save as much as 5% of our total energy consumption by the use of new and renewable energy sources.

For Egypt, which is rich in solar energy, conditions appear to be favorable for the production of what is called "solar hydrogen", a non-conventional energy form. A master plan using "solar hydrogen" is worked out for the development of our arid land as follows.

SOLAR HYDROGEN: A NON-CONVENTIONAL ENERGY SOURCE

Alternative sources of energy are available in Egypt, but are relatively undeveloped technologically or have not been utilized fully. Some are renewable[2] as shown in Figure 3; others are depletable but relatively untapped such as geothermal heat or synthetic fuels from coal.

Nuclear energy, as presently produced, depletes uranium more rapidly than energy from breeder reactors. Nuclear fusion is still in the stage of experimental research but it represents a hope for the future[3].

At present, petroleum and natural gas serve not only as energy resources but also as conventional energy carriers.

The replacement of these resources by large stationary power plants based on energy from other resources, creates a greatly expanded need for alternative energy carriers. In other words, in proposing and developing a non-conventional energy plan, we should be considering a system that consists of new primary energy resources along with an alternative energy carrier as shown in Figure 4.

What system of energy conversion that uses inexhaustible or renewable resources can provide us with our energy needs for agricultural development? Energy supplied should be:

- In a suitable media that is compatible with existing infrastructures.

- Competitive in cost.

- Pollution-free.

- Available at the time it is needed.

Prolonged consideration shows that the solar-system which will be described can meet most of the requirements stated above for a new energy option which is particularly suitable for Egypt. Needless to say, the amount of solar radiation which reaches the earth in one hour is sufficient to cover the needs of energy for the whole world. Table 1 is a manifestation of the magnitude of the solar energy received worldwide and is expressed in equivalent of millions of barrels of oil.[5] The energy of this immense source could be utilized by converting it into a more useful form according to the first law of thermodynamics as stated in Figure 5.

The production of electricity by means of solar cells is a possibility, but transport of such electricity is costly and it cannot be stored. The most effective way to store solar energy, therefore, is to convert it into hydrogen. Hence, solar energy is used as a primary source to split water via one of the proposed schemes which were reported earlier[6] and are presented in Figure 6.

FORMULATION OF A MASTER PLAN

The general scheme presented in this paper is geared towards desert development in Egypt. It utilizes, in principle, solar energy and saline water as inputs to the system in a number of interacting ways as outlined in Figure 7. The outputs from this system are identified as follows:

(1) Hydrogen (i) as a fuel, produced in a liquefied form [LH_2], or stored in the form of metal hydrides, and (ii) as a feed stock for fertilizers and chemicals, which are manufactured by chemically combining H_2 with atmospheric N_2 to form NH_3, hence the down-stream fertilizers. In addition, H_2 is combined with CO_2 (from air or sea water) to form methanol (CH_3OH).

(2) Water and mineral salts which include fresh water, valuable mineral salts (M_gCl_2) and chilled water for air cooling and refrigeration.

The most crucial step in implementing our master plan is the production of solar hydrogen which is addressed next.

PRODUCTION OF SOLAR HYDROGEN

An examination of the tremendous volume of literature published in this field[7,8,9,10], indicates that the most promising methods for producing hydrogen using solar energy in Egypt are water electrolysis and thermochemical cycles.

a) Water Electrolysis is based on the passage of DC through water that has been made electrically conductive by adding excess H^+

or OH^-. Thus O_2 is liberated at the anode and H_2 at the cathode. The state of the art offers very reliable electrolyzers. The question is, how can we supply the electric current using solar energy as a source? Two methods are suggested in this regard:

1. Direct energy conversion using photovoltaic cells, which is considered a very promising technology, because it produces electricity directly from sunlight without intermediate gas, steam, or mechanical cycle. In other words, it is not Carnot-limited in efficiency. Photovoltaic electricity is produced by modular arrays of identical modules that can be assembled to meet almost any power demand.

2. A solar concentrating system which is then used for generating steam for a conventional power generation plant.

   Figure 8 illustrates the production of hydrogen by water electrolysis using solar energy as the source.

b) <u>Thermochemical Production of Hydrogen</u>
Hydrogen production has been subject to Carnot-cycle limitations in the efficiency of the thermal operation of generating steam. Such limitations led scientists to propose thermochemical cycles. The basic idea involved in these cycles is to free hydrogen from water through chemical reactions and under the influence of heat only.[7,8,11,12] More than 200 cycles have been proposed and more than 1000 have been calculated via computer; however only a few (30 calculated) have been worked out in more detail and only three have led to demonstration plants.[12] The basic reaction schemes are depicted in Figure 9.

CHEMICAL FERTILIZERS FROM SOLAR HYDROGEN

Once produced, solar hydrogen is envisaged to be utilized in the manufacture of ammonia, which is the basic raw material for the production of many of the down-stream fertilizer products as shown in Figure 10. Such an approach is rather interesting and very promising for supplying ammonium fertilizers without using petroleum hydrocarbons, as was reported before by the author in 1984.[13]

THE APPROACH TO IMPLEMENTATION

The application of solar hydrogen for desert development is a multifunction operation in which hydrogen is first used as a source of energy to be stored and pipelined similarly to natural gas. Hydrogen can also be used as a clean fuel in the desalination of water to provide fresh water for agricultural purposes. Finally, hydrogen will serve as a raw material for the production of chemical fertilizers. This unique property of solar hydrogen serves practically all aspects of desert development.

Now, if the productivity of agricultural crops is denoted by A, cultivated land is L, water used in irrigation is W, energy consumed is E, fertilizers is F; then it can be stated that:

$$A = f_1 \ (L, W, E, F)$$

For a fixed land area, L is constant,

$$A = f_2 \ (W, E, F)$$

If the implementation of the use of solar hydrogen is considered as explained above, it becomes clear that these three variables affecting productivity A can be substituted by solar hydrogen, SH. Thus:

$$A = f_3 \ (SH)$$

As an added advantage, the use of solar hydrogen in the agricultural sector eliminates environmental pollution as shown in Figure 11.

The implementation of our proposed scheme to use solar hydrogen can proceed via three options depending on cost analysis. These options are:

- P.V. cells / Electrolysis
- Solar concentrators (Thermal)/ Power gener./Electrolysis
- Solar concentrators (Thermal)/ Thermochemical cycles

Details about the efficiencies of these processes and the land area required for the production of 100 MW H_2 for each case are found in Table 2.

CONCLUSIONS

Solar energy is a clean, non-depletable energy source and has attracted much attention as an alternative to fossil and nuclear energy for future use in a high-grade energy market. The major problem with solar energy is the cost of collecting it, as well as the ecological pollution associated with this collection. The surface of the earth receives about 550-1400 Wm^{-2} of radiation, which can be collected without concentrators at less than 520 K. Another feature of solar energy is its diurnal nature which, from an economic standpoint, means that capital tied up in solar collectors is utilized less than 50% of the time, and, in practice, not more than 25%. This, in effect, quadruples the cost of solar energy over continuous energy sources of comparable capital outlay.

Nevertheless, the utilization of renewable energy sources in Egypt (both solar and wind) as non-conventional energy forms for the development of arid areas is technically feasible. The major barriers to large scale use are the development of the existing technology and the reduction in capital cost of the plants.

To increase the share of non-fossil primary energy for Egypt in the energy supply in general, and in desert development in particular, the introduction of a new chemical carrier such as solar hydrogen is of vital importance. It must be emphasized that the significant use of solar hydrogen for desert development will require integrated systems of facilities for production, storage and distribution, and utilization.[4] Similarly to electrical energy, hydrogen may be produced and used on site, or distributed. Unlike electricity, it is an energy storage medium itself.

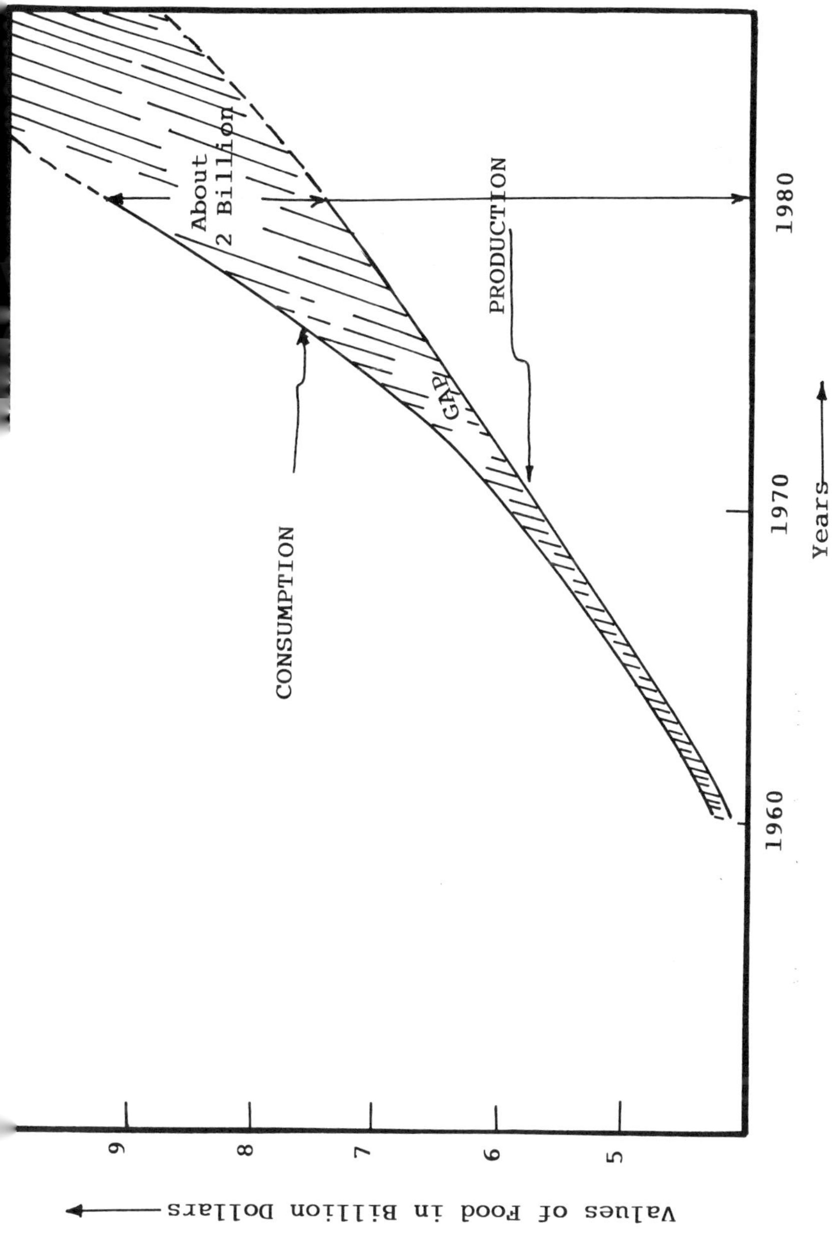

FIGURE 1: Illustration of Food Gap in Egypt
(Data are taken from Reference 1)

Developed Countries Developing Countries

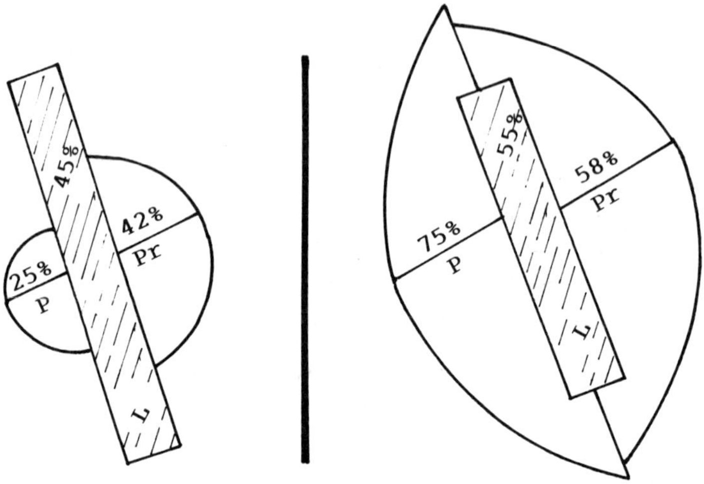

Population (P) - Productivity (Pr) Relationships, Compared with Land (L)

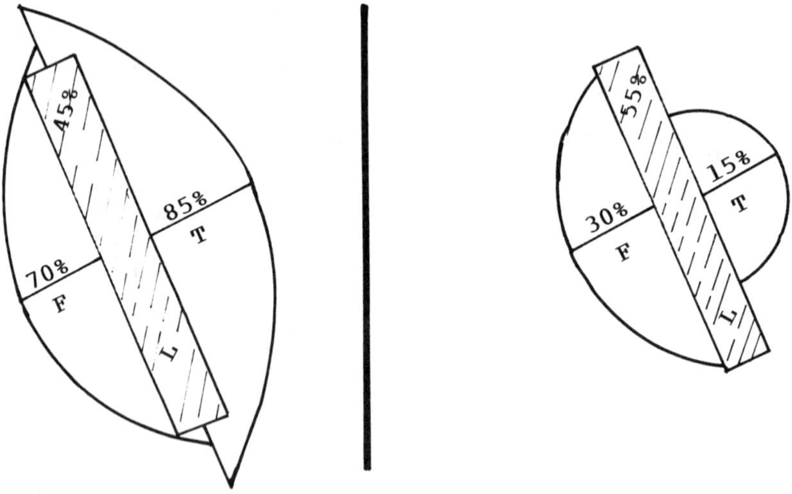

Fertilizer (F) - Tractor (T) Relationships, Compared with Land (L)

FIGURE 2: Comparative Agricultural Indicators In Developed and Developing Countries

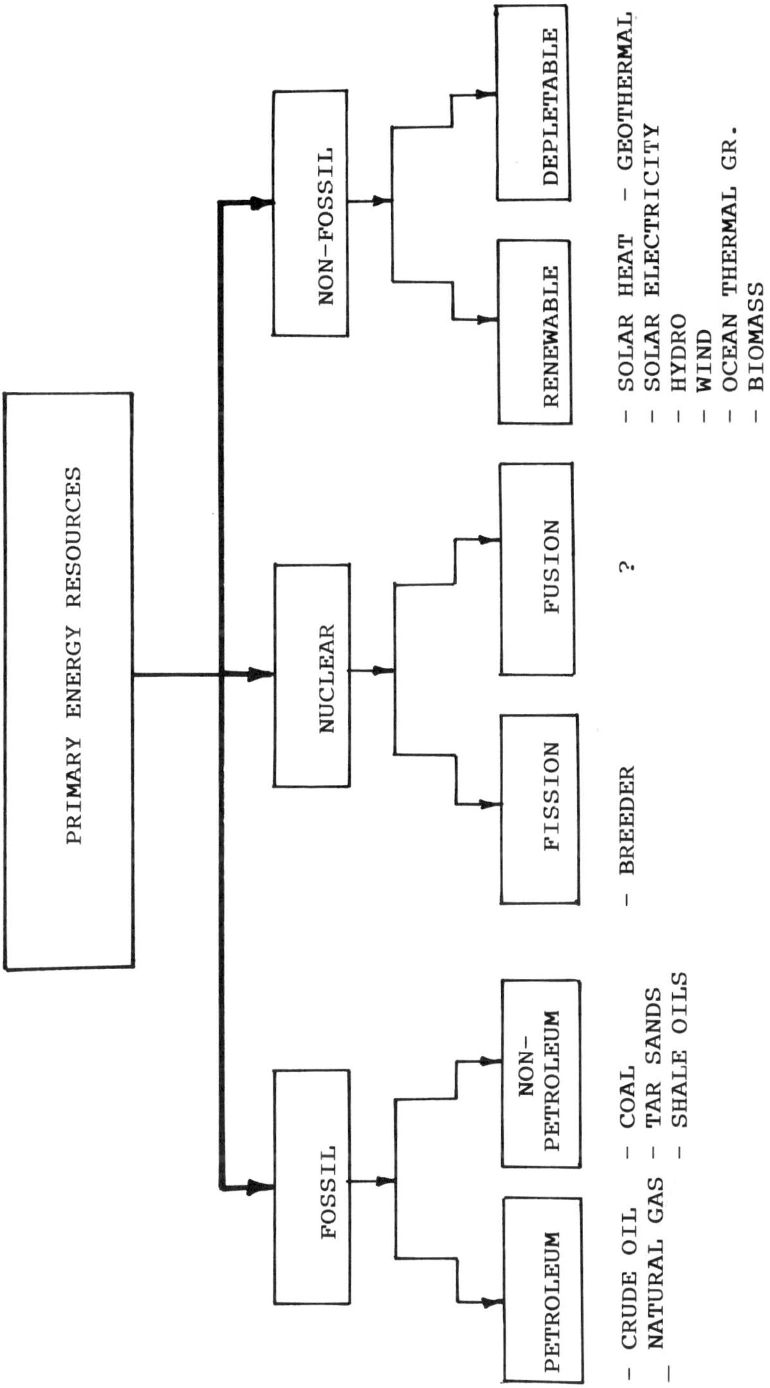

FIGURE 3: Classification of Primary Energy Resources

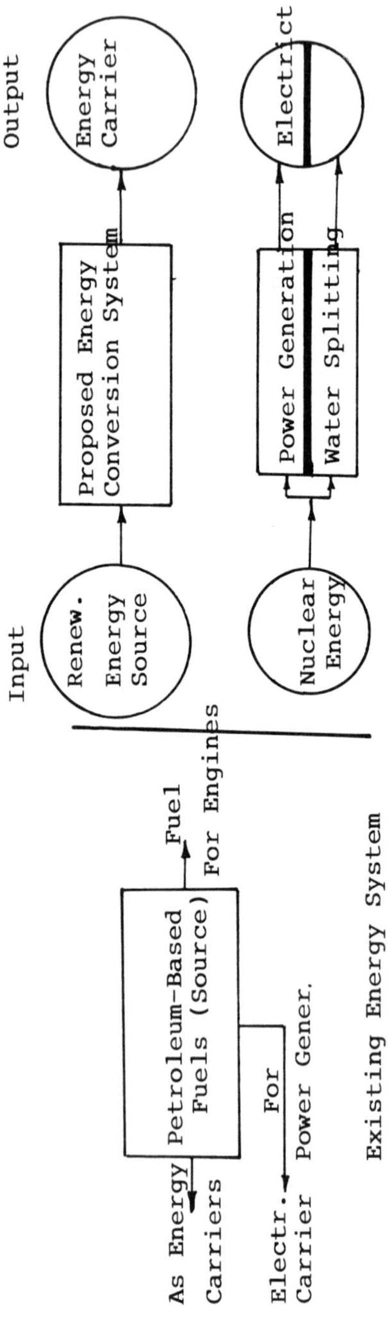

FIGURE 4: Existing and Proposed Energy System

TABLE 1. Calculations of global equivalent of solar energy (in bbl of oil).

Continent (Country)	Area (10^{12} m^2)	Solar In-solation (Kj/cm^2)	Annual P, ($Kw-hr/m^2$)	Annual Kw-hrs (10^{12})	Equivalent of oil (million bbl/year)
Asia	44	550	1,528	67.232	39,400
Africa	30	725	2,014	60.420	35,400
Australia	7.7	750	2,083	16.039	9,400
Europe	10	450	1,250	12.500	7,330
U.S.A.	9.4	560	1,556	14.626	8,570
Canada	10	325	903	9.030	5,290
S. America	17.6	630	1,750	30.800	18,060
Total (World)	128.7	570 AVG	1,583 AVG	203.732	123,450

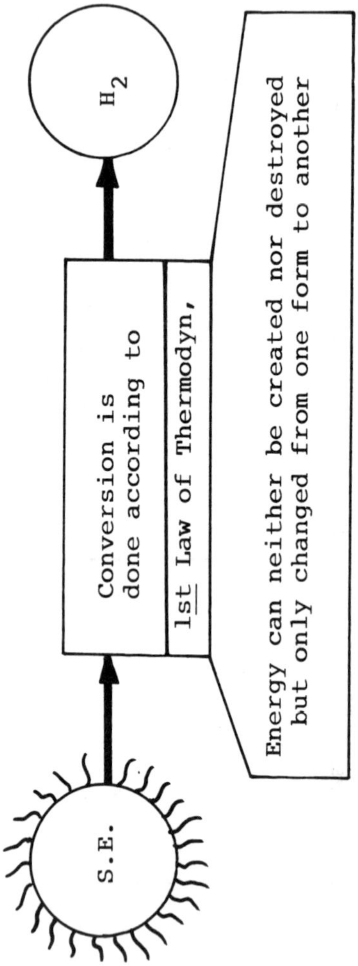

FIGURE 5: Application of 1st Law of Thermodynamics

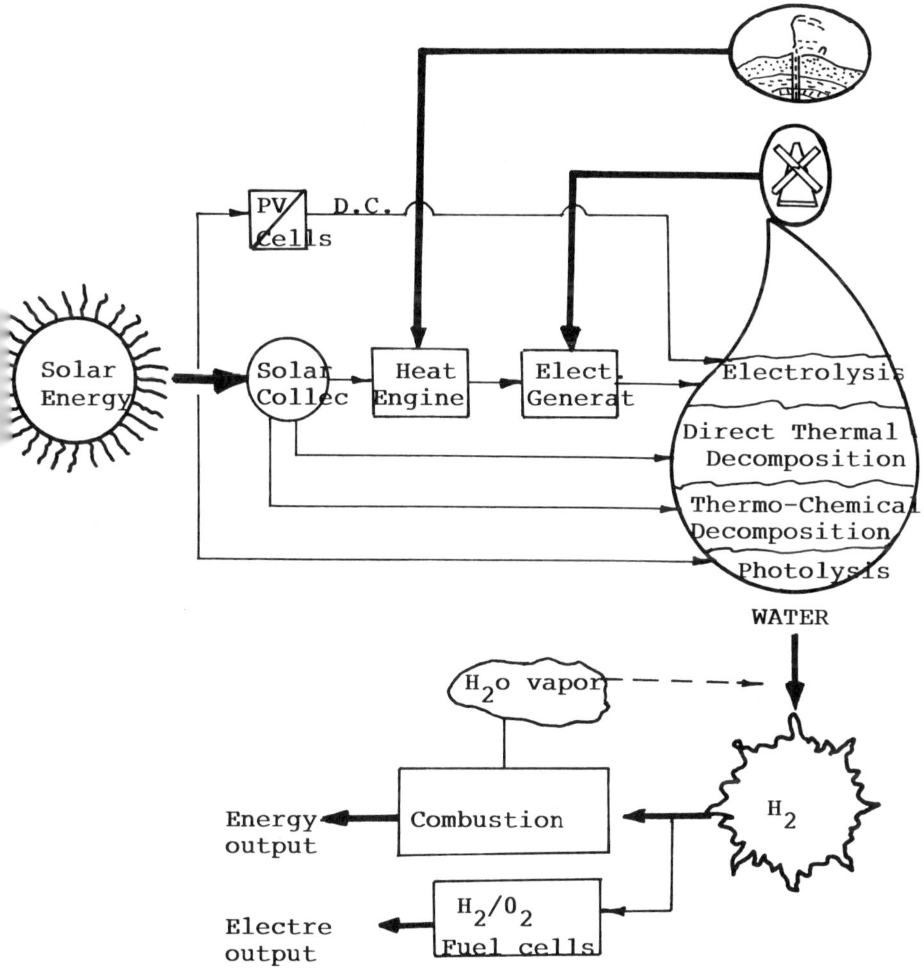

FIGURE 6: H_2-Water Cycle

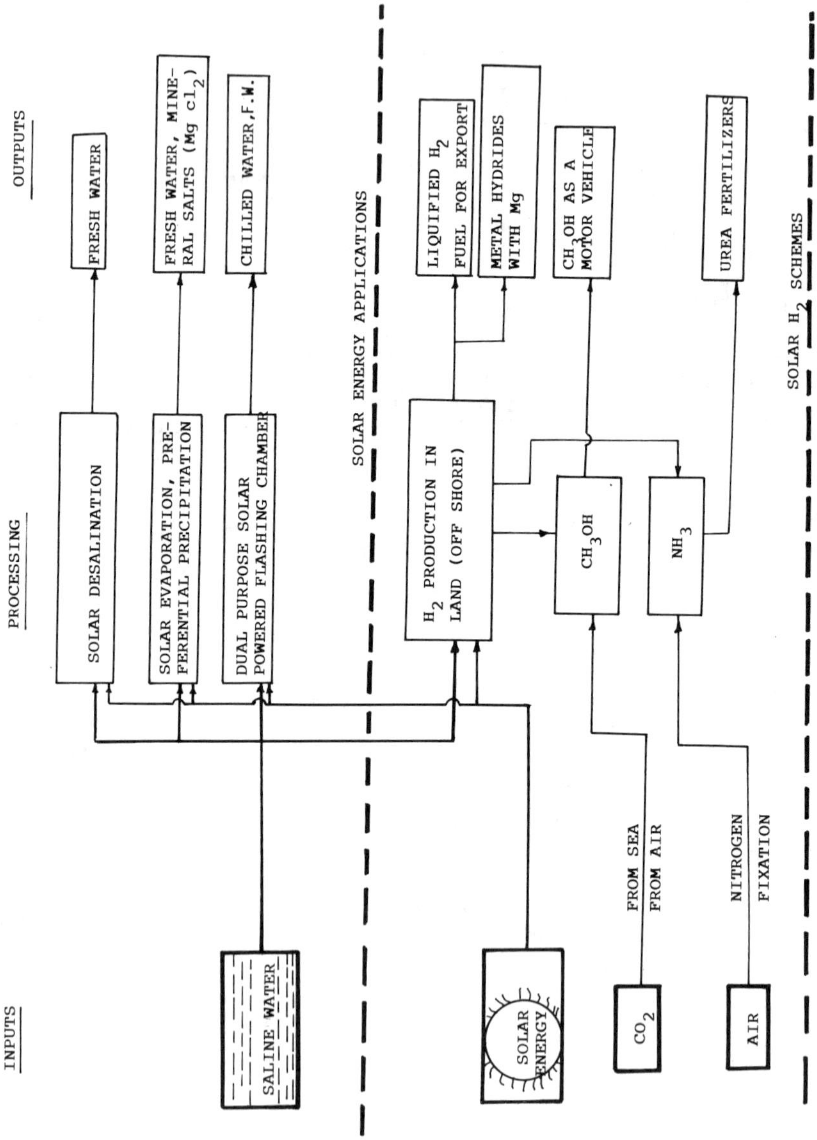

FIGURE 7: Proposed Scheme Utilizing Solar Energy and Sea Water

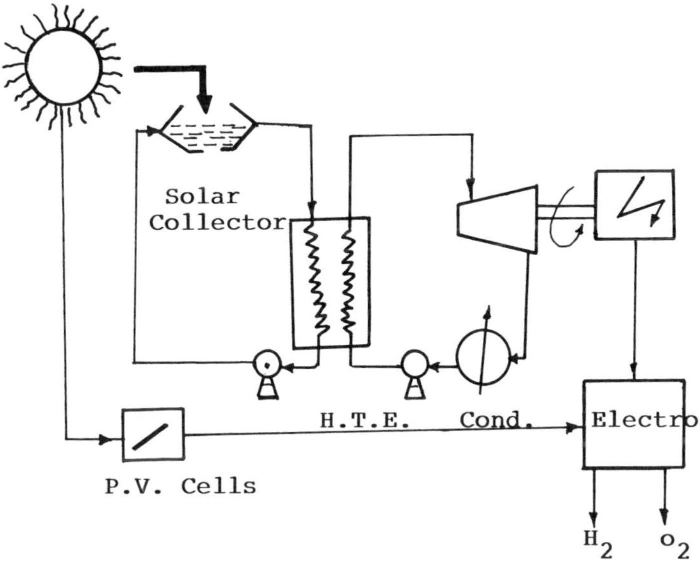

FIGURE 8: Electrolytic H_2 Production

1st: General Atomic Iodine-Sulfur Cycle

$$I_2 + SO_2 + 2\,H_2O \xrightarrow{370°K} 2\,HI + H_2SO_4$$

$$2\,HI \xrightarrow{800°K} H_2 + I_2$$

$$H_2SO_4 \xrightarrow{1150°K} H_2O + SO_2 + O_2$$

2nd: Westinghouse Sulfur Cycle (Hybrid)

$$SO_2 + 2\,H_2O \xrightarrow[0.17\ Volt]{300°K} H_2SO_4 + H_2\ \text{(Electroly.)}$$

$$H_2SO_4 \xrightarrow{1150°K} H_2O + SO_2 + O_2\ \text{(Thermolys.)}$$

3rd: Mark 13 Ispra

$$Br_2 + SO_2 + 2H_2O \xrightarrow[\text{(Electrolys.)}]{370°K} 2HBr + H_2SO_4\ \text{(aq.)}$$

$$2\,HBr \xrightarrow{470°K} H_2 + Br_2$$

$$H_2SO_4 \xrightarrow[\text{(Thermolys.)}]{1150°K} H_2O + SO_2 + O_2$$

FIGURE 9: Solar hydrogen by thermochemical cycles

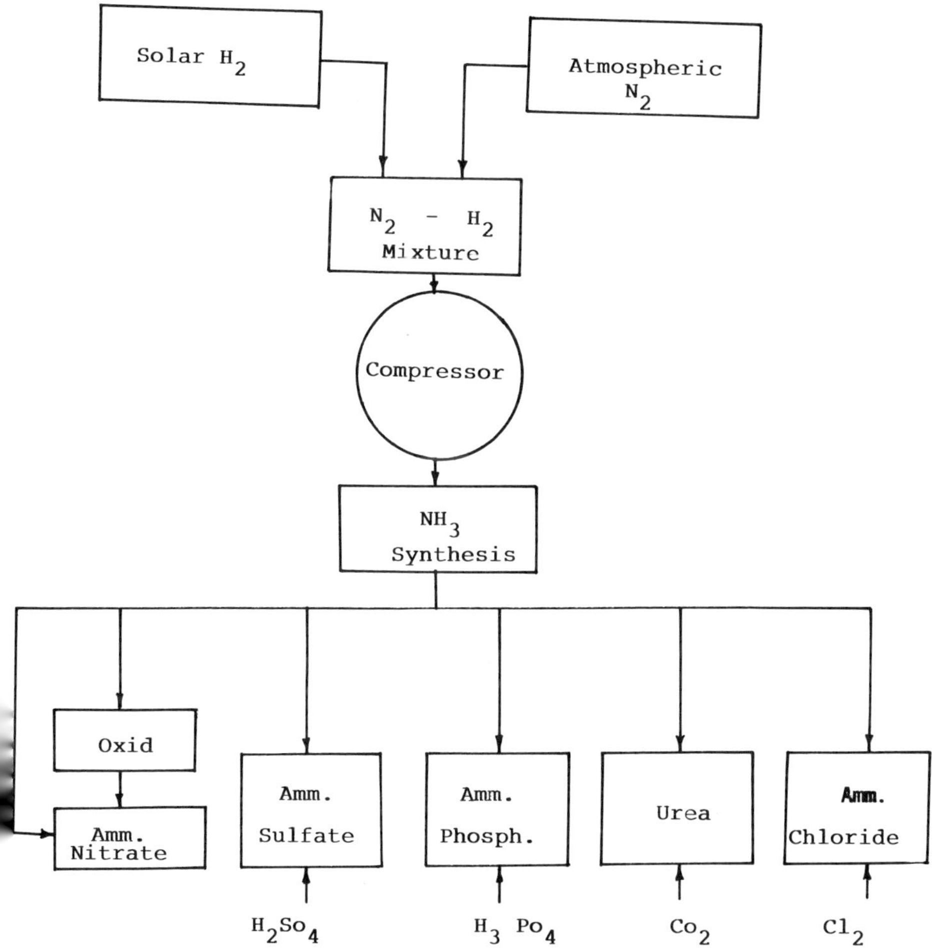

FIGURE 10: Fertilizers From Solar H_2 For Desert Development

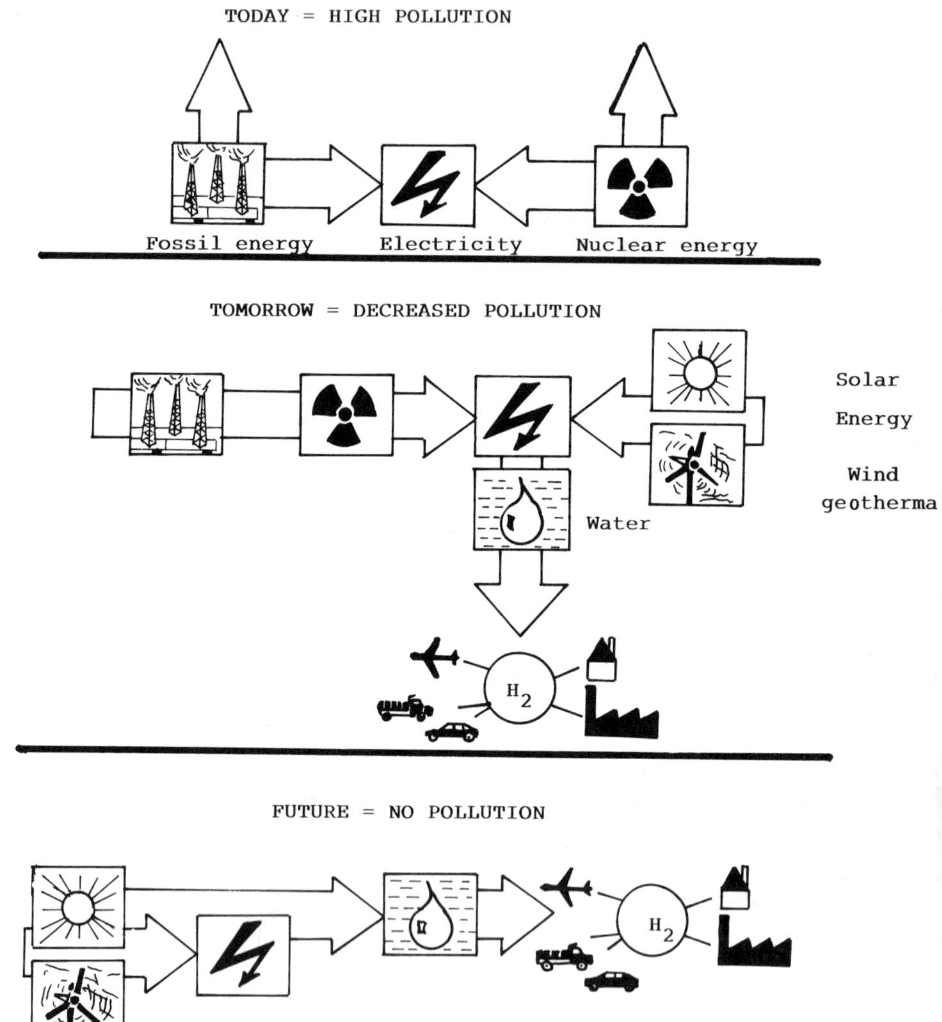

FIGURE 11: The future in using H_2 with no pollution

TABLE 2. Data for different processes producing solar hydrogen-rated capacity 100 MWH_2 or $300,000m^3H_2$ for 10 hrs/day operation.

Method \ Details	η_{PV}	η_{SC}	η_{PP}	η_{Th}	η_{E1}	η_0	Surface Area ($1000m^2$) S. collector or P.V. Cells
P.V. cells/Electroysis	0.18	--	--	--	0.8	0.144	700
Solar Concent./ Power generators Electrolysis	--	0.6	0.3	--	0.8	0.144	700
Solar concent./ Thermochemical	--	0.6	--	0.5	--	0.30	330

η_{PV} = Eff of photovoltaic cells
η_{SC} = Eff of solar concentration
η_{PP} = Eff of convent. power plant
η_{Th} = Eff of thermochemical cycles
η_{E1} = Eff of electrolysis
η_0 = Overall eff.

REFERENCES

1. A.S. El-Nokrashy, Osman Galal and J. Davenport. 1986. Applied Science and Technology, A.S.R.T, Cairo, Egypt.

2. H.K. Abdel-Aal. 1978. Energy Communication, 4 (3), 271-291.

3. L.P. Bicelli. 1986. Int. J. Hydrogen Energy, 11 (9), 555-562.

4. H.K. Abdel-Aal, Proceedings of the 2nd Arab Intern. Solar Energy Conference, Bahrain, 348-353 (1986).

5. H.K. Abdel-Aal. 1976. Energy Communications, 2 (6), 535-558.

6. H.K. Abdel-Aal. 1975. Proceedings of Mechanical Eng. Congress, Shiraz.

7. H.K. Abdel-Aal. 1984. Int. J. Hydrogen Energy, 9 (9), 767-772.

8. H.K. Abdel-Aal. 1985. Int. J. Hydrogen Energy, 10 (5), 333.

9. E. Bilgen and C. Bilgen. 1983. Int. J. Hydrogen Energy, 8 (6) 441-451.

10. W.J.D. Escher. 1983. Int. J. Hydrogen Energy, 8 (7), 479.

11. H.K. Abdel-Aal. et.al. 1974. C.O.M.P.L.E.S, 2nd Semester.

12. R. Neill. 1986. 3rd Int. Symposium on Hydrogen Produced from Renewable Energy, Hawaii.

13. H.K. Abdel-Aal. 1984. Proceedings 5th WHEC, Toronto.

OASIS ARCHITECTURE OF THE TAKLIMAKAN DESERT IN WESTERN CHINA

DONALD HANLON
School of Architecture and Urban Planning,
University of Wisconsin--Milwaukee

Abstract Results are presented of research from a recent case study concerned with integration of architecture with all aspects of community development in an isolated desert oasis village group.

The oasis of Turpan is in Xinjiang Uygur Autonomous Region of western China. It has historically been a crucial link in trade and communications between cultures of the Middle East and China, and so has been continuously inhabited for a least two thousand years. The oasis, however, has severe ecological constraints. Climate and geology play major roles in its development. Turpan, in effect, has no rainfall, yet it is a highly successful agricultural area. Energy consumption is low while productivity is high. The standard of living in this traditionally Islamic society is relatively high for China. Success is attributed in large measure to consistent perpetuation of traditional building crafts totally integrated with economic, religious and social institutions. The architecture of Turpan reflects prosperity based upon appropriate, practical responses to constraints of scarce resources in an extremely arid and hostile environment. The case study of Turpan is a valuable model for successful integrated approaches to desert development.

***

The standard television format for news, entertainment and commercials represents a common feature of contemporary life - the fragmentation of experience. Typically, only a few nanoseconds of blackness separates a tragic death in a melodrama from a sales pitch for cat food. Connections between ideas and themes are ignored; each episode of experience is a self-contained entity--a package. Even within the plots of the most successful television dramas, discrete episodes follow one another in rapid succession without articulation; there is the requisite car chase, the shooting, a brief romantic interlude, the confrontation, another car chase. Each episode within the plot is carefully choreographed and compartmentalized to fulfill the expectations of a jaded audience.

The separation of life into separate compartments of experience may have originated in the exigencies of industrial production. As factors of production, people have been required to fulfill particular roles predictably. Contemporary American society, though, has extended this format to every facet of daily life; the technique of fragmentation and packaging of experience has become the central organizing principle of modern aesthetics.

This prejudice is no less evident in the form of contemporary American cities than it is in television programming. Buildings and public spaces are packaged; connections between them are either ignored or are contrived by splicing, sewing and pinning autonomous urban experiences together.

Not surprisingly, a recurring theme of urban design criticism is the lack of cohesion between specialized or competing formal and spatial episodes in an urban landscape. Ad hoc mechanisms, such as monuments or plazas, are generally suggested as remedies.

Perhaps an approach to the problem of disjointed urban forms and spaces needs to be developed from a more systematic point of view. Instead of accepting the aesthetic of fragmentation as inevitable, another way may be discovered founded upon different principles of organization and the place of people in it. The following account of urbanization in the oasis of Turpan offers a modest example of an alternative. In this case, transparent layers of social zones from purely public through purely private are constructed systematically so as to unite ordinary social conventions with constraints of nature.

Turpan is a large oasis on the northern edge of the Taklimakan Desert in Xinjiang Autonomous Region of western China (Figure 1). The oasis shelters a cluster of villages composed predominantly of the Uygur, an Islamic people of Turkish origin. They are traditionally agricultural, though their predecessors, who occupied large, now abandoned cities nearby controlled much of the trade of the Silk Road which for many centuries linked the Levant with China. The climate is extreme; there is virtually no rainfall. Beyond the borders of the oasis lies one of the fiercest deserts of the world. In response to a harsh climate and limited resources, a tradition of sun-dried mud brick architecture has developed which closely resembles that of Persia.[1]

Life in an extremely arid environment such as that of the Taklimakan Desert is entirely dependent upon the delivery of water. Irrigation for both domestic and agricultural uses inevitably becomes the central organizing principle for human settlement. While arable land may be available on the fringe of the desert where there is virtually no rainfall, the water table may be over 500' deep and other sources of water in the mountains are too distant. There seem to be overwhelming reasons for not living in a place like Turpan, yet people have occupied it and many similar oases in the region for thousands of years. The secret of success in such an unpromising place is a Persian invention called the "kareze," or in Arabic, the "qanat." This system of horizontal wells is one of many engineering and architectural ideas that link Central Asia to Persia (Figure 2). Urbanization in much of the Middle East and Central Asia could not have occurred without the kareze. Even large cities in Iran today, such as Isphahan and Tehran, are supplied by these subterranean aqueducts

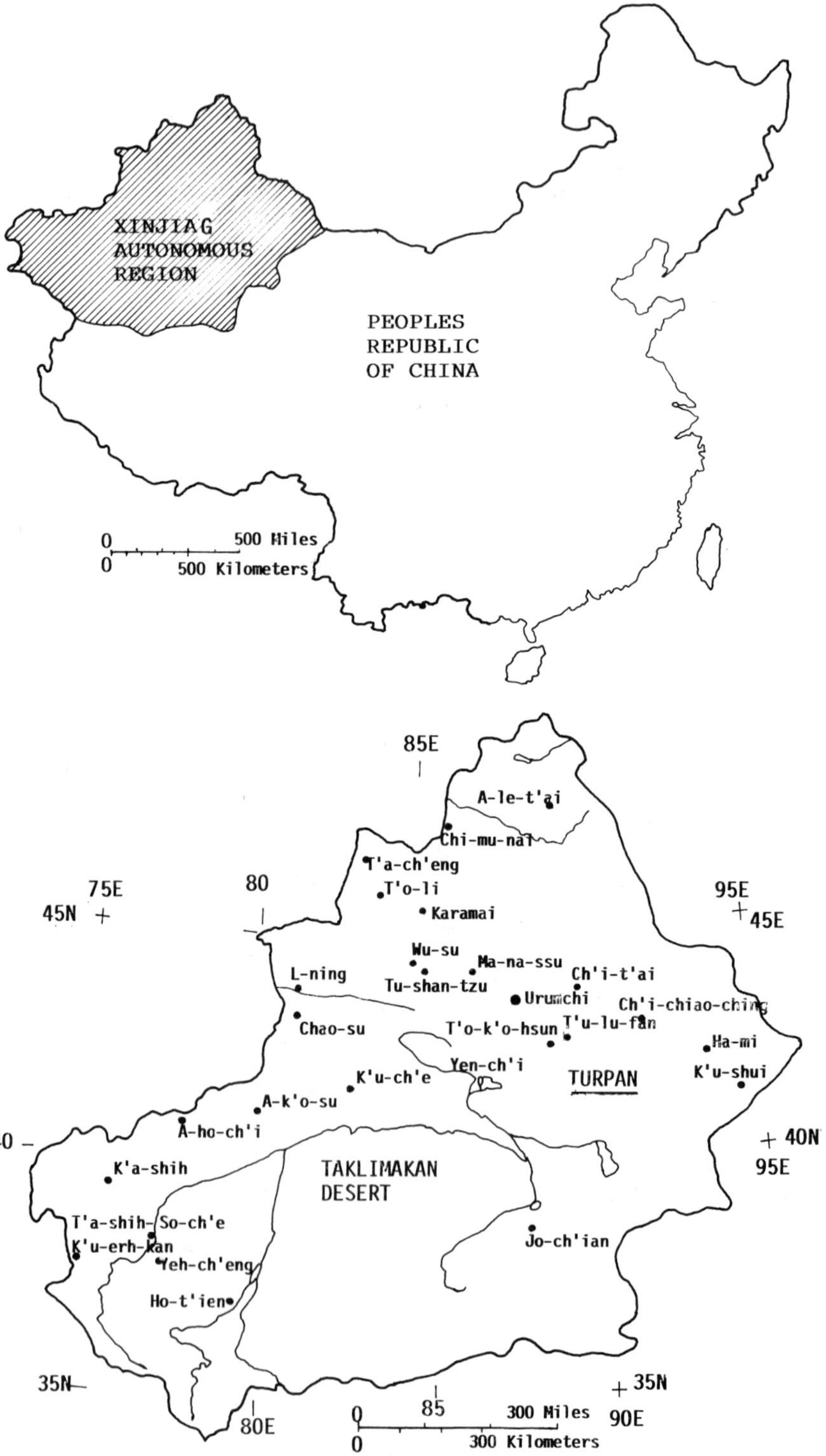

FIGURE 1: Maps of China with Taklimakan desert shown shaded on upper left

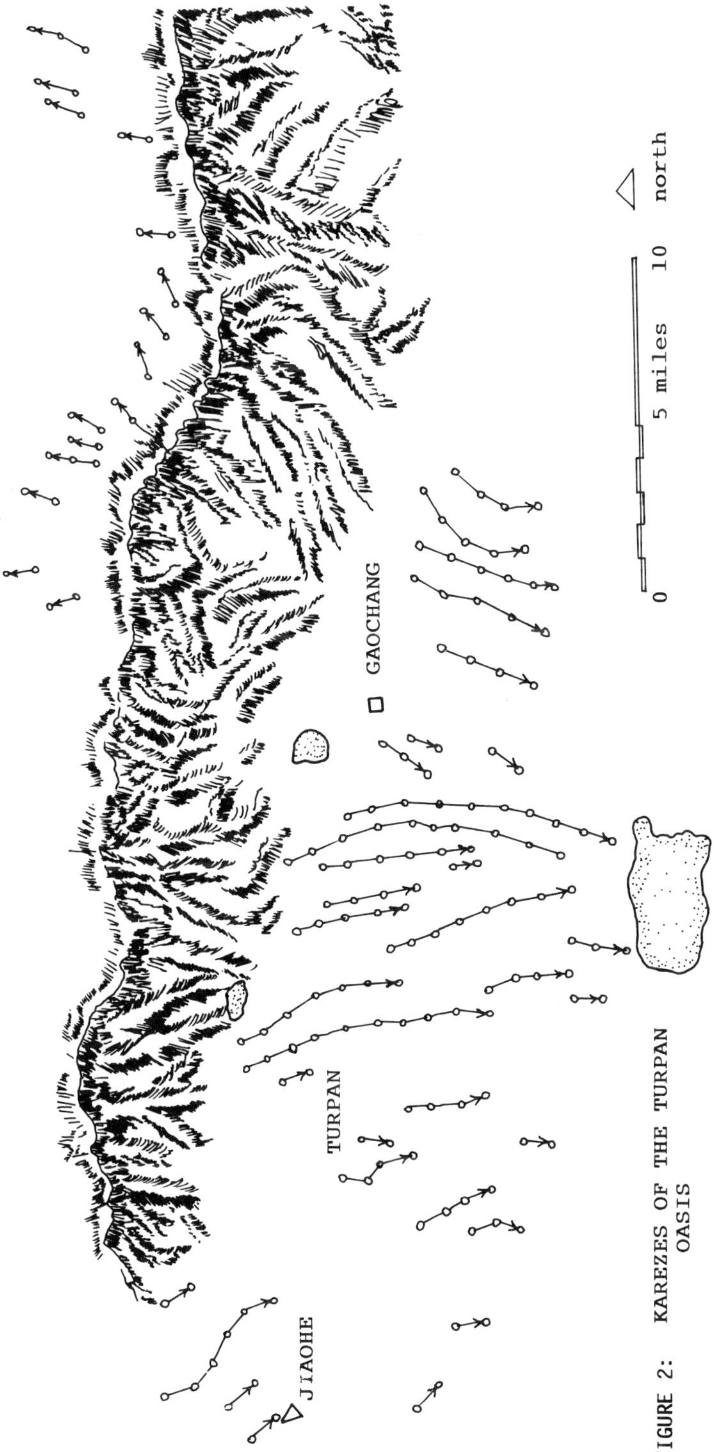

FIGURE 2: KAREZES OF THE TURPAN OASIS

hundreds of years old. The typical kareze is a tunnel varying in diameter from 4 to 10 feet and in length up to 10 miles. Its purpose is to tap water bearing soil strata deep below the base of surrounding mountains and to deliver a steady flow at ground level where it is needed in the desert valley. Karezes are constructed by first excavating vertical shafts starting with the "mother well" in the mountain foothills, then proceeding at approximately 100 yard intervals across the desert to the oasis in a straight line. While the mother well does not exceed 300' in depth, the depth of succeeding shafts are calculated so that the tunnel connecting them slopes gently enough to prevent erosion. In addition to their necessity for construction, the vertical shafts provide access to the kareze for maintenance and ventilation. Construction of kareze is entirely by hand and is extremely dangerous. In Persia, the kareze is known as "the murderer" due to frequent drownings and asphyxiations of "muquanni" who are the specialists in its construction. Understandably, intense superstition surrounds the kareze and the muquanni are never required to work. Even a child's sneeze at breakfast can be an adequate reason for its father's refusal to work that day.[2]

As water emerges from one of the hundreds of karezes around Turpan, it flows into an open tank at the edge of the oasis and joins a second system that determines, to a great extent, the pattern of urbanization. Carefully monitored channels guide the water down both sides of most streets of the villages that comprise the oasis, then into the surrounding fields. Tall, straight and fast-growing poplars are cultivated along the narrow channel banks. The union of this particular tree species with the system of channels contributes four valuable qualities toward a pleasant village life: the trees slow the water and reduce erosion thereby keeping it cleaner for domestic use; closely spaced trees form windbreaks and thus reduce the effects of blowing dust; the poplars are ideal for eventual use as posts and beams in light-weight building construction; and finally the rapid transpiration of moisture through the trees in combination with the shade they provide serves to cool and humidify the air and to form a comfortable micro-climate in the street.

The system formed from water channels and trees is an integral part of a larger system comprising streets. But the system of streets is in turn only a part of a more extensive system of spaces that have evolved to provide a clearly defined set of social conventions. This set includes the street system as well as private houses and has eight major zones of space ranging in a spectrum from totally public to totally private. Though every dwelling in Turpan is unique, each individual building being more or less complex, they all employ this system to some degree. Therefore, a clear, unequivocal spatial pattern emerges that includes all houses from the most humble to the most elaborate. The pattern suggests a traditional consensus developed over centuries by which a variety of small urban places is codified. This spatial code provides predictable social settings necessary in an isolated Islamic community.

The Uygur people, who have comprised the majority ethnic group in Turpan, tend toward a more expansive and open relationship between public spaces of the street and more private spaces of the house than may be found in most other Islamic communities. This is due to a far more visible role of women in Uygur society than one finds in an Islamic community, for example. Whereas a typical Islamic house bars any view of its domestic interior from the street, a Uygur house does not. As one walks down a street in Turpan the large wooden doors of houses are flung open to reveal much of what might be deemed semi-private domestic life such as courtyard sleeping, cooking and entertainment. Though the spectrum of spaces between purely public and purely private uses in Uygur settlements seems transparent, i.e. layered so that several spaces can be enjoyed simultaneously, the Uygur does not provide absolute privacy. The spatial system developed by the Uygur does not provide absolute privacy except in the most intimate of spaces; instead, the Uygur seem to be primarily interested in relative privacy by which various domestic activities are visible from the street but remain undisturbed because intervening layers of clearly defined space provide a series of social thresholds that cannot be crossed casually. A detailed description of the urban spatial system employed in Turpan can clarify what is meant by this transparent spatial system and relative privacy (Figure 3).

SPATIAL SYSTEM OF TURPAN

THE STREET

Streets in the residential villages that comprise Turpan are generally only wide enough for two donkey carts to pass one another comfortably with additional area available to each side for piles of bricks, logs and coal. The poplars rising directly from the narrow channels at both sides are as high or higher than the street is wide. The approximately one-to-one proportion of height ot width in the street space imparts a strong image of outdoor room. The passive air conditioning supplied by shade and transpiration reinforces the sense of enclosure since the temperature outside of the street microclimate may be as much as 15° higher. Since other spaces provide better personal socializing settings, the street is devoted to traffic (mostly animals and small tractors) and to large groups that may form for a funeral or around an itinerant knife grinder. The street is also the site for minor industries such as carpentry and the fabrication of coal dust bricks.

WATER CHANNEL AND BRIDGES

Generally along each side of the street runs a shallow water channel about four feet wide. Poplars rise on both of its rock-reinforced banks. The water runs swiftly so as to remain clean enough for domestic use. Pots, pans and clothes are

FIGURE 3: PRIVATE AND PUBLIC SPACE IN TURPAN

washed in the channel. Children occasionally bathe in it and newly woven grass mats are soaked in it. The narrow channel is spanned by small stone and mud brick bridges which are often at least as wide as they are long. These provide the only access to houses beyond, the channel thus serving as a moat. Each bridge is useful far beyond its utility as passage over water; it is the principal public socializing space of the street, a place for gossip and play. It is also the site for convening special events: marriage, a funeral or the butchering of an animal. It is not unusual to find even a bed or two on a bridge in the shade of the poplars above the cool, gurgling water. The combination of water channel/moat and bridge is the first in the series of transparent layers of relative privacy. All activity in this layer is in plain view of the street, yet dedicated solely to the life of the individual house beyond the bridge. The definite boundaries of the bridge, defined by low walls, street edge and poplars, allow it to function as a reception room (Figure 4).

FRONT YARD

Between the water channel and the front wall of a house is a narrow space, not more than five feet wide. Here, behind a double screen wall of trees, beyond the moat and accessible only via the bridge, the household can assemble for play, washing and even sleeping. Donkey carts are occasionally parked here; this zone also serves as a staging area for renovation or expansion of the house beyond, providing space for piles of sun-dried bricks and logs. The narrow yard is often the special province of children and ducks. Spatially, it is very important as it provides a buffer zone between the active atmosphere surrounding the water and the serenity of the house. This space, also, has strong features of an outdoor room; it is bound longitudinally by two high walls - one of rustling poplars, the other of impassive mud brick - as well as segmented laterally by the bridges. As the bridge links the yard with the street, so does an interesting architectural feature of the house gate also link the yard with the house. This is a small detail of the mud brick frame of the gate which serves as "feet" for the composition of the front wall. The feet are rectangular prisms of mud brick which protrude from either side of the gate opening into the yard space. In a relentlessly planar facade, the feet of the gate become very significant sculptural forms which connect the space of the yard with the mass of the house (Figure 4).

GATE

Houses in Turpan only rarely provide windows to the street. Therefore the front wall, often up to twelve feet in height, seems quite stolid. The only personal expression exhibited by the house to the street is the gate. It may be recessed, protuberant or flush with the wall; composed of massive mud brick or carefully crafted wood, most commonly a combination of the two. In an agricultural community such as Turpan, large wooden double doors are necessary; thus the gate becomes monumental. A concentration of detail in the gate in high contrast to its homogeneous flanking walls concentrates attention upon the space beyond, within the house. While the high, plastered mud walls reflect the glare of intense sunlight, the view through the gate doors, left casually ajar, reveals a deep green, cool oasis. It is not surprising to find a gate in the form of an "iwan," a vaulted niche framing a portal, which may symbolize Paradise in the Islamic architectural tradition.[3] The three preceding layers of space - street, water channel and yard - are consistently linear, running transversally to the axis of entry. The gate, reinforced by the bridge before it, presents an energetic counterpoint and signals a higher level of social threshold--a specialized outdoor reception room. But the surface one walks on has not changed at all from the street through the gate; it remains compacted earth (Figure 4).

FIGURE 4: CANAL, BRIDGE & GATE

INTERIOR STREET

Virtually every house in Turpan employs a special zone within and/or beyond the gate that acts as an intermediate space between the abruptly monumental scale of the gate and smaller scale spaces of the house interior. This takes the form of a short street. In some cases it is a vaulted extension of the gate inward toward the heart of the house; in other cases the gate leads to a narrow passage, open to the sky, bound on each side by service and storage buildings. Either vaulted or open, the interior street is a cool and shady place to lounge, entertain an infant, chat with a guest or sleep in full view of the street, but deep within the protective enclosure of the house. The open gate allows free air circulation through the passage but deep shadow provides a sense of privacy. The interior street in combination with the gate and its "feet" protruding into the outer yard all act together as a semi-public living room within which various positions may be chosen depending upon mood or occasion.

COURTYARD

The easygoing hospitality for which the people of Turpan are known can sweep even a stranger from the street through to the heart of a house. At the end of the interior street is the Paradise the gate so boldly promises. It is essentially a room paved in fired brick, enclosed by walls from which various doors, stairs and passages lead deeper into private spaces. The ceiling of the courtyard room is a grape arbor; by late summer great stalactites of the fruit hang in the cool, dappled light. The courtyard is the living room of the house throughout the warm months of the year. When the temperature of the desert exceeds 115°F and the breeze fails, even this shady outdoor room may not provide relief. Then people retire to dark subterranean chambers below. If weather permits, cooking is done in the courtyard in earthen ovens built against one wall. Often, species of trees with spreading branches are cultivated on the perimeter of the courtyard to shade the flat roofs of the various buildings that comprise the house compound. The high surrounding walls, deep gateway, trees and arbor shelter the courtyard from all but the most severe dust storms. Bed frames are pulled out of stuffy rooms in the spring so that all may sleep in the courtyard throughout the summer. A raised platform of mud-brick covered with carpets is often found along one side; this serves for sleeping as well as for entertaining guests.

LOGGIA

Closely associated with the courtyard, either on the same level or raised upon a plinth of service rooms, is a loggia. Courtyards generally do not have a single compositional focus; the loggia provides a cynosure for the central living space as well as a terminus for all the spatial layers of the house from public to semi-private. Sometimes made of brick but more often of wood, the loggia is a lofty, airy space. Its front, which opens onto the courtyard, is entirely open but for slender columns supporting a flat roof; the short ends of its rectangular plan are generally blank walls shared with neighbors, while its rear wall is pierced with windows of private rooms. The loggia's columns and railing may contribute a rare exhibition of color in an architecture dominated by brick, mud plaster and bare wood (Figure 5).

PRIVATE ROOMS

The loggia is the seventh layer of space in a traditional format for ordering social settings in domestic architecture of Turpan. Occasionally, one finds houses with a relatively small courtyard/ loggia combination and tunnel that leads beneath the principal house volume to another, more private courtyard beyond. Usually, however, the loggia signals a transition from expansive semi-public and semi-private spaces to compact private rooms, the eighth layer. An indoor living room may be large

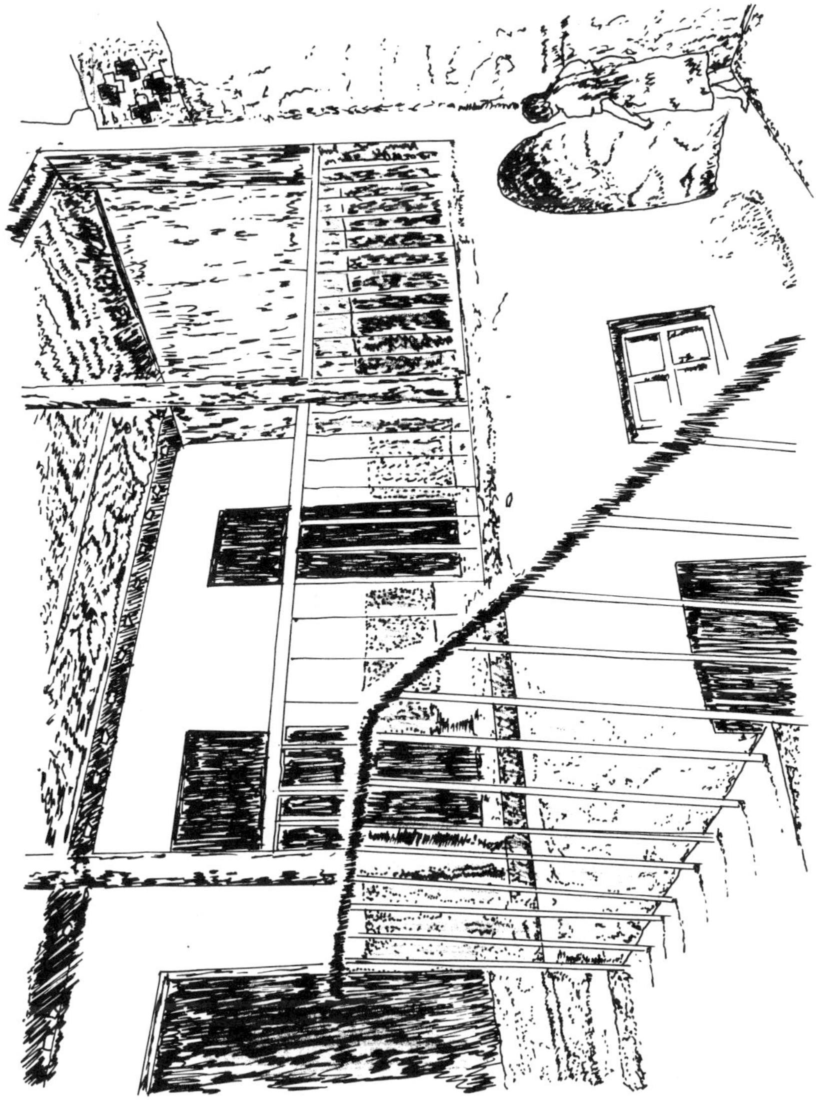

FIGURE 5: A LOGGIA

enough for ten people and well lit by windows from the courtyard but most bedrooms are small and relatively dark with only one or two windows set deeply within thick, mud-brick walls. During winter cold spells the entire extended family may assemble in a single special room equipped with its own cooking/heating oven and thick, raised floor. The walls of some rooms are about three feet thick; heat is conserved by brick barrel vaults spanning up to twelve feet.

CONCLUSION

The analysis of typical houses in the community of Turpan seeks to reveal a systematic approach to ordering architectural space in a vernacular building tradition. Well beyond picturesque and romantic associations that observers may derive from vernacular architecture, concrete lessons in design - the composition of form and space - can comprise the underlying concept of a traditional method of building. Moreover, the subtle layering and articulation of compositional elements demonstrated in the houses and streets of Turpan serves a vital social purpose while it delights the eye. Each layer of space and each formal element contributes to a clear delineation of specific social settings within an integrated whole. This is accomplished without relegating each activity to a discrete architectural compartment but, to the contrary, by utilizing the simplest of means to allow boundaries to become more implied than defined. Some degree of functional redundancy throughout a spectrum of incremental change contributes to smooth movement from one transparent zone to another. The transparency of the spatial system depends upon boundaries between spaces sometimes becoming spaces themselves rather than remaining edges. A fluid transition between public and private spaces is thus accomplished in a gracefully compelling way.

The entire system, emerging from an equitable distribution of water from the karezes, is integrated for conditioning a micro-climate at very low cost. Moreover, within the context of a traditional society, such a system of layered transparent spaces affording varying degrees of relative privacy operates as a formal popular structure. The significance of each layer and the appropriate behavior assigned to it is thoroughly inculcated by consistent repetition. Architecture thus becomes an embodiment of social conventions in everyday life as well as an extension of natural structures of the environment.

REFERENCES

1. Elizabeth Beazley and Michael Harverson. 1982. Living with the desert, (Wilts, England: Aris and Phillips Ltd.). Vernacular buildings of the Iranian plateau in relation to climate and geography are thoroughly discussed.

2. Anthony Smith. 1953. Blind white fish in Persia. (New York: E.P. Duttton), p. 80. The longest known qanat in Iran is fifty miles.

3. John D. Hoag. 1977. Islamic architecture. (New York: Harry H. Abrams), p. 405.

USING CALPAS 3 AS A TOOL TO OPTIMIZE THE DESIGN OF PASSIVE SOLAR DESERT HOUSES IN NORTH TAHRIR, EGYPT.

NADER VICTOR CHALFOUN
Zagazig University, Cairo, Egypt

Abstract Development of deserts is a task requiring coordinated applications of many areas of knowledge and experience. This paper demonstrates how the design of a suggested low-cost passive solar house in the North Tahrir region in Egypt has been optimized through the application of Calpas 3, a sophisticated computer energy design program, developed by the Berkeley Solar Group in California. A preliminary design suggested the Base Case, which was derived from two pre-studied passive solar models for the Aswan and Alexandria regions of Egypt. The two models had been developed previously by the author through conventional passive solar design techniques. In a simulation of the Base Case, the value of 11.977 KBTU/sq. ft. was obtained as the total annual space conditioning energy consumption from the source. This predicted value indicated an already energy-efficient house which compiled with the California Title-24 residential energy performance code by being well below the 38.9 KBTU/sq. ft. of Zone 15 of El-Centro, California. This California Zone has weather conditions closest to those of North Tahrir. In addition to the initial Base Case run, and through inexpensive conservation measures, a sequence of design enhancements were applied in an attempt to minimize the existing annual space conditioning energy consumption. Significant improvement was achieved over the Base Case results; the house annual heating load decreased from 249 to 8 KBTU/yr, a 96.8% saving. The house annual cooling load also decreased from 4612 to 1913 KBTU/yr, a 58.6% saving. Also, the annual source heating and cooling load decreased from 11.98 to 4.52 KBTU/sq. ft. per year, a 62.3% saving. Further computer runs can be conducted in an attempt to eliminate the house heating load --a desirable situation for a good passive solar house in a hot climate region like Tahrir, Egypt, and to decrease the house cooling load.

INTRODUCTION

The tremendous increase in the population of Egypt has caused the Egyptian government to rethink its settlement policies by planning for new communities in the remote arid areas while preserving the Nile Valley for agriculture development. Accordingly, the need to increase the productivity of arid lands has become a national priority.

The development of deserts and remote arid regions presents a challenge demanding coordinated application of many areas of knowledge and experience. A few of the areas that should be included in an integrated approach are agriculture and irrigation techniques, soil improvement, ground water, animal husbandry, and community planning including architecture and renewable energy. The last is the focal point of this particular study.

In Egypt, as well as most arid parts of the world, attractive, low-cost, energy efficient houses are needed for new desert settlements. Under the desert's harsh climatic conditions, buildings can be passively designed to respond to environments which would otherwise be intolerable. Such desert houses, which need to be thermally comfortable, should also offer a quality of life attractive enough to encourage migration of families from overcrowded cities and urban towns.

Within the context of the general objectives mentioned above, this paper demonstrates design guidelines for a low-cost, passive-solar desert house which could be replicated in the establishment of viable desert communities in the North Tahrir region in Egypt. The study also shows the advantages of computer application techniques which are currently becoming a common engineering tool for design professionals working on energy simulation in buildings.

SELECTION OF THE BASE CASE

The optimization process described below begins with the selection of a base case. The criteria for selecting this base case were established by comparing the heat loads of two pre-studied passive solar houses designated for the Aswan and Alexandria regions in Egypt (Figure 1), and selecting the one that has the best thermal performance corresponding to the North Tahrir climate. The two models were previously developed by the author through conventional passive solar design techniques.[3] The design process was based on biological evaluation, microclimatic modifications, and recommendation of envelope materials for the region under study. Finally, the thermal performance of each of the concluded designs was predicted through a tedious hand method calculation. Figure 2 shows the applied criteria for selecting the best shape and orientation of the Alexandria and Aswan designs, and Figure 3 illustrates the heat gain/loss calculation of each, as well as the architectural drawings.

The Calpas 3 computer program was then used to assess the thermal performance of the two above-mentioned houses. The North Tahrir climatic file was used however, rather than the Aswan or the Alexandria weather data. Calpas 3 is a sophisticated computer energy analysis simulation program for houses and small commercial-building design. It was developed by the Berkeley Solar Group of the California Energy Commission (CEC) to comply with the energy code for New Residential Building Standards, Title-24. It calculates hourly the air, surface, and mass temperatures throughout the building, as well as heat transfer among components, the contribution of natural energy to comfort levels, and the mechanical heating or cooling loads required to maintain indoor

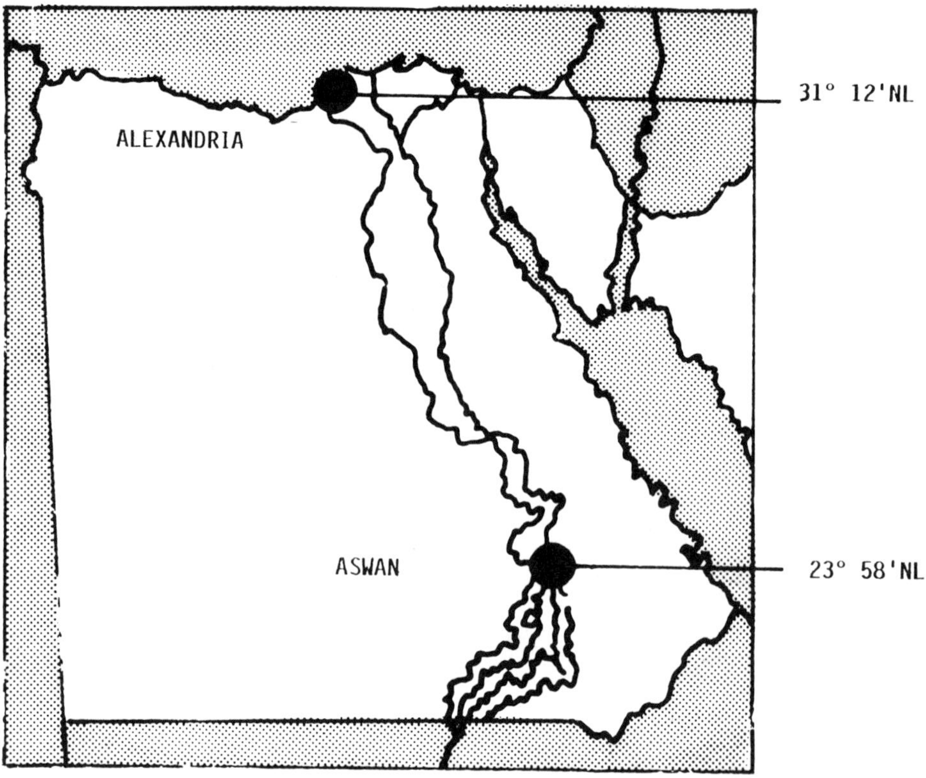

FIGURE 1: Geographical location of the two selected regions in Egypt; Alexandria and Aswan

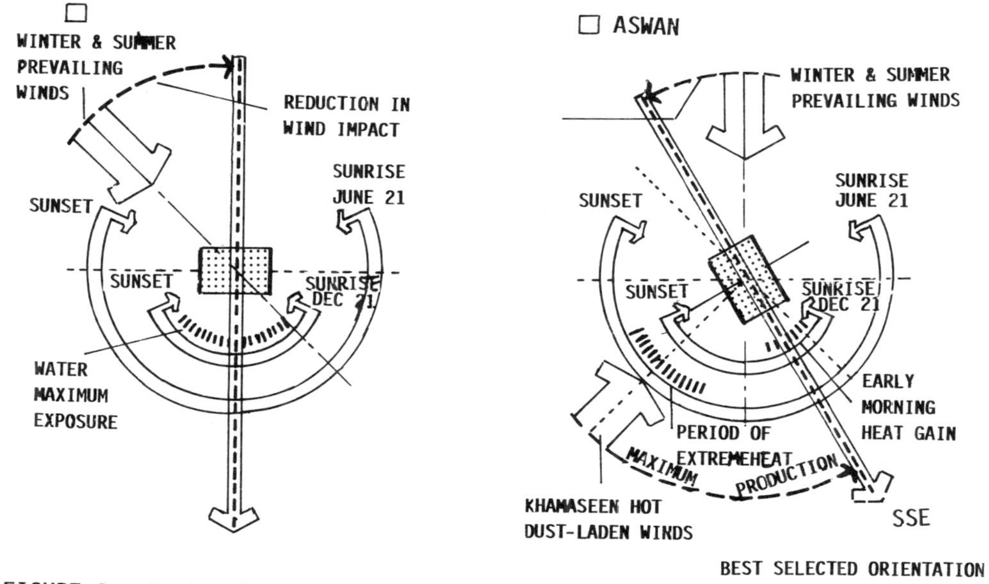

FIGURE 2: Best selected building orientation in Alexandria and Aswan corresponding to sun and wind[3]

ENVELOPE HEAT GAIN/LOSS CALCUALTIONS, AND DEVELOPED DRAWINGS: ASWAN

Zone: 1

1-CONDUCTIVE GAIN / LOSS

	AREA	(U)	LOAD
Windcw 1	(1X4X4)	X 0.9	= 14.4
Window 2	(2X3X3)	X 0.9	= 16.2
Door	(1X3X8)	X 0.4	= 9.6
Wall R12	(142)	X 0.08	= 11.3
Roof	(20X16)	X 0.06	= 19.2
Perimeter	20	X 0.55	= 11.0

(BTU/hr/°F) TOTAL			= 81.7

2-CONVECTIVE GAIN/LOSS

Infiltration (BTU/hr/°F)= 60.0

Conductive & Convective =141.7

3-TOTAL PER DAY

141.7 X 24 hrs = 3400 BTU/DD/°F

4-LOAD PER SQUARE FOOT

3400/(20X16)=10.6 BTU/DD/ft²/°F

5-INTERNAL GAIN

From occupants	600 BTU/hr
From electric lights	600 BTU/hr
From equipment	1800 BTU/hr

TOTAL	3000 BTU/hr

Zone: 2

1-CONDUCTIVE GAIN / LOSS

	AREA	(U)	LOAD
Window 1	(1X4X6)	X 0.9	= 21.6
Windcw 2	(1X4X4)	X 0.9	= 14.4

Wall R5	(160)	X 0.2	= 32.0
Roof	(20X16)	X 0.06	= 19.2
Perimeter	20	X 0.55	= 11.0

(BTU/hr/°F) TOTAL			= 98.2

2-CONVECTIVE GAIN/LOSS

Infiltration (BTU/hr/°F)= 40.0

Conductive & Convective =138.2

3-TOTAL PER DAY

138.2 X 24 hrs = 3316 BTU/DD /°F

4-LOAD PER SQUARE FOOT

3316/(20X16)=10.3 BTU/DD/ft²/°F

5-INTERNAL GAIN

From occupants	600 BTU/hr
From electric lights	600 BTU/hr
From equipment	200 BTU/hr

TOTAL	1400 BTU/hr

6-DEVELOPED DRAWINGS:

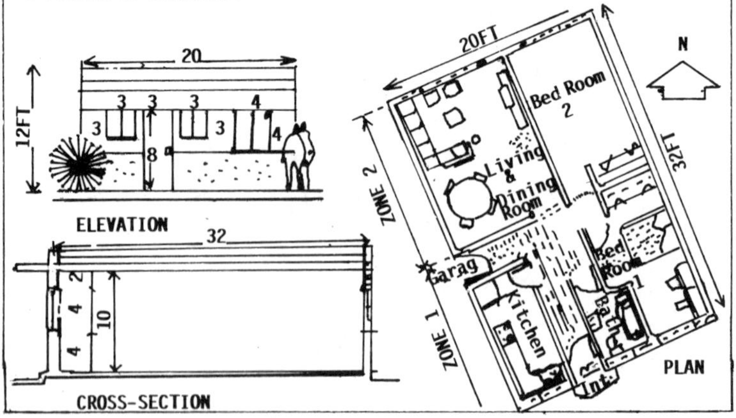

FIGURE 3: Envelope energy heat gain/loss calcaulation and architectural drawings for the Aswan region

ENVELOPE HEAT GAIN/LOSS CALCULATIONS, AND DEVELOPED DRAWINGS: ALEXANDRIA

Zone: 1

1-CONDUCTIVE GAIN / LOSS

	AREA	(U)	LOAD
Window 1	(1X4X4)	X 0.52	= 8.3
Window 2	(2X3X3)	X 0.52	= 9.3
Door	(1X3X8)	X 0.4	= 9.6
Wall R12	(262)	X 0.08	= 20.9
Roof	(10X32)	X 0.06	= 19.2
Perimeter	32	X 0.55	= 11.0
(BTU/hr/°F) TOTAL			= 84.9

2-CONVECTIVE GAIN/LOSS

Infiltration (BTU/hr/°F)= 60.0

Conductive & Convective =144.9

3-TOTAL PER DAY

144.9 X 24 hrs = 3477 BTU/DD/°F

4-LOAD PER SQUARE FOOT

$3086/(10X32)=10.8$ BTU/DD/ft^2/°F

5-INTERNAL GAIN

From occupants	600 BTU/hr
From electric lights	600 BTU/hr
From equipment	1800 BTU/hr
TOTAL	3000 BTU/hr

Zone: 2

1-CONDUCTIVE GAIN / LOSS

	AREA	(U)	LOAD
Window 1	(2X6X4)	X 0.52	= 24.9

Wall R5	(208)	X 0.2	= 41.6
Wall R24	(2 X32)	X 0.04	= 2.5
Roof	(10X32)	X 0.06	= 19.2

(BTU/hr/°F) TOTAL			= 88.2

2-CONVECTIVE GAIN/LOSS

Infiltration (BTU/hr/°F)= 40.0

Conductive & Convective =128.2

3-TOTAL PER DAY

128.2 X 24 hrs = 3076 BTU/DD/°F

4-LOAD PER SQUARE FOOT

$3076/(10X32)=9.6$ BTU/DD/ft^2/°F

5-INTERNAL GAIN

From occupants	600 BTU/hr
From Electric lights	600 BTU/hr
From equipment	200 BTU/hr
TOTAL	1400 BTU/hr

6-DEVELOPED DRAWINGS:

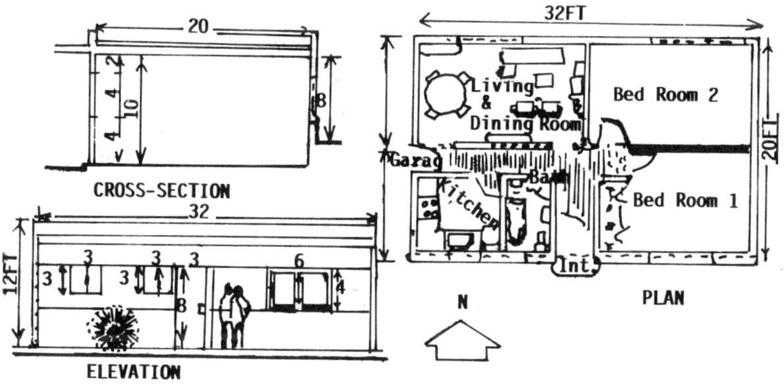

FIGURE 3 (Cont.): Envlope energy heat gain/loss calculation for the Alexandria region

temperatures specified by the designer.

Figure 4 shows that, after running the Calpas 3 program, the Aswan house yielded its best performance when its orientation was due south. Its total heating and cooling load was 4861 KBTU/yr. the least of the simulated models, and it was therefore selected as the Base Case for North Tahrir region. The total annual summary of the electric and/or fuel energy consumed by each of the space-conditioning subsystems after adjustments to approximate generation and transmission losses was 11.97 KBTU/sq. ft. per year. This value is well below the California Title 24 source energy budget for Zone 15 (38.9 KBTU/sq. ft. per year) of El-Centro, California. The weather conditions of El-Centro are similar to those of North Tahrir.

The results, as presented above, indicate an energy-efficient base-case design. Figure 5 illustrates the architectural drawings of the selected Base Case.

BASE CASE DESCRIPTION

The selected Base Case represents a low-cost attached, two bedroom, single story dwelling located on an open site in the North Tahrir region in Egypt, 30.9° north, and 30.1° west. Each unit is 640 square feet in floor area, 10 feet in height, rectangular in shape with its longitudinal axis elongated true north-south, and is designed to accommodate a family of four, two adults and two children. All walls, roofs, windows and finished materials are from types that are relatively inexpensive and commonly used in the construction of low-cost housing in Egypt; i.e. red brick walls, a flat concrete roof, an ordinary concrete slab on grade, single-glazed wood-frame windows, and solid-crop soft-wood doors. Shading is provided from the south only through a continuous horizontal concrete overhang of two feet and outside woodscreen shutters (Figure 5).

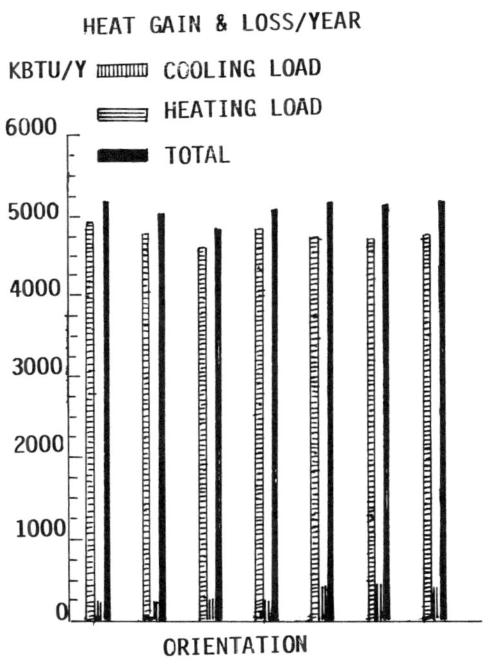

FIGURE 4: Selection of the base case

ASWAN HOUSE:

1- 22.5 D. east from south
2- 10.0 D. east from south
3- 00.0 D. due south
4- 10.0 D. west from south

ALEXANDRIA HOUSE:

5- 10.0 D. east from south
6- 00.0 D. due south
7- 10.0 D. west from south

FIGURE 5: Architectural drawings of the base case

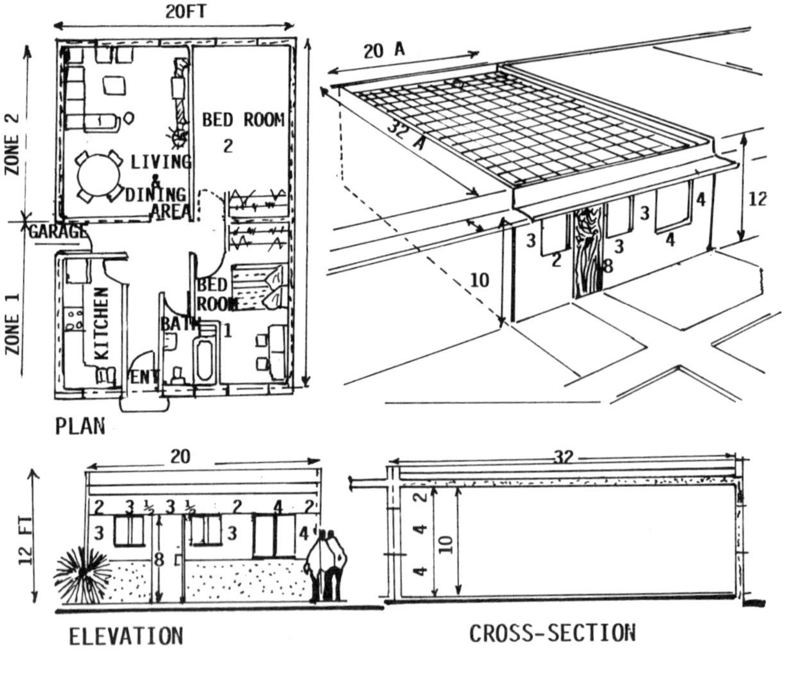

MODELLING TECHNIQUES

In order to model the Base Case design described above, an input file must first be created. Creating an input file using the Calpas 3 program is a process of inserting argument values to a corresponding group of commands listed in the program manual.[2] Most commands have default argument values which Calpas 3 will use if none has been supplied by the user.

The input file which describes the Base Case is given in Figure 3. Since the selected location for the project is an open site in the desert of North Tahrir, a 0.5 wind factor was chosen. The surrounding landscape was assumed to be dark soil with 0.2 ground reflectivity. The space heat capacity is 1960 Btu/F for the contents of a "normal" house. The concrete roof topped by cement tiles has a U-value of 0.06 Bth/hr. sq. ft. F and has an absorptivity of 0.3.[1] The south wall is a thermal mass having a U-value of 0.13 whereas the U-value of the north wall is 0.2. The absorptivity of each wall is 0.3 for the applied exterior white plaster. The U-values and absorptivities of the attached east and west walls are zero. The house slab is on grade and has a perimeter loss U-value of 0.55 for the provided R-7 slabedge insulation, while the absorptivity is zero. All interior walls are made of brick with 24 BTU/cu. ft. F volumetric heat capacity. All windows are single glazed with 1.07 U-value. The south windows are equipped with outside woodscreen shutters having a 0.2 shading coefficient (SC), and are designed to be fully closed in summer and open in winter. Therefore, the solar gain factors are assumed to be 1.0 (100%) during the five winter months from November through March, and 0.2 (20% only) during the summer months from April through October. Also, a continuous horizontal concrete overhang which is two feet deep is provided from the south. The applied inside shutters on the north and south fenestration are operable light-colored curtains having a transmissivity of 0.8. No mechanical means for ventilation is used. However, 10% of the window area is designated for natural ventilation, as defaulted by the program. The values of the winter and summer thermostat settings are derived from the upper and lower levels of the developed comfort zone for Egypt as illustrated in Figure 6.[3,5] Calpas 3 simulation of the summer and winter operating modes is controlled by the indoor ambient temperature of the house rather than by seasons. All of the mechanical heating and cooling equipment are assumed to be operated by electricity.

The input file which describes the Base Case to the Calpas 3 program is illustrated in Figure 7.

PARAMETRIC STUDY AND OPTIMIZATION METHODOLOGY

After running the Base Case, an annual cooling load of 4612 KBTU/yr, and a heating load of 249 KBTU/yr were found (see Figure 8). An annual source for total heating and cooling loads of 11.977 was obtained, which, as previously explained, complies very well with the California Title-24 performance code indicating an already energy-efficient design. The Base Case, thus, indicates that it requires no further modifications. However, for optimization purposes, and in an attempt to decrease or eliminate the mechanical loads, a series of design enhancements were applied using relatively inexpensive architectural conservation measurements.

Nine additional computer runs were conducted. Each run comprised a different conservation measure which was selected to comply with low-income house construction, as well as the cultural traditions of the region. Examples of these architectural measures are: adding outside woodscreen shutters to the north facade, increasing the glass area while replacing it with double glazed windows, berming parts of some walls, optimizing the south horizontal overhang depth, and improving the absorptivity of the roof by covering it with traditional reed panels.

THE BIOCLIMATIC CHART SHOWING THE PARAMETERS OF THE OPTIMUM AND DESIRABLE COMFORT ZONES FOR THE INHABITANTS OF EGYPT.

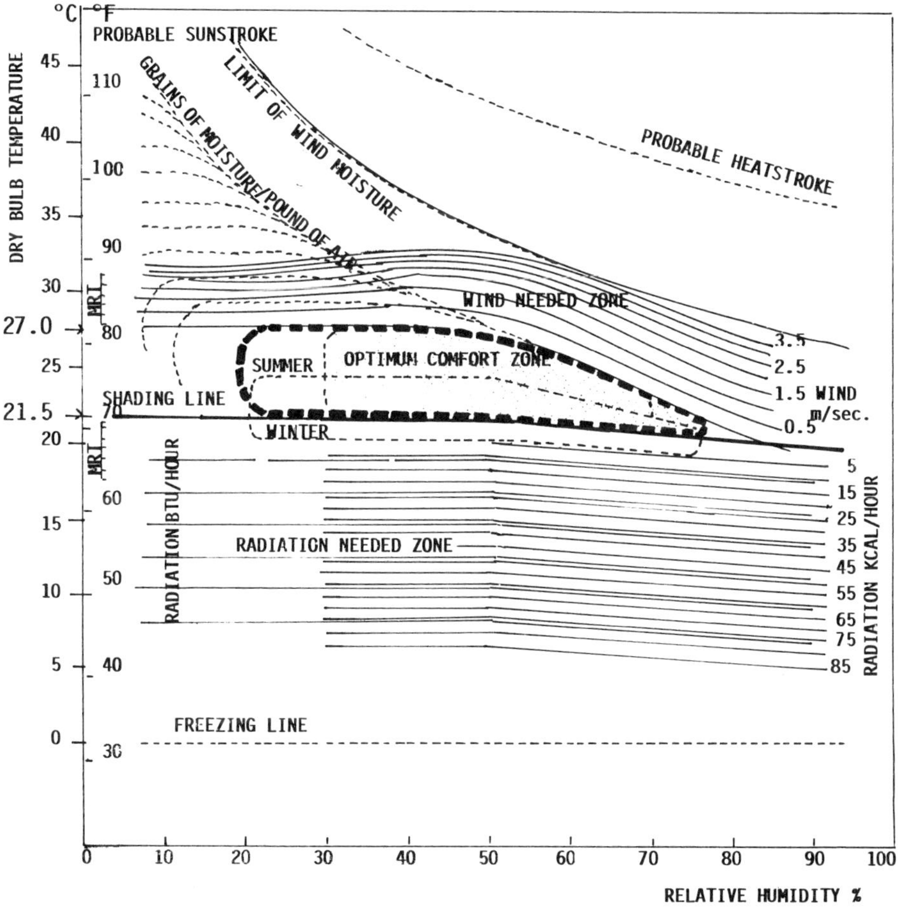

O THE OPTIMUM COMFORT ZONE IS AT 50% RELATIVE HUMIDITY WITH AIR TEMPERATURE BETWEEN 21.5°C AND 27°C.

FIGURE 6: The comfort zone in Egypt[3]

FIGURE 7: CALPAS 3 INPUT FILE OF THE BASE CASE

```
TAHR10 ATTACHED ONE-STORY HOUSE ORIENT:S  NADER  CALPAS3  V3.12  LICENSES PC0201

OPEN  SITE  IN  NORTH  TAHRIR REGION Weather: NATAHRIR. EG1 (North Tahrir Egypt MKW)

Line
----  ------------------------------------------------------------------
      *TITLE       TAHRO1 ATTACHED ONESTORY HOUSE ORIENT:S NADER
      SITE         LOCATION-OPEN SITE IN NORTH TAHRIR REGION
      AZMSOUTH     0 ; basecase
      GREFLECT     JANGR=0.2  FEBGR=0.2  MARGR=0.2  APRGR=0.2 &
                   MAYGR=0.2  JUNGR=0.2  JULGR=0.2  AUGGR=0.2 &
                   SEPGR=0.2  OCTGR=0.2  NOVGR=0.2  DECGR=0.2
      HOUT         2.64; average outside file coefficient,  3 'aph,  light air
      *HOUSE       FLRAREA=640 VOL=6400 CAIR=1960; firarea & 1.5 + 1000 &
                   heat capacity of the contents of a normal house
      ROOF         AREA=640 AZM=0 TILT=0 UVAL=0.06 ABSRP=0.3; cem. tiles
      WALL         NAME-SOUTH  AREA=142  AZM=0  TILT=90  UVAL=0.13 &
                    ABSRP=0.3    INSIDE=EXWALL  ;  val.0fABSRP  for  white plastr
      WALL         NAME=NORTH  AREA=160  AZM=180  TILT=90  UVAL=0.2 &
                   ABSRP=0.3
      WALL         NAME=EAST   AREA=320  AZM=-90  TILT=90  UVAL=0.0 &
                   ABSRP=0.0; this is an attached wall
      WALL         NAME=WEST   AREA=320  AZM=+90  TILT=90  UVAL=0.0 &
                   ABSRP=0.0; this is an attached wall
      WALL         NAME=DOOR   AREA=24   AZM=0  UVAL=0.4  ABSRP=0.5 ; &
                   solid core soft wood door
      WALL         NAME=SLABLOSS AREA=40 AZM=0 UVAL=0.55  ABSRP=0.0 ; &
                   perimeter slabloss with R-7 slabedge insul. 18 in. deep
      SLAB         AREA=640  THKNS=10  MATERIAL=CONC120  HTAHS=1.3 &
                   ASURF=0.05 ; vinyilile, loss  to  ground  by  perimeter loss
      INTWALL      AREA=1480 THKNS=3  MATERIAL=BRICKMN   HTAHS=1.5 &
                   ASURF=0.29; walls covered by cement and gypsum plaster
```

FIGURE 7: (Cont.)

```
Line
       -------------------------------------------------------------------
EXWALL                THKNS=20   MATERIAL=BRICKCMN HTAHS=1.5 &
            RSURF=0.29; wall covered by stucco plaster
GLASS       NAME-SOUTHBEDRM  AREA=16.0  AM=0.0  TILT=90  NGLZ=1 &
             UVAL=1.07  GLSTYP=1 XRFLCT=0.14 RSHTR=0.0 TRSHTR=0.8 &
            SCFWNTR=0  SCFSMR=1
SGDISTWNTR  AIR=0.4  SLB=0.6  IW=0.0
SGDISTSMR   AIR=0.7  SLB=0.3  IW=0.0
GLSGREFLECT JANGR=0.2  FEBGR=0.2  MARGR-0.2   APRGR=0.2 &
            MAYGR=0.2  JUNGR=0.2  JULGR=0.2   AUGGR=0.2 &
            SEPGR=0.2  OCTGR=0.2  NOVGR=0.2   DECGR=0.2
SGFACTORS   JANSGF=1.0  FEBSGF=1.0  MARSGF=1.0  APRSGF=0.2 &
            MAYSGF=0.2  JUNSGF=0.2  JULSGF=0.2  AUGSGF=0.2 &
            SEPSGF=0.2  OCTSGF=0.2  NOVSGF=1.0  DECSGF=1.0; &
            outside woodscreen shutters are used in summer & SC=0.2
SHADING     WHEIGHT=4.0  WWIDTH=4.0  OHDEPTH=2.0 &
            OHWD  =2.0  OHLX  =50.0  OHRX  =50.0  ONFLAP=0.0 &
            FLDEPTH=0.0  FLTX=0.0  FLWD=0.0  FLWBX=0.0 &
            FRDEPTH=0.0  FRTX=0.0  FRWD=0.0  FRWBX=0.0
GLASS       NAME-SOUTHBATH  AREA=9.0  AZM=0.0  TILT=90  NGLZ=1 &
            UVAL=1.07 GLSTYP=1 XRFLCT=0.14  RSHTR=0.0  TRSHTR=0.8 &
            SCFWNTR=0  SCFSMR=1
SGDISTWNTR  AIR=0.4  SLB=0.6  IW=0.0
SGDISTSMR   AIR=0.7  SLB=0.3  IW=0.0
GLSGREFLECT JANGR=0.2  FEBGR=0.2  MARGR=0.2  APRGR=0.2 &
            MAYGR=0.2  JUNGR=0.2  JULGR=0.2  AUGGR=0.2 &
            SEPGR=0.2  OCTGR=0.2  NOVGR=0.2  DECGR=0.2
SGFACTORS   JANSGF=1.0  FEBSGF=1.0  MARSGF=1.0  APRSGF=0.2 &
            MAYSGF=0.2  JUNSGF=0.2  JULSGF=0.2  AUGSGF=0.2 &
            SEPSGF=0.2  OCTSGF=0.2  NOVSGF=1.0  DECSGF=1.0; &
            outside woodscreen shutters are used in summer & SC=0.2
```

FIGURE 7: (Cont.)

```
----------------------------------------------------------------------
SHADING      WHEIGHT=3.0  WWIDTH=3.0  OHDEPTH=2.0 &
             OHWD=2.0  OHLX=50.0  OHRX=50.0  OHFLAP=0.0 &
             FLDEPTH=0.0  FLTX=0.0  FLWD=0.0  FLWBX=0.0 &
             FRDEPTH=0.0  FRTX=0.0  FRWD=0.0  FRWBX=0.0
GLASS        NAME-SOUTHKITCH  AREA=9.0  AZM=0.0  TILT=90  NGLZ=1 &
             UVAL=1.07 GLSTYP=1 IRFLCT=0.14  RSHTR=0.0  TRSHTR=0.8 &
             SCFWNTR=0  SCFSMR=1
SGDISTWNTR   AIR=0.4  SLB=0.6  IW=0.0
SGDISTSMR    AIR=0.7  SLB=0.3  IW=0.0
GLSGREFLECT JANGR=0.2  FEBGR=0.2  MARGR=0.2  APRGR=0.2 &
             MAYGR=0.2  JUNGR=0.2  JULGR=0.2  AUGGR=0.2 &
             SEPGR=0.2  OCTGR=0.2  NOVGR=0.2  DECGR=0.2
SGFACTORS    JANSGF=1.0  FEBSGF=1.0  MARSGF=1.0  APRSGF=0.2 &
             MAYSGF=0.2  JUNSGF=0.2  JULSGF=0.2  AUGSGF=0.2 &
             SEPSGF=0.2  OCTSGF=0.2  NOVSGF=1.0  DECSGF=1.0; &
             outside woodscreen shutters are used in summer & SC=0.2
SHADING      WHEIGHT=3.0  WWIDTH=3.0  OHDEPTH=2.0 &
             OHWD=2.0  OHLX=50.0  OHRX=50.0  OHFLAP=0.0 &
             FLDEPTH=0.0  FLTX=0.0  FLWD=0.0  FLWBX=0.0 &
             FRDEPTH=0.0  FRTX=0.0  FRWD=0.0  FRWBX=0.0
GLASS        NAME=NORTHGLASS  AREA=40.0  AZM=180  TILT=90  NGLZ=1 &
             UVAL=1.07 GLSTYP=1 IRFLCT=0.14 RSHTR=0.0  TRSHTR=0.8 &
             SCFWNTR=0  SCFSMR=1
SGDISTWNTR   AIR=1.0
SGDISTSMR    AIR=1.0
GLSGREFLECT JANGR=0.2  FEBGR=0.2  MARGR=0.2  APRGR=0.2 &
             MAYGR=0.2  JUNGR=0.2  JULGR=0.2  AUGGR=0.2 &
             SEPGR=0.2  OCTGR=0.2  NOVGR=0.2  DECGR=0.2
```

FIGURE 7: (Cont.)

```
Line
----  ---------------------------------------------------------------------
  85  SGFACTORS   JANSGF=1.0  FEBSGF=1.0  MARSGF=1.0  APRSGF=1.0 &
  86              MAYSGF=1.0  JUNSGF=1.0  JULSGF=1.0  AUGSGF=1.0 &
  87              SEPSGF=1.0  OCTSGF=1.0  NOVSGF=1.0  DECSGF=1.0; &
  88              no outside shutters in winter or summer & SC=1.0
  89  INFIL       ACBASE=0.75
  90  INTGAIN     INTGAIN=20  SCHED=RES; value of intgain KWhr/day
  91  VENT        TYPE-NATURAL  HDIFF=2  STACKEFF=0.0
  92  TSTATSWNTR  THEAT=67  TDSRD=77  TCOOL=81  THEATNIGHT=60 &
  93              TIMEDOWN=24  TIMEUP=6  LGTRANGE=4
  94  TSTATSSMR   THEAT=67  TDSRD=71  TCOOL=81  THEATNIGHT=60 &
  95              TIMEDOWN=24  TIMEUP=8  LGTRANGE=4
  96  DBMEMORY    7
  97  GROUNDTEMP  GROUNDTEMP=DBBAR
  98  WINDFACTOR  0.5; adjusted for an open site
  99  CHNGSEASON  TYPE=TEMP TEMP=67; program operates in summer mode &
 100              whenever the moving mean drybulb temp. (DBBAR) is >67
 101  DAYTIMES    WDBEG=8  WDEND=18  SDBEG=7  SDEND=19
 102  OPCOST      ELPRICE=0.095  ACCOP=2  HEATING=ELECTRIC  HTCOP=1
 103  SOLARCALC   FREQ=MONTHLY
 104  *END
```

Table 1 summarized the nine conservation measures, their heating, cooling, and total loads, and the percent saving of each from the Base Case according to rank order of saving.

RESULTS

Significant improvement over the Base Case run was achieved. It is clear from Table 1 that more than 60% from the Base Case total heating and cooling load was eliminated. The house annual heating load decreased from 249 KBTU/yr to 8 KBTU/year, a 96.8% saving. Also, the house annual cooling load decreased from 4612 KBTU/year, to 1913 KBTU/year, a 58.6% saving. The annual source total heating and cooling load decreased from 11.977 KBTU/sq. ft. per year, to 4.522 KBTU/sq. ft. per year, a 62.2% saving. Finally, the house annual operating cost decreased from $0.11/sq. ft. per year, to $0.04/sq. ft. per year, a 63.4% saving.

These results are also shown in Figures 8 and 9 where the Calpas 3 sample output sheets are illustrated for the Base Case, Run 1, and the last run, No. 10. A summary of the monthly house energy balance is also shown, as well as monthly temperatures.

CONCLUSION

This presentation demonstrates an approach for designing guidelines for low-cost energy-efficient attached houses. It is proposed that such houses be developed and replicated for the establishment of viable settlements in the newly reclaimed desert areas of the North Tahrir region in Egypt. Although the study highlights some successful passive solar design techniques for desert hoses, most importantly, it demonstrates the advantage of using computer energy modeling and simulation programs, such as Calpas 3, in the optimization process of such designs. It was shown that, with

TABLE 1. Summary fo the nine conservation measures and the energy consumption results

RUN	Measurement & Architectural Description	Source Energy (KBTU/sf) Cooling	Heating	Total	% (+) Saving of base	$$.$ per S.F.
1	Base Case: Orientation = South	4,612	249	4,861	100%	0.11
2	North shutters: (C) Outside woodscreen shutters are added to the north. SC=0.2	3,415	249	3,664	-24.6%	0.09
3	Double glazing: (H) on the north and south NGLZ=2 XRFLCT=0.18 UVAL=06 TRSHTR=0.75 Vertical fins deleted.	3,433	93	3,526	-27.5%	0.08
4	Bermed Walls: (H&C) 4 ft. above the ground berming from the north slab perimeter is 20 ft (not 40), R-26 is the value of soil resist.	3,379	43	3,422	-29.6%	0.08
5	South glass area: (H) increased by 22 sq. ft. subtracted from S. wall	2,939	23	2,962	-39.1%	0.06
6	Roof Absorptivity: (C) Roof covered by Reed Panels, area between is vented, Absorp. is 0.02%	2,211	36	2,247	-53.8%	0.05

TABLE 1: (Cont.)

RUN	Measurement & Architectural Description	Source Energy (KBTU/sf) Cooling	Heating	Total	% (+) Saving of base	$$.$ per S.F.
7	South glass area: (H) Additional 19 sq. ft. is added to overcome the heating problem due to applying reed panels.	2,058	22	2,080	-57.2%	0.05
8	Shutter transmis.: (C) decreased by applying dark color close weave drapes (type H ASHRAE) south Bedroom & Kitch. north Bedroom & Living SC=0.42 TRSHTR=0.48	1,932	22	1,954	-59.8%	0.04
9	Shading: (C) overhang depth is optimized to 2.75	1,911	23	1,934	-60.2%	0.04
10	Reflectors: (H) South Glass ground reflectivity increased to 0.7 in Dec. and Jan. month by adding 3 feet concrete footpath.	1,913	8	1,921	-60.5%	0.04

(C) Indicates a cooling strategy
(H) Indicates a heating strategy
(H&C) Indicates a heating and cooling strategy

FIGURE 3. CALPAS 3 SAMPLE OUTPUT FOR THE BASE CASE.

```
TAHR10 ATTACHED ONE-STORY HOUSE ORIENT:S  NADER CALPAS3 V3.12 License: PC0201
OPEN SITE IN NORTH TAHRIR REGION  Weather:  NTAHRIR.EG1 (North Tahrir Egypt MKW)

S U M M A R Y   Run period: JAN-01 - DEC-31  Conditioned floor area: 640 sf

SPACE CONDITIONING LOADS          Run totals               Peaks
========================      -------------------        ----------
                                  kBtu   kBtu/sf              kBtuh
                              ----------  -------        ----------
 House
    Cooling                       4612     7.206              9.646
    Heating                        249     0.389             19.078

ENERGY CONSUMPTION                Run totals               Peaks
==================            -----------------------------  ----------
                                       Prop line  Source
                              kWh;  kBtu  kBtu/sf  kBtu/sf  kW,  kBtuh
                              ------------------------------  ----------
 Electricity

    House cooling                  676     3.603    10.809        1.413
    House heating                   73     0.389     1.168        5.590

    Total                          749     3.992    11.977
                                          ------------------

    Building total                         3.992    11.977
                                          ==================

OPERATING COSTS
===============
 Electricity @ 0.095 $/kWh          $      71
 Fuel        @ 0 $/kBtu             $       0
                                    --------
 Total                              $      71  ($0.11/sf)
                                    ========

Note:  CALPAS3  is  the  property  of  and is licensed by Berkeley Solar
Group.
```

FIGURE 8: (Cont.)

MONTHLY HOUSE ENERGY BALANCE (kBtu; + into house)

MON	GAINS & LOSSES COND	SHCND	INFIL	SLR	INT	TRANSFERS STRG	RB+SS	VENT	COOL	HEAT
JAN	-2203.1		-1127.9	1276.0	2116.1	-168		-64.790	0	178.95
FEB	-1945.7		-1021.4	1219.0	1911.3	-91.4		-67.885	0	0
MAR	-1906.8		-1029.6	1091.1	2116.1	32.4		-304.22	0	0
APR	-1047.5		-667.74	522.55	2047.8	-109		-715.50	-25.893	0
MAY	-426.20		-401.67	569.67	2116.1	-1.06		-1432.3	-424.18	0
JUN	-54.608		-234.65	595.13	2047.8	-57.0		-1158.5	-1136.7	0
JUL	-8.259		-216.93	575.74	2116.1	33.1		-1145.8	-1354.7	0
AUG	-97.643		-247.52	478.83	2116.1	8.49		-1222.2	-1036.7	0
SEP	-369.16		-344.95	419.58	2047.8	79.0		-1422.6	-413.08	0
OCT	-667.98		-461.94	448.08	2116.1	-71.4		-1168.8	-190.37	0
NOV	-1553.6		-828.38	1147.6	2047.8	121		-909.71	-30.259	0
DEC	-2210.0		-1110.4	1182.7	2116.1	239		-297.82	0	70.171
TOT	-12491		-7693.0	9525.9	24915	15.4		-9910.1	-4612.0	249.12

FIGURE 8: (Cont.)

M O N T H L Y   C O N D I T I O N S (Units as shown)

	TEMPERATURES (F)						WTHR (F; Btus/sf)				PEAKS (kBtuh)			
MON	THL	THH	THM	TSL	TSH	TSM	DBL	DBH	DBM	SGL	HSCL/DY	HSHT/DY	SSCL/DY	SSHT/DY
JAN	69	73	72				45	63	54	1055	0	19.1 3		
FEB	72	76	74				46	67	57	1376	0	0		
MAR	73	77	75				49	70	59	1651	0	0		
APR	73	78	76				53	76	65	2125	-3.37 30	0		
MAY	74	79	77				58	83	71	2306	-9.12 17	0		
JUN	76	80	78				63	86	75	2351	-9.65 24	0		
JUL	78	81	80				67	87	77	2376	-9.26 10	0		
AUG	78	81	80				66	87	76	2209	-8.01 18	0		
SEP	75	80	78				60	86	73	1858	-6.19 3	0		
OCT	75	80	77				60	82	70	1430	-6.20 9	0		
NOV	75	79	77				54	75	64	1109	-2.67 10	0		
DEC	72	76	74				49	66	57	936	0	14.8 31		
TOT	74	78	77				56	77	66	1733	-9.65	19.1		

FIGURE 9: CALPAS 3 SAMPLE OUT-PUT FOR RUN 10

```
TAHR10 ATTACHED ONE-STORY HOUSE ORIENT:S  NADER CALPAS3 V3.12 License:PC0201
OPEN SITE IN NORTH TAHRIR REGION  Weather:  NTAHRIR.EG1 (North Tahrir Egypt MKW)

S U M M A R Y   Run period: JAN-01 - DEC-31  Conditioned floor area: 640 sf

SPACE CONDITIONING LOADS            Run totals               Peaks
========================       -------------------         ----------
                                   kBtu   kBtu/sf               kBtuh
                               ----------  -------         ----------
  House
     Cooling                       1913     2.988               8.075
     Heating                          8     0.013               5.152

ENERGY CONSUMPTION                  Run totals               Peaks
==================            -----------------------------  ----------
                                     Prop line  Source
                              kWh;  kBtu   kBtu/sf  kBtu/sf  kW,  kBtuh
                              ------------------------------  ----------
  Electricity

     House cooling                  280     1.494    4.483        1.183
     House heating                    2     0.013    0.040        1.510

     Total                          283     1.507    4.522
                                          ------------------
     Building total                         1.507    4.522
                                          ==================
OPERATING COSTS
===============

  Electricity @ 0.095 $/kWh                $     27
  Fuel        @ 0 $/kBtu                   $      0
                                          --------
  Total                                    $     27  ($0.04/sf)
                                          ========
```

Note: CALPAS3 is the property of and is licensed by Berkeley Solar Group

FIGURE 9: (Cont.)

M O N T H L Y   H O U S E   E N E R G Y   B A L A N C E (kBtu; + into house)

	GAINS & LOSSES						TRANSFERS			
MON	COND	SHCND	INFIL	SLR	INT	STRG	RB+SS	VENT	COOL	HEAT
JAN	-2363.8		-1283.6	2431.9	2116.1	-149		-753.32	0	8.486
FEB	-2034.8		-1112.8	1794.7	1911.3	-22.7		-534.81	0	0
MAR	-1939.0		-1064.4	1420.2	2116.1	30.4		-564.15	0	0
APR	-1039.3		-601.66	200.73	2047.8	-58.9		-544.97	-1.156	0
MAY	-448.13		-294.73	165.82	2116.1	47.4		-1372.5	-215.68	0
JUN	-145.15		-140.32	148.69	2047.8	-117		-1214.6	-575.68	0
JUL	-168.31		-149.74	146.01	2116.1	69.4		-1496.1	-519.73	0
AUG	-179.99		-158.55	126.90	2116.1	17.8		-1506.1	-417.38	0
SEP	-294.56		-219.23	125.83	2047.8	94.9		-1681.6	-77.409	0
OCT	-353.76		-356.69	204.27	2116.1	-88.2		-1282.5	-34.669	0
NOV	-1459.6		-813.20	1667.0	2047.8	30.4		-1405.0	-69.465	0
DEC	-2252.4		-1215.6	2267.1	2116.1	149		-1068.7	-1.714	0
TOT	-12879		-7410.5	10699	24915	3.17		-13424	-1912.6	8.486

FIGURE 9: (Cont.)

M O N T H L Y   C O N D I T I O N S

(Units as shown)

MON	TEMPERATURES (F) THL	THH	THM	TSL	TSH	TSM	WTHR (F; Btus/sf) DBL	DBH	DBM	SGL	PEAKS (kBtuh) HSCL/DY	HSHT/DY	SSCL/DY	SSHT/DY
JAN	72	76	74				45	63	54	1055	0	5.15 3		
FEB	73	77	76				46	67	57	1376	0	0		
MAR	74	77	76				49	70	59	1651	0	0		
APR	73	76	74				53	76	65	2125	-0.87 30	0		
MAY	73	78	75				58	83	71	2306	-7.62 17	0		
JUN	75	79	77				63	86	75	2351	-8.08 24	0		
JUL	77	81	79				67	87	77	2376	-7.07 19	0		
AUG	76	80	78				66	87	76	2209	-6.37 13	0		
SEP	73	79	76				60	86	73	1858	-4.20 3	0		
OCT	74	78	76				60	82	70	1430	-4.31 9	0		
NOV	75	79	77				54	75	64	1109	-5.29 22	0		
DEC	74	77	76				49	66	57	936	-1.71 6	0		
TOT	74	78	76				56	78	66	1733	-8.08	5.15		

as few as nine runs, each simulating a different inexpensive architectural conservation measure, substantial savings in both the annual heating and cooling loads were achieved. Further computer runs with additional design enhancements can be easily conducted in an attempt to eliminate the heating and cooling loads, a desirable situation for a good passive solar house in remote regions with hot climates.

REFERENCES

1. The American Society of Heating, Refrigeration and Air Conditioning Engineers. 1981. Ashrae Handbook of Fundamentals, New York.

2. Berkeley Solar Group. 1984. Calpas 3 Program User Manual. Berkeley, California.

3. Nader Chalfoun. 1985. A passive solar design methodology for housing in Egypt. Master's Thesis, University of Arizona, Tucson.

4. Nader Chalfoun and Mohammed Hamdy. 1986. Optimization of a passive solar desert house design using Calpas 3. Paper, The Second National Conference on Micro-Computer Applications for Conservation and Renewable Energy, October 21-23, Tucson, Arizona.

5. P.O. Fanger. 1970. Thermal comfort, analysis and applications in environmental engineering. Danish Technical Press, Copenhagen, Denmark.

THE POTENTIAL OF A DESERT ECONOMY

PINCAS JAWETZ
UNITAR Consultant on Technology and Policy
P.O. Box 6297, New York, N.Y. 10150-6297, USA

Abstract The presentation is based on a recent report that uses the information contained in the United Nations Institute for Training and Research (UNITAR) four volumes - "Alternative Strategies for Desert Development Management".

When reviewing the large quantity of material gathered for the 1977 Sacramento Conference it becomes clear that as a generality there exists a vast potential for economic development of arid lands. As such it is stressed that deserts should not be regarded as a curse but rather as a resource to be developed - perhaps even as an ultimate resource. If we believe that the earth is entering a cooling period, then areas that enjoy more sun will increase in importance in the decades to come. Desert areas are the great reserve in reserves of unoccupied land today and land must be increasingly reviewed as the most essential resource as the population of the globe continues to grow rapidly. Such growth will probably continue for a number of decades to come. Africa's population is expected to triple by the year 2025.

In most desert areas, water is the limiting factor for the development of the region and it will be important to consider the income effect of each unit of water consumed. This may lead to the conclusion that agriculture for the production of low value staple foods which produce a low value-added per unit of water used, may have the lowest development effect. Thus, in situations of limited water availability careful planning will be required in order to maximize national income on the basis of the water available.

Unfortunately there is no ready-made formula for determining the optimum proportionate effort between industrial and agricultural development. Historically, mining and/or agriculture have been the pioneers, but a more serious consideration of the potential industrial contribution to development of arid areas could possibly indicate the need for a stronger role for this type of activity. The same is true for many aspects of a service economy - from tourism and retirement settlements to information storage.

* * *

The presentation is based on a recent report that uses the information contained in four volumes of the United Nations Institute for Training and Research (UNITAR) called "Alternative Strategies for Desert Development and Management"[1] These volumes contain material gathered via a UNITAR conference organized jointly with the Government of the State of California at Sacramento, California, 31 May - 10 June 1977. The material was put together in view of a United Nations General Assembly Special Session on Africa, and in preparation for the organization in 1988 of the Second UNITAR Conference on Development and Management of Arid Lands. A first version of this paper was presented to a workshop sponsored by the Club of Rome on "The African Predicament in the Global Society: Towards the Management of Drought, Desertification and Famine," Lusaka, Zambia, 24-27 June 1986.

A review of the large quantity of material gathered for the Sacramento conference makes it clear that, as a generality, there exists vast potential for economic development of arid lands. As such, it is stressed that deserts should not be regarded as a curse but rather as a resource to be developed -- perhaps even as an ultimate resource. If we believe that the earth is entering a cooling period, then areas that enjoy greater sun will increase in importance in the decades to come. Desert areas are the great reserve of unoccupied land today. As the population of the globe continues to grow rapidly, and probably will do so for a number of decades to come, land must be viewed as the most essential resource. Africa's population is expected to triple by the year 2025.

Desertification is the process of desert creation and is considered a nearly irreversible process. Desert reclamation is viewed as a process of retrieving the land from the desert's grasp. For most people, desert reclamation is declared a success when green vegetation is being sprouted in the desert, even if this is not the optimum use of the particular desert location or a best use of the available resources. Water, capital, modern technology, and an educated population can, however, usually bring about development of desert areas in harmony with long- term stability and with benefits to mankind as a whole.

In most desert areas, water is the limiting factor for the development of the region and it will be important to consider the income effect of each unit of water consumed. This may lead to the conclusion that agriculture for the production of low-value staple foods which produce a low value-added per unit of water used, may have the lowest development effect. Thus, in situations of limited water availability, careful planning will be required in order to maximize national income on the basis of the water available.

The advantages of self-sufficiency in staple foods and the need to reduce the consequences of recurring droughts are obvious. However, mining and industrial development also have their advantages. Furthermore, they are endeavors which can propel a nation into the twentieth century (if that is considered desirable--and it is considered so by many). Mining and industrial development tend to promote a higher level of education and can accelerate economic progress and social change. Rates of capital accumulation and foreign earnings can more rapidly be increased. Unfortunately, there is no ready-made formula for determining the optimum proportionate effort between industrial and agricultural development. Historically, mining and/or agriculture have been the pioneers, but a more serious consideration of the potential contribution of industrial development to development of arid areas could possibly indicate the need for a stronger role for this type of activity.

Seacoasts comprising some 20,000 miles in arid and semi-arid areas pose many advantages. The revolution in sea transport during recent decades has greatly lowered shipping costs, thereby opening these areas to importation of new materials and export of local products. Saudi Arabia might, for example, cheaply import iron ore from as far away as Brazil, use cheap natural gas in the direct reduction process, and export the resulting iron at competitive prices. Desert seacoasts also have various marine resources and can provide cooling water for power and industrial plants. In comparison, landlocked desert nations are at a considerable disadvantage and will often suffer from severe limitations to their industrial growth capacity.

The development and expansion of agricultural production in arid and semi-arid zones is intimately related to the availability and use of water in irrigated and rainfed cultivation, and to range utilization. Various problems arise in determining overall strategies to overcome the difficult and complex obstacles to agricultural development. Should, for example, priority of resource use be given to rainfed agriculture and/or to developing new irrigated areas? Should new irrigation projects be constructed or should the existing ones be improved? What priority should be given to rangelands or woodlands? What level of technology is appropriate? However, behind all these problems loom the social and cultural barriers of archaic systems of land tenure and rigid water rights, and lack of development capital and of technological skills needed to absorb new technologies.

Several methodologies exist for strategy development. These are known and accepted methodologies in developed countries and should be applied with care until more experience is gained in less developed areas. However, the disciplined thinking that they require can only increase the effectiveness of planning procedures. Some technologies could lead to combined uses, multiple purposes and sustained yields. Considering the wide disparities among the nations with arid regions it is apparent that there will be no single overall methodology or technology level appropriate to all of these regions and there will possibly be great differences in approach within the region itself.

Several technologies are worth stressing here. For desert nations with the ability to make large capital investments, a major priority would be that of affecting a significant increase in the efficiency of agricultural water use. This would include employing drip irrigation systems and, under exceptional circumstances, controlled environment agriculture. For most arid-zone nations the installation of drainage systems still remains at the top of the priority list. The resultant salinity and waterlogging because of poor drainage is causing a yield reduction of 30-50% in those areas which have not already gone completely out of production. Drainage and better irrigation technology, combined with some increase in overall levels of agro-technology, could probably result in doubling agricultural production in the presently irrigated areas of the arid and semi- arid regions.

One important asset is the large amount of underground water which could be exploited at economic cost in many regions under the desert surface. While great caution has to be taken in verifying the estimates (especially since most of these resources are non-renewable fossil waters accumulated over the last 30,000 years), these resources could undoubtedly have tremendous impact on those regions where they exist. Groundwater development has several advantages over use of surface resources. Large capital investments in dams and canals are not necessary as wells can be drilled where needed and at a rate commensurate with capital availability.

In many semi-arid regions, rain-fed agricultural development will warrant highest priority. In most developing countries in the semi-arid zone, this is the mainstay of the economy because it produces a large proportion of the food and gives employment to a vast majority of the population. Hence any incremental increase in productivity in this activity would have a far-reaching effect on the nation's economy. Conflicts in this sector relate to producing food for a subsistence economy and alternatively growing more commercial crops (e.g., cotton, peanuts, etc.) for a cash economy. While most researchers consider food self-sufficiency as being of great importance, others say it is difficult to see how essential agricultural inputs needed to raise productivity can be financed without some amount of cash earnings. A second problem concerns the proportion and integration of rain-fed agriculture with irrigated agriculture and the possibility of combining these with range- and cattle-raising into a coordinated production system. There is a great need for better crop rotation, for the introduction of valuable nitrogen-fixing legumes, and for new grain varieties which would produce accumulated yields over a period of several years. (A variety developed only for its "drought resistance" could give less accumulated yield over the same period.) Good grain storage facilities are vital to this strategy.

Range management will probably always remain one of the main activities for utilizing arid and semi-arid land resources. In many areas it is the only way to use these resources. However, more attention will have to be paid to proper grazing controls and management as well as integration with other aspects of the economy.

An innovative idea is that of wildlife production in place of conventional cattle herds. Proponents of this scheme view cattle as being responsible for the widespread desertification we are witnessing and claim that wildlife raising for meat, hides, etc. would be more in harmony with "nature's" intent. While a case could be made for the need for further experimentation and even for field trials, it is difficult to believe that simply by changing the instrument of man's exploitation of his environment we will at one stroke eliminate mismanagement and all those abuses for which man alone and not the animal is responsible.

Industrial development in arid and semi-arid areas has historically been associated with mining activities. Deserts seem to have accumulated large deposits of minerals and salts which have been preserved by the natural aridity and whose locations are relatively easily discovered.

Unfortunately, most resources have been mined and exported as raw materials or, at best, as semi-finished products. The mineral resources could be used today as basic materials for more sophisticated manufactured products and become a far more significant source of foreign currency income for the arid region. However, extensive transportation, power, and water infrastructure is required. To this could be added the need for skilled labor and management with a strong inclination for entrepreneurial undertaking.

Deserts do have many advantages for industrial development. For example, land is readily available at a low price and the mild winters could possibly ease construction costs. Industries may be able to easily and safely dispose of their waste products by simple evaporation of industrial effluents which in other areas could contaminate surface or groundwater. Low air humidities are often useful and solar energy could be utilized for drying processes resulting in considerable reduction of energy costs.

A shortage of water is not as limiting for industry as it is for agriculture. Income per unit of water is much higher for industrial products and water is proportionately a lower cost factor in product price. Industry can afford to pay more for water so that deeper or more distant water resources can be used. There is a great deal of active research on minimizing water consumption in industry and there already exists a wide spectrum of "dry" technologies, especially for mining of minerals and metals in desert areas. The recycling of industrial waste water seems to be at a practical stage and water of low quality could be used often. Agricultural products and wastes and agro-industrial crops could be a suitable basis for initiating industrial development. Urbanization also requires less water than agriculture or growing crops. Developing small-scale industries in order to provide occupations and income to an urban population presents further advantages.

The human factor remains, as in other fields of endeavor, the most critical element. In many areas there is a lack of sufficiently trained personnel. Many desert areas are hampered by lower educational levels or do not have suitable institutions to utilize that portion of the population which has been educated or adequately trained. Obviously this is a principle direction for future aid.

REFERENCE

1. UNITAR, 'Alternative Strategies for Desert Development and Management, Proceedings of the Sacramento, California, conference held May 31 - June 10, 1977. Published by Pergamon Press in 1982.

Volume I - Energy and Minerals - pages 1-254
Volume II - Agriculture - pages 255-481
Volume III - Water - pages 482-946
Volume IV - Desert Management - pages 947-1396

Annex I to this paper presents a short review of activities at the United Nations following the 1986 General Assembly Special Session on Africa.

ANNEX I

THE UNITED NATIONS PROGRAMME OF ACTION FOR AFRICAN ECONOMIC RECOVERY AND DEVELOPMENT 1986-1990 following the 27 May - 1 June 1986 U.S. General Assembly Special Session on Africa's Critical Economic Situation

Following Africa's Priority Programme for Economic Recovery 1986- 1990 (APPER), adopted by the Assembly of Heads of State and Government of the Organization of African Unity (O.A.U.) at its session held in Addis Ababa 18-20 July 1985, and presented before the preparatory committee of the U.N. General Assembly Special Session by O.A.U. and the U.N. Economic Commission for Africa (E.C.A.) on behalf of the Governments and peoples of all Africa, and remembering the recent food shortages, the main emphasis of the document is on the agricultural sector.

The document summarizes the O.A.U. estimates that a five-year plan for all of Africa will require about $US 57.4 billion for agricultural development, $US 60.1 billion for agricultural support industries, $US 3.4 billion for measures to fight drought and the growth of deserts, and $US 7 billion for human resources development including education and training. Above estimates amount to a total of $US 128.1 billion and the O.A.U. document suggested that $US 82.5 billion can be provided from Africa's own funds. This leaves a gap of $US 45.6 billion or about a further request of $US 9 billion per year. The above gap assumes that the payment of interest and repayment of capital for the present debt of the African states can be stalled for the period of the quinquennium.

In response to the O.A.U. estimates, the donor countries, though noting with admiration the comprehensive resolve of the African countries to break away from the clutches of under-development with notable exceptions such as Canada, the Netherlands and Denmark, rejected the idea of a moratorium on present debt and also were not ready to go beyond the general promise "to make every effort to provide sufficient resources to support and supplement the African development effort."

Having described the content of the document, and now looking over relevant outside material one can nevertheless predict that the document may have the needed impact on opening the purse- strings at the development banks.

Looking back at the activities of the World Bank as an example: The traditional criteria for lending was a robust economic rate of return and the expectations that loans will "pay-off" in a relatively short period of time. These criteria are obviously inappropriate when dealing with most of Africa and more help and time are required than those criteria allow. Given the right policy environment, and if you wish the needed excuse, the Bank can nevertheless be flexible. A case in point was when aware of past limitations of policy, when lending only for the importation of capital goods, this was changed and relaxed to permit financing of local and recurrent expenditures.

Furthermore, while lending for agriculture was only around 6 percent of the total World Bank's lending in the 1960s it grew 8.75 times in the 1970s and reached 30 percent of much larger total Bank's lendings in 1985. This development was not smooth and occurred indeed in three stages: (1) the Bank started lending for capital goods such as tractors and large projects such as irrigation structures for Latin America; then stage (2) raising farmer's productivity by making services and inputs such as fertilizers available especially in Asia. Eventually, finally the Bank was ready to stage (3) increasing equity and raising the productivity of low-income producers in Sub-Sahara

Africa. Upon reaching this third stage the Bank made quite a few mistakes. Not realizing the impact of fewer irrigation facilities then in Asia, little high-yielding technology, and underdeveloped institutional framework for investment, the Bank, without first developing pre-conditions for successful investment in rural Africa, made loans which failed thus adding to the debt burden of these countries without contributing to their growth. A most important impact of the new U.N. Programme of Action and the impact of the African commitment in the preamble, will be to revive the Bank's interest in lending to Africa.

The main argument in African development is the need for an integrative development involving agriculture, forestry, renewable energy, rural development, and education. Also, as F.A.O. has shown agricultural development that was very successful in other parts of the world turned out to be disastrous when applied to inhospitable African conditions. Thus the great achievement of the "green revolution" in wheat and rice in other parts of the world turned out of little use to the different ecological conditions in Africa and to the sorghum and millet eaters in the indigenous population. To achieve a successful marriage of advanced technology with the local conditions a local network of technological institutions will have to be established in order to help direct the development process. Wherever the time does not allow such a slower pace, working with the local small farmer will enable the funding organizations to achieve at least first tangible results and the present Programme of Action avails the lenders with such an avenue of experimentation.

The Programme of Action calls for establishing of efficient regional networks of crop protection agencies and an eventual increased use of fertilizers, improved seeds and pesticides, development of mechanization and the use of modern farm and processing machinery, improving and expanding storage capacity, distribution, and marketing systems. Also, the development of industries for production of agricultural tools and equipment, small-scale irrigation equipment, and agricultural inputs; the processing of raw materials and the development of agro-related industry and the local production of essential consumption products.

To facilitate the trade in these products, the infrastructure and communication and transportation networks will have to be improved.

The above allows for many investment possibilities and the international community in its response to the African appeal agreed to strive to ensure that no individual bilateral donor, during the period 1986-1990, become a net recipient of official capital flows from African countries that undertake adjustment programs - read those countries that loosen internal regulations enough to allow for production incentives to their local agricultural private sector. If this can be seen as an attempt at a program of reinvesting debt payment and debt servicing payment this could become a partial start for the funding mechanism of the proposed Programme.

At most positive statement to the General Assembly was made in the opening session by Dr. Edgard Pisani, Chairman of the Preparatory Committee for the special session who said:

> "It was crucially important that Governments, organizations and information media unit to provide international public opinion with an image of Africa not only as a continent of drought, famine and disease, but also <u>stressing its resources</u>, its vitality, its determination, its youth and its importance in the global balance."

In the course of the meetings of the Preparatory Committee, Dr. Pisani went beyond the question of food and agriculture and expressed the hope that in order to enhance the development potential for Africa, the documents will include also in-depth looks at such topics as energy and industrialization.

The Foreign Minister of Cameroon, H.E. William Eteki Mboumoua, speaking before the General Assembly said that of all Africa's resources, its human resources had been the least rationally utilized. Programs are needed to create new perspectives of human resource development.

In general, Africa's education and employment policies, institutions and methodologies had not allowed for the most effective development of those resources. The content of educational programs was far from responding to the needs of accelerated development. Efforts to promote technical training, to co-ordinate formal and informal education and to reorient education to national development needs had not produced the results that had been hoped for. Africa was suffering from a shortage of skilled labour and technical personnel, and a high rate of unemployment and underemployment even among those who had followed a course of education.

EXPORT CROPPING AND DESERT DEVELOPMENT

JIMMYE S. HILLMAN
Director, International Programs,
University of Arizona, Tucson, Arizona, U.S.A.

Abstract Export cropping has been blamed for many problems in countries when incomes are low, where agriculture is a large sector in national employment, and where real costs of food are relatively high. There is a conflict in policy objectives as to how resources in the agricultural sector should be used in developing countries: Should countries use advanced technology to specialize in high-value crops for export, thus solving an exchange problem; or should resources be devoted to food production for self sufficiency?

Desert countries and irrigated agriculture are not excluded from this economic/political dilemma. In fact, because specialty crops grown under irrigated conditions tend to be high value by necessity, the problem is compounded.

Research which has examined this question in tropical and other contexts will be explored, and the agriculture of northwest Mexico will be examined as to problems of producing and exporting specialty crops.

The self-sufficiency in food argument will also be explored.

Can agricultural specialty exports lead economic growth?

These questions will be examined by looking at the research literature.

Tentative findings suggest that "export cropping" is a buzz word term that has led to some misunderstandings about economic growth, food self-sufficiency and political economy.

INTRODUCTION

The debate surrounding the efficacy and the equity considerations of export cropping, like all great debates on political economy, continues to draw major international attention. Item: Sudan's Prime Minister Sadek El-Mahdi said at the December 1986 annual meeting of the International Fund for Agricultural Development (IFAD) in Rome that his country has a food surplus and will begin exporting. Acknowledging that this might seem "ridiculous" for a nation in which relief officials say two million people face starvation, El-Mahdi said that enough food would be stored to ensure food security and some would be exported to help people in worse situations. Thus, the paradox and the conundrum about agricultural exports from poor countries continue.[1] The situation in arid environments is a special case to which some attention will be devoted after an elaboration of the general problem.

During the past few years, a rather large body of literature has emerged. It is comforting to note, however, that the topics in the debate remain nearly the same as outlined then: definitional problems and clarification of rhetoric; the economics of resource use, or comparative advantage; food, self-sufficiency and attendant nutritional problems; and growth and trade issues associated with cash and export cropping. The conclusions of most of the research are similar, and moreover, there is nothing about arid countries that negate those conclusions.

Critics of cash and export cropping are concerned, correctly, about many indicators of slow economic development, endemic hunger and malnutrition. Their prognoses as to cause, and their prescriptions as to cure run counter, however, to the results of most researchers who have studied these problems in depth. Books such as <u>Food First</u>[2]; <u>How the Other Half Dies</u>[3]; <u>Feeding the Few</u>[4] and other titles, to be sure, have called attention to the problems and have forced traditional aid organizations and many countries to reexamine their development plans and methods. Fortunately in her new book, World Hunger; Twelve Myths,[5] Frances Lappé (1986) appears more moderate in her observations on economic development as it relates to issues such as hunger, free trade and technical assistance.

Fundamental solutions to the growth paradox of developing countries have not, however, resulted from the rhetoric and the suggestions of the "food firsters". Critics of commercial agriculture often ignore research and practical evidence which demonstrate the benefits to poor countries when they follow growth models that emphasize cash cropping, trade and comparative advantage. Moreover, they forget the disadvantages, the costs and the risks associated with a predominantly self-sufficient food policy.

One further introductory comment should be made: It is most difficult to generalize about the complex problems surrounding cash and export crops on the one hand and self-sufficiency on the other. What is true of one country may not be applicable to another. For example, tropical development schemes may not work in arid areas.

Before turning to substantive issues, a word about terminology: cash cropping, export cropping, or the old colonial term "commodity scheme", are all treated here in the same vein. Such crops are those which are produced for sale, be they food or nonfood. Export crops are particular types of cash crops.

EXPORT CROPPING AND NUTRITION

In recent years, considerable research has been done and literature published on the nutritional impact of export cropping. The International Food Policy Research Institute (IFPRI) has made this subject the focus of one of its major efforts. It has undertaken research to assess the effects on nutrition, as well as on income, the transition from semi-subsistence to cash cropping in several countries in order to, among other things, avoid negative outcomes and to enhance positive nutrition and income effects. One recent working paper from IFPRI (Von Braun and Kennedy)[6] provides a basic set of information and analysis which will be used in this section. In addition, the seminal thinking and expression of Pinstrup-Andersen (1985)[7] on the subject was of great help.

As already noted, there is danger in generalizing about the problem: the critical question in most cases seems not to be whether increased money income and foreign exchange offer potential for increased food security and for improved nutrition but, instead, whether national policies and decisions of policy makers will direct that such exchange be used to increase food production, and food imports, thereby maintaining or increasing food availability.

A review of the published research on the effects that cash and export cropping have on household food consumption and nutrition, shows a mixed picture. This is not a surprising conclusion, given the wide variety of cash crops and the conditions of their production. Design of the studies, for example: their cross-sectional, static comparisons, frequently ignore <u>ex-ante</u> cash cropping household conditions. In comparing a cash-crop household with a food-crop household one might conclude that the food-crop household is better nourished, without having determined whether the nutritional status of the cash-crop household has changed since it adopted cash-cropping.

In 78 countries studied by Von Braun and Kennedy (1986),[6] only 16 devote less than 10 percent of their cultivated area to major cash crops (excluding basic staple crops), and 28 have more than 30 percent of their cropped area in such crops. This represents an underestimation of cash cropping, however, because, for statistical reasons, cash crops are identified by types of crops such as sugar, rubber, cocoa, tea, coffee, vegetable oils, etc., while in practice many staple crops are often also sold for cash.

Most analyses of the problem show that agricultural production patterns for cash crops in developing countries change over time, and that cash cropping is much less stable in low-income countries than in countries where incomes are higher. Further, research reviewed for this paper questions the assumption that cash cropping expands at the expense of staple food crop production. Static comparison often supports the claim that countries with high percentages of their land in cash crops produce less food per capita, but this is superficial in that it tells us nothing of food availability, a condition largely influenced by trade policies of such countries. Analyses demonstrate, moreover, that countries which manage their cash and export crop sector well will also manage their staple food production sector well, and vice versa. Policies for input supply, product marketing and rural infrastructures have an impact on both sectors and, hence, influence the growth outcome and success in both.

Additional factors are important in the nutrition-export cropping argument. For example, the complex issue of cereal grain-production-for-livestock-feed involves a broad set of factors which are not easily analyzed. A continuation of recent animal feeding practices in middle income countries might be disadvantageous to the poor in developing countries, but in most cases the nation's problem is one of poverty and is affected only marginally by the fact that 40 percent of the world's cereal harvest is fed to animals. The food-feed controversy, moreover, must not be confused with other livestock policy issues where, for example, livestock is an indispensable source of income with many opportunities for development.

In sum, while it is evident that food consumption and nutrition are affected by cash cropping, the results are not as bleak as some food-firsters lead us to believe. As Pinstrup-Andersen points out:

--"The use of scarce resources for the production of agricultural commodities for export by countries with widespread nutritional problems cannot be assumed always to have negative nutrition implications. However, it would be equally simplistic to assume that it would necessarily and automatically result in positive nutrition impact. First, using comparative advantage as a guideline for the production patterns in the agricultural sector helps to assure the most efficient utilization of scarce resources across countries. But it tells us nothing about the distribution of the associated economic benefits vis-a-vis nutritional status. Furthermore, inefficienciences in agricultural marketing, uneven distribution of the power to extract economic benefits from agricultural production and marketing, and existing government policies and regulations, all exercise great influence over the benefit distribution. Thus, while production along the lines suggested by comparative advantages may assure that any negative nutrition effects from the production of export crops could be compensated for, it does not assure that population groups experiencing negative effects will, in fact be compensated."[7]

Finally, as most analysts have concluded, and as was stated at the outset, rather than rejecting export cropping on the basis of rhetorical and unsubstantiated assertions that economic and nutritional conditions associated with it are unacceptable, governments should invent more equitable and beneficial policies. If export crop production results in more efficient resource use in a country, the government of that country should start from there and design and implement policies which would assure equity measures: i.e., beneficial income distribution and increased food availability, along with efficiency. A high level of long-term food security is a desirable objective. In the long run, a simple avoidance of negative nutrition effects is not sufficient -- other export cropping objectives must be fulfilled.

For example, export cropping and resultant income patterns might alter nutrition modes through changes in consumer preference and spending behavior: that is, through modifying the desire for different foods and the need to acquire them; and through changes in food distribution within households. More could be said on export cropping and nutrition, but it is a large subject, and will not be covered in detail here.

RESOURCE ALLOCATION, SELF-SUFFICIENCY AND GROWTH

Resource allocation, like comparative advantage, has been the grist for economists since the original economic development and trade models of Adam Smith and David Ricardo in The Wealth of Nations (1937 ed.)[8] and in The Principles of Political Economy and Taxation (1911 ed.).[9] The essence of the theory is that regions are differently endowed with resources, including human capital. These differences also extend to asset ownership, taste-endowment and demand-creating attributes. Production and specialization, including the exploitation of economics of scale, lead to increased levels of world output, and an exchange of this increased output represents the gains from trade.

The dynamic forces of trade promote and facilitate the complex processes of economic development and add important dividends, such as access to world food supplies. Further, trade encourages technological innovation and raises worker skills. Increases in the range of goods that can be purchased stimulates new wants and can create new areas of production or higher productivity. Only exports can enable low income nations to pay for those goods and services necessary for development but which they lack the capacity to produce. Finally, exports provide an important source of direct government revenue in many countries, as in Malaysia where in recent years 30 percent of federal funding has been derived from export taxes. Curtailing exports and trade would mean seriously undermining the fiscal positions of nations which are without alternative sources of taxes. In sum, trade, through the execution of the principle of comparative advantage, brings not only status gains to a country, but also acts as a stimulus to growth.

Table 1 contains information presented on another occasion around the turn of the last decade but which illustrates the practical logic of comparative advantage. Recent data do not change the conclusions (see Table 1a). Even with technical and economic qualifications to the contrary, the table demonstrates that developing countries (LDCs) would experience a foreign exchange loss (and farmers a loss in gross and net revenues) if they switch land out of traditional export crops into the domestic production of food which is now imported.

Caution must be advised, however, in recommending an abrupt shifting from domestic food production into further concentration on export cropping. Very important in such decisions will be the potential benefits from an increase in foreign exchange, which is a function of both productivity increases in export products and the elasticity of demand for such products in the importing countries. More logical and more helpful in the long run to the LDC is the raising of technical productivity of domestically-consumed foodstuffs. The potential for such increased productivity, it is clear, will not be great without major increases in research and development expenditures.

Turning to self-sufficiency in food, many countries view it as the only alternative to cash and export cropping, and they place it at the heart of their national development and their food and agricultural policies. Provisional self-sufficiency estimates by FAO for 103 countries (only 5% of the world population was not included) show that 3 billion of the world's population of 3.7 billion at that time lived in countries, each of which was at least 95% self-sufficient in kilocalories.[10] Somewhat unexpectedly, developed countries were less self-sufficient than developing countries, a fact which should not be lost on the export-cropping critics.

The most frequently articulated reasons for self-sufficiency involve conservation of scarce foreign exchange, protection of an economy against instability arising from fluctuating import prices, and insulation of a country from presumed political vulnerability associated with dependence on imports. Self-sufficiency, in short, often tempers comparative advantage because governments are responding to uncertainty in food supplies, depleted foreign exchange reserves, and poor income distribution with measures of direct intervention.

Demands for self-sufficiency rarely address the appropriate degree of self-sufficiency, nor do they consider the costs involved in attaining self-sufficiency. But these factors are frequently the crux of the matter. Large-scale subsidization of agriculture is an expensive business for a country with 50 percent or more of its labor force on farms. Also, production costs in farming can rise rapidly when output is stimulated in regions badly suited to such farming. Variability in output can be more severe when self-sufficiency levels are higher than warranted by natural resource endowments. Moreover, imports of production inputs and investment goods for agriculture can put almost as much strain on the payments position as would importing the foodstuff itself. Despite self-sufficiency and export-cropping rhetoric and genuine aspirations, these factors suggest that continued expansion in the absolute level of food imports by developing countries might be required due to the constraints on domestic budgets.[11]

Whether more or less self-sufficiency would be desirable becomes an empirical question for each country, demanding more objective analysis. First of all, the costs of a self-sufficiency policy must be given more emphasis. Achieving improvements in income distribution involves both income sacrifices and transfers from urban consumers through higher food prices -- thus, the urban poor suffer. While government policy discriminates against food production in some countries by taxation of inputs and outputs, it is questionable whether strong policies of self-sufficiency can be pushed in most countries before effective research and development programs are established for domestic food production and distribution. On the other hand, it must also be noted that current export levels in many developing countries do not reflect free trade, and distorted policies may in some cases overemphasize exports (e.g. by subsidies) or restrain export development (e.g. by export tax schemes).

How to achieve increased efficiency and development in a developing country's agriculture is the problem which concerns both export croppers and those interested in food self-sufficiency. Johnson (1980)[12] has given a general policy recipe which stimulates (a) adequate incentives for farmers through prices of products and imports that have a reasonable relationship to the real alternative values in the international market; (b) the development and availability of useful technology that will make possible higher levels of productivity of agricultural resources, both from land and labor; (c) adequate and reliable supplies of modern inputs for farmers; (d) access to national and international markets for farmers; (e) a significant improvement and expansion of irrigation; and (f) economic, political, and social systems which permit the farm population to share in economic growth. The importance of the latter factor, equity, has been emphasized by Schuh (1978),[13] who asserts that farm and rural people in developed countries are poor and inefficient because their governments have seriously underinvested in them, particularly in terms of health and education. Instead, governments have followed inappropriate policies which keep farm prices low by decree and which intervene in product markets to correct problems rooted in factor markets.

Ultimately, food security is one of the real reasons countries turn toward self-sufficiency schemes rather than let comparative advantage and free trade run their course. In Egypt, for example, like many developing countries, the major policy issue is whether to strive for self-sufficiency or rely on international trade for food. Egyptian policy currently favors the latter. A popular definition of food security in Egypt is to strive for a balanced agricultural trade--to have cotton and other agricultural exports pay for the imports of food.[14]

A policy of export-led growth, in addition to dealing with problems of increasing productivity, employment, and equity in the agricultural sector, also must deal with the problems of economic uncertainty and the instability in food supplies. The principal controversy surrounding food security relate primarily to man's policy role, not to the limitations of man and nature to produce enough food.

Time and space limitations do not permit adequate treatment of trade issues as they affect the export cropping debate. Suffice to say that the LDCs are linked by trade to growth in the developed countries. They are, moreover, almost unavoidably propelled toward agricultural production for export in order to gain foreign exchange for their own industrial growth. Numerous cases could be cited. Zaire, for example, has been trying to increase its exports of farm products by returning control of expropriated plantations to their former owners and by encouraging farm workers to return to the land.[15]

Other countries have attempted to boost export earnings while at the same time stabilizing prices and collecting revenue by establishing government-controlled marketing boards. This would be an optimum strategy if it worked, but selfish interests usually reduce its effectiveness. For example, a study in Jamaica[16] found that the implicit taxation of export enterprises imposed high social costs on Jamaica in terms of efficiency and equity. Farmers in export industries via such taxation and price policy have subsidized select interest groups which has led to a worsening of income distribution. The loss of foreign exchange resulting from this type of trade policy has compromised Jamaica's capacity to import and has had an impact on all groups in its society.

On another front, unfortunately, preferential treatment, tariff elimination and free entry for LDC products have suffered greatly in recent years as more and more developed countries are withdrawing from former commitments to liberal trade policies. Witness the recent (late 1986) withdrawal by the United States of preferential status for products from several developing countries. The new round of GATT trade negotiations will be vital to developing countries as to the exportation of primary commodities.

ARID ZONE EXPORT CROPPING

While the generic case of cash and export cropping might apply to most situations, there are differences among countries and regions. Some unique characteristics of desert agriculture set it apart with respect to economic development and trade. The one overriding characteristic is the dependence of agriculture, in all its aspects, on supplemental water resources, i.e., in addition to natural rainfall. Added to water considerations is the soil factor; desert soils, for example, often contain sodium chloride and a high soil pH which limit plant growth. Finally, climatic conditions characterize desert regions which distinguish the production and potential marketing possibilities for desert products.

In addition to such natural factors, the requisite infrastructure of desert agriculture such as dams, pumps, and canals is unique. Water resource planning is fundamental in every case. While the capital costs that are attendant to such construction and water delivery are different, making debt load and cost of development of desert agriculture quite high, the resulting agricultural production and distribution possibilities are enormous. A few examples are obvious: Egyptian agriculture, the San Joaquin Valley in California, the Arizona-Sonora-Mexican desert, the Niger Valley and so on. Almost year-round production and distribution possibilities place these regions in a competitive position not enjoyed by typical temperate-zone agriculture. Only by overcoming high and additional storage costs for off-season delivery can temperate-zones compete market- wise with the warm deserts.[17,18]

Very little literature specifically treats export cropping in arid zones in contradistinction to that in other climes. We are, of course, well aware of the Israeli case in desert agriculture and market development for such products as citrus, cut-flowers and avocadoes. In the southwestern United States, the success story of the Mexican states of Sinaloa and Sonora is familiar in which some 300-500 million dollars worth of irrigated vegetables are shipped annually to the United States and Canada. In both of these cases, Israel and Mexico, they are grain importers. Why don't they use their resources for the production of staple foods? This question is at the heart of our inquiry.

For comparative purposes, consider some countries in arid zones -- mostly Africa, and some which are designated as "other representative countries". Data are from FAO Production Yearbooks and the research of Von Braun and Kennedy.[6] It can be seen from Table 2 (particularly the last columns) that, with the exception of Egypt, the arid countries do not plant cash crops in nearly the same proportion as the other countries shown. Many conclusions could be drawn but one is notable: cash cropping has not been established in these arid zones. There must be many reasons as to why a country such as Sudan has not produced more for cash and export markets.

Other data were analyzed in order to compare arid and non-arid experience in export cropping. Studying those data to see if there had been differential gains in various countries over the period 1968-82, one can find no outstanding examples where one set of countries show unique experiences. In fact, most countries tend to have progressed toward more staple food crop production, with some such as Niger, Somalia and Sudan showing greater gains than the average. (See column Total Cash Crops as Percent of Basic Food Crops, Table 3. In the table, a decrease over time of total cash crops percentages means a move toward more self sufficiency; i.e., more staple food production.) Like other statistics, where one country stands out there are usually extenuating circumstances to explain the situation.

Table 4 is presented primarily to corroborate the general evidence of the other two tables. The important column here is Growth in Per Capita Food Production Per Year 1968-82. Again, desert countries, with the exception of Jordan, do not stand out as being very different.

Looking simply at the agricultural aspects of domestic growth situations one can, likewise, find little to differentiate desert environments as such. Many LDCs whose principal exports are agricultural products must nevertheless import much of their basic food needs, such as wheat, corn, rice and soybeans. In other words, nations that produce 20 to 50 percent of their gross domestic product (GDP) in the agricultural sector are importing significant amounts of food and are thereby spending scarce foreign exchange on this rather than capital imports that could assist them in economic development. In Table 5, the arid countries are well mixed with others as to both percent of GDP derived from agriculture and the percentage of total imports that are food imports. Egypt stands out in the latter category.

Finally, achieving higher income growth in desert countries will take about the same prescription as that for other developing countries

-- and it is not something that one or two countries, or sets of countries, such as the United States and Europe can handle alone. It will, for example, be important to assure that foreign exchange requirements for debt service do not preclude the importation of agricultural commodities in which those countries have a comparative advantage, do not preclude the investment in and importation of capital goods which are necessary for economic growth, and do not result in pressures to adopt radical import substitution development policies.

In sum, the cash and export cropping problems of desert economics, with the exceptions noted herein, are generic to all developing countries and must be dealt with using a sound set of principles of economics, political economy, and social justice and equity.

TABLE 1: Foreign exchange value of selected crops to developing countries

	1963	1975
	Export Unit Values per Hectare (US$)[a]	
Coffee	320	769
Cocoa	127	436
Tea	1,001	1,197
Tobacco	719	1,250
Sugar	352	1,462[b]
	Import Unit Values per Hectare (US$)[c]	
Wheat	73	210
Rice	83	283
Barley	48	145
Maize	58	169

Source: FAO trade and production yearbooks.

[a] Average export unit value multiplied by average yield in developing market.

[b] $1,462 in 1972-74.

[c] Average import unit value multiplied by average yield in developing market economies.

TABLE 1a. 1983 Export/import values per hectare

	Export Unit Values per Hectare US $
Coffee	1,253.5
Cocoa	520.0
Tea	1,364.2
Tobacco	3,150.0
Sugar	16,450.4
	Import Unit Values per Hectare US $
Wheat	369.0
Rice	1,074.1
Barley	99.0
Maize	313.2

TABLE 2: Area planted in cash crops as share of total crop area by region, country, and income group, 1982

Region/ Country	Sugar	Oil seeds	Fibers	Vegeta- bles	Coffee, Tea, Cacao	Other	Total Cash Crops Percent	Total Cash Crops Area	Total Nonfood Percent	Total Nonfood Area
								(1,000 hectares)		(1,000 hectares)
Typical arid countries:										
Egypt	4.6	3.2	20.2	12.7	0.0	6.6	47.2	1,165	20.2	498
Ethiopia	0.1	1.2	1.0	0.1	5.2	0.1	7.7	1,072	6.2	864
Jordan	0.0	0.0	0.7	7.3	0.0	9.2	17.2	71	0.7	3
Mali	0.1	9.7	5.1	0.1	0.1	0.1	15.2	312	5.1	105
Mauritania	0.0	1.0	0.0	0.4	0.0	1.4	2.8	6	0.0	0
Niger	0.1	5.2	0.1	0.1	0.0	0.0	5.5	201	0.1	3
Somalia	0.8	0.3	1.1	0.0	0.0	1.4	3.6	40	1.1	12
Sudan	0.2	7.7	2.9	0.2	0.0	0.5	11.6	1,439	2.9	363
Syria	0.4	0.3	2.8	4.1	0.0	8.8	16.5	955	2.8	162
Other representative countries:										
Brazil	4.1	12.0	6.0	0.3	3.2	12.4	38.0	28,314	9.1	6,764
Costa Rica	8.0	0.0	1.5	0.7	15.3	40.2	65.6	416	16.7	106
Ghana	0.1	3.6	0.4	2.7	40.1	20.2	67.1	1,858	40.5	1,122
India	1.9	0.3	0.8	0.3	0.4	2.0	19.6	33,250	7.1	12,069
Indonesia	1.3	5.5	1.3	1.3	3.2	24.0	36.5	7.158	4.5	880
Jamaica	16.5	0.7	0.6	2.2	3.4	19.2	42.6	113	4.0	11
Mozambique	1.3	6.7	6.0	0.1	0.7	6.8	21.6	664	6.7	206
Pakistan	4.7	2.5	11.5	0.5	0.0	1.7	20.9	4,234	11.5	2,326
Peru	1.4	0.2	2.6	0.8	4.4	4.5	14.0	491	7.1	248
Turkey	1.4	2.1	2.9	2.2	0.2	10.4	19.2	5,237	3.1	852
Uganda	1.0	2.3	7.8	0.2	3.4	25.0	39.6	2,315	11.2	654
Vietnam	1.6	3.6	0.8	1.0	1.3	6.4	14.9	912	2.1	129
Zaire	0.2	8.2	3.0	0.3	4.2	4.5	20.3	1,297	7.2	459

Source: Food and Agriculture Organization of the United Nations, Production Yearbook, 1982 (Rome: FAO, 1983)

TABLE 3. Changes in the share of cash cropland in total cropland, 1968-82

Region/ County	Sugar	Oil seeds	Fibers	Vegeta- ble	Coffee Tea, Cacao	Other	Total Cash Crops	Nonfood	As Percent of Basic Food Crops: Total Cash Crops	As Percent of Basic Food Crops: Total Nonfood
					(percen/year)					
					Typical arid countries:					
Egypt	5.2	11.8	-0.8	3.8	0.0	4.0	1.7	-0.8	-0.2	-2.7
Ethiopia	0.5	-0.9	-3.1	2.0	1.0	-13.7	-0.7	0.4	0.4	1.4
Jordan	0.0	-5.6	-4.6	-1.9	0.0	1.4	-0.4	-4.6	4.3	-0.0
Mali	8.0	-3.2	1.9	0.0	63.6	-2.4	-1.7	1.9	-3.0	0.6
Mauritania	0.0	0.8	0.0	22.4	0.0	2.7	2.9	2.7	4.6	4.5
Niger	6.3	-9.4	15.0	3.4	0.0	-2.9	-9.3	-15.0	-9.4	-15.1
Somalia	5.8	-12.6	0.4	0.0	0.0	1.4	-0.6	0.4	-2.3	-1.3
Sudan	6.4	6.8	-2.6	2.1	0.0	4.1	2.9	-2.6	-2.1	-7.4
Syria	9.3	4.3	-4.5	6.5	0.0	5.8	2.9	-4.5	-0.4	-7.6
					Other representative countries:					
Brazil	1.7	11.0	-3.5	-3.7	-3.3	2.2	1.9	-3.4	2.3	-3.0
Costa Rica	1.0	0.0	9.7	1.7	-3.6	7.4	2.2	-3.0	2.6	-2.6
Ghana	-4.8	0.9	13.2	2.5	-2.2	7.8	-0.6	-2.1	-0.1	-1.6
India	1.5	1.0	0.2	1.9	1.4	-1.1	0.5	0.2	0.4	0.1
Indonesia	9.6	0.6	0.4	2.5	3.3	-0.7	0.1	2.4	0.5	2.9
Jamaica	-3.0	8.6	-1.4	3.7	-2.5	1.3	-0.8	-2.3	-2.1	-3.6
Mozambique	-0.7	-1.8	-7.8	0.1	0.4	-0.2	-3.5	-7.3	-5.2	-8.9
Pakistan	3.3	-1.5	0.9	5.1	0.0	5.8	1.4	0.9	0.9	0.4
Peru	-3.2	3.8	-4.2	-3.6	0.1	-0.8	-1.8	-2.1	1.2	1.0
Turkey	8.8	3.6	-1.7	2.0	6.8	1.1	1.2	-1.4	1.1	-1.5
Uganda	0.7	-7.4	11.6	3.0	-4.6	2.9	-2.6	-9.3	-0.0	-6.9
Vietnam	20.1	2.7	3.9	1.5	4.9	-1.1	1.6	4.5	0.2	3.0
Zaire	1.5	2.2	-3.1	1.7	1.5	0.2	0.8	-0.3	-0.8	-1.8

Source: Food and Agriculture Organization of the United Nations, Production Yearbook, 1982 (Rome: FAO, 1983)

As reported in TABLE 5. (Reference No. 6)

TABLE 4. Importance of cash cropping in food production and food trade in developing countries

Region/ Country	Cash Cropping: Share of Total Cropland Use, 1982	Cash Cropping: Growth of Share 1968-82	Growth in Per Capita Food Production Per Year, 1968-82	Share of Total Agriculture Exports: Basic Food Imports (1979-82 Averages)	Share of Total Agriculture Exports: Basic Food Exports (1979-82 Averages)
		(percent/year)			
		Typical arid lands:			
Egypt	47.2	1.7	-1.5	2.3	0.0
Ethiopia	7.7	-0.7	0.6	0.1	0.0
Jordan	17.2	-0.4	-9.1	0.8	0.1
Mali	15.2	-1.7	-1.7	0.2	0.0
Mauritania	2.8	2.9	-4.5	0.9	0.0
Niger	5.5	-9.3	1.4	0.5	0.2
Somalia	3.6	-0.6	-1.9	0.6	0.0
Sudan	11.6	2.9	2.2	0.1	0.2
Syria	16.5	2.9	3.3	0.6	0.2
		Other representative countries:			
Brazil	38.0	1.9	-0.2	0.2	0.1
Costa Rica	65.6	2.2	2.1	0.1	0.0
Ghana	67.1	-0.6	-4.2	0.1	0.0
India	19.6	0.5	0.0	0.3	0.1
Indonesia	36.5	0.1	2.1	0.3	0.0
Jamaica	42.6	-0.8	2.6	0.7	0.0
Mozambique	21.6	-3.5	-3.8	0.6	0.0
Pakistan	20.9	1.4	0.8	0.4	0.5
Peru	14.0	-1.8	-2.1	1.1	0.0
Turkey	19.2	1.2	1.4	0.1	0.2
Uganda	39.6	-2.6	-3.2	0.0	0.0
Vietnam	14.9	1.6	0.6	1.8	0.2
Zaire	20.3	0.8	-0.2	0.5	0.0

Source: Food Agriculture Organization of the United Nations, Production Yearbook, 1982 (Rome: FAO, 1983)

As reported in TABLE 7. (Reference No. 6).

TABLE 5. High percentage of agricultural share of GDP and high food importers

	Percent of GDP in Agriculture (1982)	Percent Food of Total Imports (1981)
Bangladesh	47	20
Burkina Faso (U. Volta)	41	25
Liberia	36	22
Sudan	36	19
Niger	31	23
Dominican Republic	27	18
Ivory Coast	26	20
North Yemen	26	28
Jordan	23	17
Togo	23	26
El Salvador	22	17
Senegal	22	28
Nicaragua	21	18
Egypt	20	34

Source: World Bank, *World Development Report 1984*.

REFERENCES

1. Jimmye S. Hillman. 1981. The role of export cropping in less developed countries. American Journal of Agricultural Economics 63: 375-383.

2. Frances M. Lappé, and Joseph Collins, with Cary Fowler. 1977. Food First: Beyond the Myth of Scarcity Boston: Houghton Mifflin Co.

3. Susan George. 1976. How the Other Half Dies: The Real Reasons for World Hunger, New York: Penguin Books.

4. ____1979. Feeding the Few: Corporate Control of Food. Washington, D.C.: Institute for Policy Studies.

5. Frances M. Lappé and Joseph Collins. 1986. World Hunger: Twelve Myths New York: Grove Press, Inc.

6. Joachim Von Braun and Eileen Kennedy. April 1986. Commercialization of subsistence agriculture: income and nutritional effects in developing countries. Working Paper No. 1, Washington, D.C.: International Food Policy Research Institute (IFPRI).

7. Per Pinstrup-Andersen. 1985. The impact of export crop production on human nutrition. In Margaret Biswas and Per Pinstrup-Andersen (eds.) Nutrition and Development, International Food Policy Research Institute (IFPRI), United Nations University, New York: Oxford University Press. Chap. 3: 44-45.

8. Adam Smith. 1937. The Wealth of Nations. ed. Edwin Cannan (ed.) New York: New York. Random House, Inc.

9. David Ricardo. 1911. Ernest Rhys (ed.) In The Principles of Political Economy and Taxation, New York: E.P. Dutton & Co., Inc.

10. James P. O'Hagan. 1976. National self-sufficiency in food. Food Policy 2: 355-61.

11. Timothy Josling. 1980. International trade and world food production. In D. Gale Johnson (ed.) The Politics of Food. Chicago: Chicago Council on Foreign Relations: 36-59.

12. D. Gale Johnson. (ed.) 1980. The Politics of Food. Chicago: Chicago Council on Foreign Relations.

13. G. Edward Schuh. 1978. Approaches to 'basic needs' and to 'equity' that distort incentives in agriculture. In Theodore W. Schultze (ed.) Distortions of Agriculture Incentives. Bloomington: Indiana University Press: 307-27.

14. MASIGROUP. June 4, 1980. Assessment of the agricultural inputs supply system and policies in Egypt. Report submitted to the Agency for International Development.

15. Foreign Agricultural Service (FAS). March 14, 1977. U.S. Department of Agriculture, Foreign Agriculture 15: 10.

16. Steven K. Pollard, and Douglas H. Graham. Undated. Price policy and agricultural export performance in Jamaica. Ohio State University Paper, ESO No. 953.

17. Karima Korayem. April 1982. Distributing disposable income and the impact of eliminating food subsidies in Egypt. Cairo Papers in Social Science, Vol. 5, Mono. 2, American University in Cairo.

18. Earl L. Sullivan (ed.) September 1984. Impact of development assistance on Egypt. Cairo Papers in Social Science, Vol. 7, Mono. 3, American University in Cairo.

19. E. Boyd Wennegren, et. al. 1986. Solving World Hunger: The U.S. Stake. Cabin John, MD.: Seven Locks Press.

PUBLIC POLICY IMPLICATIONS OF AGRICULTURAL MARKET PERFORMANCE UNDER DROUGHT CONDITIONS

MARK SPEECE
Central Washington University, Ellensburg, Washington

THOMAS E. GILLARD-BYERS
Department of Agricultural Economics, Washington State University, Pullman, Washington

BAKHEIT AZRAG
Central Washington University, Ellensburg, Washington, U.S.A.

Abstract The Western Sudan Agricultural Research Project (WSARP)*, Washington State University, has been active in Kordofan for the past five years. In addition to technical work, the project has carried out much socio-economic research and has collected a wealth of data on markets and marketing in Kordofan. During the 1983 to 1985 cropping seasons, Kordofan suffered severe droughts, and we were able to observe how this drought influenced market performance, as well as the results of government marketing policy during the adverse conditions. The impact of drought upon agricultural production is well known, and Kordofan's food production certainly declined drastically, contributing to food shortages in the province. Drought impact upon market performance and structure is not usually as well understood, but the food shortages in Kordofan were partially the result of drought-induced marketing failures. Annual price fluctuations and price volatility are marketing constraints characteristic of

* Supported by the Government of Sudan (Agricultural Research Corporation), the U.S. Agency for International Development, and the World Bank.

subsistence agriculture in arid regions under normal conditions. They generally place consumers and small producers at a disadvantage to start with. Drought conditions greatly increase the seasonal price differential and the price volatility. Because subsistence farmers are primarily risk avoiders rather than profit maximizers, the increased uncertainty in prices leads to holding product off the market, even if the producer has it available. High market risk also induces small merchants to stay local rather than move product, so that the market does not move food very efficiently from surplus areas to shortage areas. In addition to these influences upon market performance, drought also causes a shift in many marketing channel functions. In particular, functions such as financing, assembly/bulking, transportation, storage, and risk-bearing tend to shift to producers and small operators. These people become more responsible for moving product at the same time that they assume more risk, but they are precisely the people in the marketing channels least able to finance additional functional responsibility. Again, the result is that product is often held off the market.

* * *

A significant proportion of agriculturally productive land in the developing world lies in arid and semi-arid regions. Agricultural systems in these areas must contend with harsh environmental conditions, including recurrent droughts. The impact of drought upon agricultural production is fairly well known, and the famines which are often triggered by drought have come to world attention in the last two decades. The impact of drought upon market performance and structure is not as well understood. Market failures, however, may seriously contribute to food shortages during famine. A recent popular article explicitly notes this, and observes that people often starve even when abundant food is available in the market.[1] The food is simply too expensive for most people to afford.

Policy makers cannot ignore the lesson that such situations can provide. Typical responses to market failure have usually been to step in and impose government control on the private market. Such policies have rarely been successful in correcting market failures, however, because they discourage rather than encourage market activity. The mechanisms behind the poor performance of the market must be analyzed, and policy must be formulated to allow the market to function more efficiently under drought conditions. Behind the high prices lie some kind of supply problem; policy must encourage the flow of product to the market. The production deficit during drought is usually substantial enough, without misguided policies which often hinder rather than improve performance of drought-stressed markets.[2]

Data which allows preliminary examination of these issues was gathered in Kordofan, Sudan, as part of the Western Sudan Agricultural Research Project. WSARP operates a socioeconomic section at the Kadugli research station in South Kordofan. It has gathered extensive data on the organization of productive activities in general, and specifically on how markets function in the region. Life in Kordofan is intimately tied to the environmental conditions under which the inhabitants must live. The north is largely desert, while in the south conditions improve enough that South Kordofan can be classified as merely semi-arid. As in other such dry regions, life is precarious, even under conditions which are regarded as "good" by the inhabitants.

There is almost no surface water to serve as an irrigation source; crops depend on rainfall from the summer rainy season. Only one crop per season can be raised in Kordofan. This one crop must provide food for the entire year until the new crop is harvested in the next season. In good years, the inhabitants call the several months before the new harvest the "hunger period." In years of drought, when many crops fail, the hunger period lengthens and many inhabitants face starvation. Several of the five years in which WSARP operated in Kordofan were drought years, and therefore the project was able to gain a first-hand view of production and marketing under drought conditions.

Crop production in Kordofan certainly fell drastically, which led to serious food shortages in the province. Western Sudan, including Kordofan, has traditionally been self-sufficient in sorghum, the main staple food. By 1984, however, it was a net food-deficit area.[3] Market performance and structure were also affected by drought conditions. For example, the impact of seasonal price differentials and of price volatility was greatly increased. Overall, prices were high in 1984/1985, but the seasonal price differential was not greater than in normal years. Post-harvest prices, however, did not show the usual sharp drop in 1984/1985. The patterns of price volatility were even more important. Price fluctuations on the order of 100 percent within a few days were not at all uncommon for most agricultural commodities, regardless of whether they were produced in Kordofan or in the less affected Nile provinces. Volatility increased as prices increased throughout the season. This has serious implications for the movement of product to market, because high price volatility greatly increases the risk on market transactions.

Subsistence farmers in Kordofan are risk avoiders rather than profit maximizers. They prefer enterprises with low yield and low risk to activities which may be more profitable but carry higher risk. Even if prices are high generally, and even if they have surplus in storage, such small producers may not sell. Prices may drop by the time they move the commodity to the market, and they would either have to sell at the low price or carry the commodity back to their village. Small shopkeepers and traders are very risk conscious as well. They regard themselves as farmers first, with only supplementary commercial activities, and they have the same mentality as other farmers. Even some slightly larger scale merchants who regard trading as their primary occupation recognize the high price risk and choose not to move product. The high prices characteristic of production shortfalls during drought may be attractive to large merchants, who are more nearly profit maximizers. Small operators, however, do not respond very well to high prices if the high prices bring increased risk.

Channel structure also adapted to drought conditions in ways which inhibited the flow of agricultural product. In any marketing channel, certain functions must be performed to move goods from producer to consumer. One important set of functions typically involves such things as buying/selling, which entails identifying, locating, and negotiating with parties interested in selling/buying products. A second main set has to do with the actual physical movement of product. Agricultural commodities must be assembled, transported, stored, graded/sorted, and broken down into lot sizes which customers/consumers are willing to buy. A third major set might be classified as facilitating functions. Financing/credit provisions must be made, risk must be assumed on transactions, and market information must be available. Typical channels are made up of various levels of wholesalers and retailers, who assume responsibility for performance of the functions. Performance of any function entails a certain cost, and there must be an associated compensation if the function is to be performed at all.

Drought conditions changed the cost and compensation relationships of channel functions, with the result that functions were shifted to the lower ends of the channels. Large operators restricted their offering of credit-financing. Shopkeepers, then, in addition to facing greater price risk, had to extend more credit, further increasing their risk exposure. Village level shopkeepers also became more responsible for functions associated with the physical handling of products, such as buying/selling, assembly, and storage. Farmers were less willing to take their crop to distant markets themselves and face price risk, so the shopkeepers had to purchase a relatively greater share of the marketed crop. (Of course, in absolute terms, there is less marketed crop in drought years.) Responsibility for the transportation function shifted to them, because fewer merchants came to villages from the larger markets when the prospect of a good buy on crops was diminished.

Smaller operators are much less able to finance additional functions in the first place. Furthermore, the increased responsibility during the drought did not bring increased compensation to shopkeepers. Analysis of price series data showed average markups of 20 percent or less on sorghum sales by small merchants in Kadugli area villages in 1985. WSARP data from 1980 shows a markup of 40 percent on sorghum by small merchants. Rather than taking advantage of the drought situation, a charge which is often leveled at merchants, it appears that they cut their markups. With diminished markups, but increased functional responsibility, one can safely conclude that profits suffered. The profit margins documented in 1985 were fairly low, on the order of a few percent. The fact that small operators do not receive increased compensation in return for shouldering additional costs makes them reluctant to take on these additional functions during drought. As a result, product flow is impeded.

Policy makers must address such issues, rather than simply impose inflexible constraints (such as price controls) which only further discourage market activity. Price differentials and rapidly fluctuating prices can be partially traced to physical infrastructure deficiencies. Another key factor behind them is uncertainty. There is uncertainty about production levels, about the proportion of product which will reach the market, about the climate's impact on product movement, and about the ability of the physical infrastructure to move product. These problems become even more severe when climatic shocks impose changes in channel structure on the marketing system. During the drought, small merchants, shopkeepers, and producers need stable prices more than ever. They need access to technical and marketing expertise to cope with the expanded scope of their operations. They require better access to financing, and they should receive additional compensation for their increased functional responsibility. Policies which can deliver these aspects could greatly ease the marketing constraints associated with drought and facilitate the movement of agricultural commodities from areas of relative surplus to areas of relative deficit.

Many of the requirements in marketing are beyond the scope of the private sector to provide. Some government role is certainly necessary to solve physical infrastructure deficiencies, for example. Improved roads and storage facilities are greatly needed to allow lower level channel members to stay in the market during bad years. Large operators can deal with the high cost of poor roads and inadequate storage by spreading the cost over a large volume. When functions are shifted to small operators, they must either take losses or raise prices to maintain their margins. A network of all-weather roads which are easily passable in the rainy season would contribute significantly to a reduction in the erratic market behavior during drought. Such projects must concentrate on feeder roads and a regional network, rather than on trunk roads to connect major Kordofan cities with Khartoum. Road development which builds trunk roads only has contributed to the decline of lower level market centers. When local and regional trade declines because of fewer markets, this reduces the options available to small operators, and thereby increases the risk they must face.

Storage constraints reduce the ability of mid-level channel members to overcome the predictable transportation problems. Storage is not a major problem to small farmers who utilize traditional methods and sustain quite low losses. Traditional methods are not adequate for large volumes, however, and mechanized farmers and larger merchants face considerable storage loss. Crops held for the entire year commonly lose fifty percent or more of their volume. A merchant in Kordofan who holds grain for six to nine months must see prices rise at least 72 percent to break even. Storage loss, then, coupled with high transportation costs which coincide with the "hunger period", contributes substantially to the very high prices faced by consumers in the summer months. Small operators do not need to move product to avoid storage loss, while larger operators must cover losses with higher prices.

Storage facilities must also be constructed mainly by the public sector. There are few merchants who specialize in sorting grain; most prefer to rent storage space, usually from the government. Merchants feel that they lack expertise to manage storage facilities, and they are reluctant to invest in them. Provincial governments should be granted ownership of government-financed facilities, so that storage policy remains closely attuned to the needs of local markets. Management contracts should be worked out so that private business operates the facilities for the provincial government. Business expertise is readily available in the private sector, but is in short supply in government. Facilities should be run on a profit-making basis, and management contracts should be granted and renewed based on acceptable profit performance. Over time, such a plan will increase private sector experience with storage management, and will eventually lead to increased private sector investment in storage facilities.

Charges for use of the modern storage facilities should be similar to the current cost structure which grain merchants face. The facility must remain profitable because Sudan, especially the provincial government, simply does not have the financial resources to operate yet another subsidy program. In addition, cumulative storage losses currently limit the ability of merchants to hold grain off the market during shortage periods, because losses may rise faster than prices. With better facilities, losses will be lower, and fees must be correspondingly higher to discourage hoarding. Modern facilities will also moderate the seasonal price differential, because more of the stored grain will actually be available instead of lost.

Farmers need mechanisms which eliminate the need to sell at harvest to pay debts or to resupply the family's stocks. They would then be able to hold their crop until prices had risen. Selling later would also shift the sale to a time when more family labor was available. At harvest time, labor demands are high. The farmer must utilize channels within the village which offer lower return because there is no free time to assume more marketing functions. With more time, farmers could move their own crops in many cases, and sell in alternative higher return channels. Such an adjustment in temporal sales patterns would reduce immediate post-harvest supply of grain and increase availability later on, thereby reducing the seasonal price differential.

Such programs will probably have to come through some sort of village-level credit cooperative, since large institutions will not loan to small farmers. Such co-ops could grant short-term consumer loans which would hold farmers until a few months after harvest when prices had stabilized at higher levels. They could also grant production investment loans which do not come due at harvest. Local loan decisions and community pressure on borrowers greatly decreases the default rate, and loan cost will be lower than for large institutions. As with storage, however, credit should not be subsidized, but should reflect the cost of operating a credit co-op. Such co-ops need income generating activities in order to raise loan capital. WSARP has had considerable success at introducing village co-op owned ox-carts into several villages around Kadugli. Ox-carts meet a critical need for transportation capable of carrying relatively large loads. They free labor during labor shortage periods, and they help small farmers bypass very expensive local hauling methods (up to several £S per sack for relatively short distances). The village co-ops have generated considerable income for village projects through these ox-carts.

High price differentials and volatility would be further moderated through a floor and ceiling price set by the provincial government. The floor should not be an administered price imposed upon private traders, but rather a trigger price at which the government would step in to buy grain for a regional stockpile. It should be determined at harvest time and should be set only slightly above the prevailing market price. Neither should the ceiling be fixed or imposed upon private merchants. It should be set only slightly below prevailing prices during the hunger period. Sales would be made out of the stockpile accumulated during the harvest. The objective is to moderate the seasonal price differential, rather than to eliminate it, and to accumulate modest stockpiles for emergencies (drought years).

The program must be self supporting. Realistic prices would insure that a modest stockpile could be accumulated, but the government would not be forced to buy major portions of the harvest because the private market would continue to operate. Subsidized prices which drove private merchants out of the business would suggest that the government program would be operating at a loss. As noted above, provincial governments do not have sufficient resources to finance subsidized programs. Nor would consumers be very well served if the government held a virtual monopoly over channels for a particular crop. Prices must be set so that merchants who compete with government channels remain profitable, or no grain will be stored by the private marketing channels. However, competition from the government will force merchants to reduce costs to remain profitable. In particular, there will be greater incentive to improve storage capabilities, which contribute the major cost involved in taking advantage of temporal price fluctuations.

It should be clear from the preceding discussion that marketing considerations play a critical role in the food shortage problems in drought affected areas. Technical advances in such areas as drought-resistant crops will not by themselves be entirely successful in solving the problems, because they address only production issues. Price performance and market structure also play key roles in determining food availability. This discussion of Kordofan can serve as a case study on the integration of marketing into the analysis of what is happening in subsistence agricultural areas which come under drought stress. For those regions which are quite similar to Kordofan in climate, production systems, and marketing structures, much of the analysis and many of the recommendations will be directly applicable. The greater value of this study, however, is probably the presentation of a framework on how to approach some of the key issues. Kordofan and many other arid and semi-arid areas have considerable agricultural potential. Policies which improve the ability of the marketing system to cope with drought will aid greatly in realizing that potential.

REFERENCES

1. T.A. Bass. 1986. Famine: an early warning System. Science Digest (May): 57-62; 82-84.

2. M. Speece and T.E. Gillard-Byers. (Forthcoming). Government market intervention in Kordofan, Sudan. Northeast African Studies.

3. B.C. D'Silva. 1985. Sudan: Policy reforms and prospects for agricultural recovery after the drought (ERS Staff Report No. AGES850909). Washington, D.C.: International Economics Division, Economic Research Service, U.S. Dept. of Agriculture.

AGRICULTURAL MARKETING IN SOUTH TAHRIR AREA*

NABIL T. HABASHY
Ministry of Agriculture, Agricultural Economic Institute, Egypt,

Abstract A study of marketing in the desert communities of South Tahrir in Egypt is discussed. Included are the market structure; marketing performance of different commodities, especially horticultural crops, peanuts, and livestock products; margins and marketing costs.

The main constraints to marketing from the farmer's point of view are poor roads; high transportation costs; unsuitable handling which causes post-harvest losses; absence of collective power for bargaining; lack of information about prices prevailing in wholesale and retail markets; and bureaucracy faced in dealing with public sector companies. Solutions for overcoming these difficulties are suggested.

INTRODUCTION

Evidence indicated that many problems occur in the marketing of agricultural products in South Tahrir. No marketing studies had been conducted in this area; accordingly this study aimed to cover the following:

1. To determine the prevailing marketing pattern in the area.

2. To estimate marketing costs in relation to production costs and margins for farmer, wholesaler, and retailer.

---

* The research was done by the author for the Desert Development Center, American University in Cairo, and was sponsored by the Ford Foundation.

3. To identify specific marketing problems facing the farmers in the area.

4. To make suggestions for improvement of marketing in the South Tahrir area.

The study also aimed to gain some information about socio-economic factors such as the role of women in marketing and the kind and nature of relationships between the farmer and middlemen.

The study focussed on the marketing of peanuts, horticultural crops and milk. Because of limited time and financing, this study gives an overview of marketing of these crops in the area. It should be followed by further in-depth studies to tackle specific marketing problems appearing in the area.

Interviews based on a structured questionnaire were conducted to fulfill the objectives of the study. In view of the limited number of observations possible in some cases, the verbal approach was used. Some simple statistical analyses were applied when feasible.

SAMPLE DESCRIPTION

A random sample was drawn from a list frame of producers of different crops in two regions of the area, South Tahrir and El Tahadi. Smallholders and graduates are presented whenever possible; some wholesalers and retailers were interviewed. Agricultural companies in the area were also investigated to identify marketing problems and attitudes toward solutions. The sample was composed of 68 farmers who grow peanuts, almost 7% of the peanut producers in the two regions. For horticultural crops, a sample of 63 farmers was randomly selected of which 38 were vegetable producers and 25 were citrus producers. The sample includes both graduates and smallholders in the two regions.

For livestock, a list of livestock owners which focussed on milk producers was consulted and a sample of 17 farmers was selected (about 7% of the milk producers). The number of producers in the sample includes seven smallholders in the southern area and ten graduates in Tahadi region. One cooperative society owning milk cattle was investigated, namely Salah El Din farm. Wholesalers and retailers were chosen by convenience because of the absence of a list frame.

PEANUT MARKETING

Marketing of peanuts is subject to regulations of the government cooperative system. The farmer is required to submit a certain quota of peanuts to the government at a fixed price which is lower than the market price. The government price for the last season of 1986 was L.E. 40 per ardab while the market price ranged between L.E. 50-55 per ardab.

DISTRIBUTION BETWEEN DIFFERENT USES

1. The quantity marketed in the two regions constituted about 69% of the total production; the quantity marketed in the South Tahrir (ST) region represented 64.5% while it was 77.6% in the Tahadi (T) region. The quantity stored for later sale was about 12% of total production. It was 16.5% in (S) region and 3.9% in (T) region. The quantity used for home consumption or gifts was about 10% of the total production, the percentages being almost the same in the two regions. The rest of the crop which is almost 9% is kept as seed. The price increases gained as a result of storing was 20% to 37% more than the government price.

2. Although the obligatory quota to be delivered is about 60% of the crop, some farmers deliver more than the quota required which implies that those farmers prefer to market through the cooperative system. This was especially true

for smallholders in (ST) region. Seventy-nine percent indicated that access to marketing is a main reason for selecting the crop. Only 13% of the answers indicated that net revenue is a main criterion for selecting peanuts.

MARKETING COSTS

Marketing costs for the farmers include labor needed for harvesting, sorting and packing, and transportation costs either to the collecting centers of the village bank or to the wholesaler. In addition to commissions (fees paid to the cooperative society), total marketing costs per ardab for (ST) region ranged between L.E. 9-14 and for (T) region L.E. 6-10 per ardab. The difference may be attributed to variations in the costs of the sorting operations, packing and transportation costs. These may differ according to the distance of the farm from the collecting center or from wholesaler. Eighty percent of the sample used hired labor for marketing operations. About 1.16 man days per ardab were hired labor while .42 man days were from family labor. The relationship between marketing costs for ardab of peanuts and quantity marketed show a highly significant relationship which implies the existence of economics of scale in costs of marketing operations as a result of the increase in the quantities marketed:

$$Y = \underset{(6.43)}{21.4 - 2X} \qquad R^2 = .58$$

Y is the quantity marketed per ardab and X is the marketing costs in L.E. per ardab.

MARKETING OF HORTICULTURAL CROPS IN SOUTH TAHRIR

For most vegetable crops, particularly green peas, the farmer in the Tahrir area sells his crop on the field either to local traders or wholesalers from Rod El Farag Market of Cairo. However, a small number of farmers market their crops by themselves either by shipping it to some wholesalers in Badr district or to main wholesale markets in Rod El Farag in Cairo or to El Nozha Market in Alexandria. The price for vegetables is subject to bargaining relative to the prices prevailing in the wholesale market after deduction of marketing costs and commissions. When sales take place on the farm, which is the most common method, the farmer does not incur transportation costs, commissions or packaging costs. His main marketing operations are harvesting, sorting and packing. The middleman will provide him with the packaging needed.

A very small number of farmers sell their vegetables, particularly green peas, to exporters for higher prices and incur higher marketing costs. However, those who sell to exporters will enjoy a net margin that is higher by at least 150% than that gained from the domestic market.

A limited number of graduates sell their crop at Rod El Farag Market at about L.E. 400 per ton and pay marketing costs of about L.E. 80 per ton. This implies a significantly higher margin for those who sell their crop at Rod El Farag. However, because of the oligopolistic situation prevailing in this market, the farmer has to compete with local traders who have gained favored positions in Rod El Farag. Thus, farmers cannot sell easily in this market and may be obliged to sell to local traders. In addition, losses result through the marketing channel as a result of an unsuitable handling system.

COSTS AND MARGINS

Marketing costs and production costs for vegetable crops were higher for smallholders (SH) than for graduates (G). Graduates with larger areas realized marketing costs per ardab that were about 40% less than those of smallholders.

The gross margin for green peas was L.E. 172.4 per ton for (G) and L.E. 131 for (SH). The gross margin for onions was L.E. 30 because of government intervention which required a quota for lower prices. Gross margins were L.E. 207 and 88 per ton for cucumber and watermelon respectively.

Regarding the choice of growing green peas, 58% of the sample of (SH) stated that net revenue is the main reason for their choice. Thirty-one percent attributed their choice to access to marketing. For graduates, 58% of the sample stated that access to marketing is the main reason for their choice while 37% stated that net profit was their reason.

The relationship between marketing costs as a dependent variable and quantities marketed as an independent variable showed economies of scale. The relationship for smallholders was significant. It was:

$$Y = 99.49 - \underset{(4.58)}{2.5} X \qquad R^2 = .6$$

While for graduates it was

$$Y = 9.77 - \underset{(2.09)}{.53} X \qquad R^2 = .4$$

The average gross margin for a green pea producer who devotes his crop to different marketing channels is about L.E. 172 (G) and L.E. 131 (SH) per ton. The net margin for wholesaler and retailer was L.E. 20 and L.E. 80 per ton respectively. Almost the same relationship between margins occurs for cucumbers. However, for onions and watermelon, margins were L.E. 30, L.E. 21, and L.E. 71 per ton for farmer, wholesaler and retailer respectively. It was L.E. 88, L.E. 36, and L.E. 56 per ton for watermelon. Breaking down the consumer pound showed a share for the farmer of 42% of the retail price for (G) and 39% for (SH) for green peas. It amounts to 28%, 13.9%, and 18.9% for cucumbers, onions and watermelon respectively.

The nature of the relationship between the farmer and the middleman appears to be a good permanent relationship in 62% of the cases. Financing from middlemen is limited for vegetable producers.

MARKETING PATTERN OF FRUITS

Our investigation of fruit crops in the area shows that citrus is the main crop.

The prevailing marketing pattern for fruits is to sell the crop on the trees (Kelala system). In this case the farmer does not pay any marketing costs, not even harvesting. However, he is responsible for taking care of agricultural and pest or weed control operations in the field. The farmer follows this method of marketing his crop to avoid any risk, save marketing costs, and get rid of post-harvest losses due to handling even though he will not realize the same margin that he could get if he markets by himself.

Another method adopted in marketing is to sell to the trader at the farmgate after the farmer has harvested, sorted and packed the crop. Marketing costs in this case amount to about L.E. 29 per ton. In a third method, farmers transport their crops to Badr district where a middleman sells it for the farmer at a commission of about 7% of the selling price. He pays for all marketing costs in this case.

PRICE, MARKETING COSTS, AND MARGINS

The average farmgate price of oranges last season (1985/86) was L.E. 344 for smallholders selling Navel Oranges while it amounted to L.E. 300 per ton for graduates because they sold at a different time during the season. The farmgate price of Baladi Oranges amounted to L.E. 100 per ton. Marketing costs per ton were almost the same, L.E. 23.8 and 22% per ton for graduates and smallholders, respectively. In contrast, marketing costs of the main operations of harvesting, sorting and packing constituted only about 60% of the total marketing costs for graduates but constituted 100% for smallholders. This implies that some graduates transport their crop to other markets but it seems that they did not realize higher prices than (SH) who did not incur any marketing costs except those mentioned earlier. This verifies that graduates face an oligopolistic situation in these markets. The graduate farmers' margin for Navel Oranges was L.E. 277 while the margin for (SH) was only L.E. 175. This difference can be attributed mainly to the difference in cost of production which amounted to L.E. 67 for (G) and L.E. 146 per ton for (SH). The relationship between marketing costs as a dependent variable and quantities marketed as an independent variable showed a significant negative relationship in the case of mandarine as follows:

$$Y = 66.29 - \underset{(2.177)}{1.81} X \qquad R^2 = .38$$

In the cases of lemon and orange, the relationship was also negative but insignificant.

MARGINS OF DIFFERENT DEALERS IN THE MARKETING CHANNEL

Margins of the farmer, wholesaler and retailer differ according to the method of sale and many other factors such as location, quality of the produce, etc. In the case of oranges, the margins of the three categories were L.E. 209, 30, and 81 per ton, respectively.

In the case of mandarine, the margins were L.E. 212, 24.6, and 127.8 per ton for farmer, wholesaler and retailer respectively for (G) while it amounted to L.E. 268, 30.8, and 88.3 for the same categories for (SH). About 25% of the members of the sample are financed through middlemen of the El Wadi Company for Exporting. Farmers in the area receive about 25% of the crop price in advance. In breaking down the consumer pound, it was found that the farmer's margin constitutes about 44.7% for (G) and 32.8% for (SH) of the retail price. For mandarine, the two percentages were 47.8% and 45.4%. For Baladi Lemon, the share for the farmer is 39.3% of the consumer pound.

Women have a significant role in the marketing of citrus. About 55% of the sample stated the women participate in sorting, packing and carrying fruits from inside the farm to the outside.

PROBLEMS FACING MARKETING OF HORTICULTURAL CROPS

It seems that the relatively small quantities produced by each farmer, particularly (SH), represent a constraint toward realizing marketing efficiency, either technological efficiency reflected in the physical marketing operations such as sorting and packing, or allocative efficiency reflected by the prices obtained. Marketing problems in the area can be summarized as follows:

1. Farmers prefer to sell their crops at the farmgate with minimum marketing operations completed. Accordingly, they get a lower margin than they could realize if they sold their crops in the wholesale markets or to exporters.

2. Bad road conditions accompanied by high transportation costs and post-harvest losses due to unsuitable handling methods represent disincentives to marketing outside the area.

3. Small farmers are exploited by traders and middlemen because of the absence of collective power for bargaining; there is no link between them.

4. Lack of financial capability is a constraint affecting ability to perform marketing operations of a good quality and at the proper time.

5. Farmers lack information and a communication system about prices prevailing in the wholesale

and retail markets or about export prices.

6. Dealings with governmental companies such as Al Ahram Company is characterized by bureaucracy and delay in receiving their payments.

PROBLEMS OF DEALING WITH KAHA COMPANY

1. The farmers stated that the company deals in quantities that are larger than each can deliver. The company also expects farmers to deliver their crops to the company. For the farmers, this means high transportation costs in addition to prices that are lower than market prices and, furthermore, the company lowers their grades.

2. Other farmers stated ignorance about contracting with the company and the procedures required. In addition, no representatives from Kaha come to the area to deal with them.

3. The company prefers to deal with certain traders who are familiar with its procedures and methods of dealing with bureaucracy, and who can provide large quantities.

MARKETING OF MILK

MARKETING MARGINS

Graduates produce more milk for commercial purposes than do the smallholders. The cost of production of one kg milk by the producers and some companies owning milk cows, is estimated at 25.35 piasters. The farmers sell their milk at about 40-45 piasters/kg. Costs for milking and feeding operations are estimated by the company owning the milking herd in Tahadi at about 5 piasters/kg whereas farmers in this region estimate these costs to be about 3.2 piasters/kg. The farmers in the south area smallholders and most have family labor and do not hire labor for milking or feeding operations.

The total cost of producing milk in the Tahadi Region appears to be between 32-39 piasters per kg and in the southern region 25-30 piasters per kg (SH). Thus, the margin per kg is about 5-8 piasters in Tahadi Region and 9.3 piasters in the southern region. If margins for farmers, wholesalers and retailers are compared, the margin for wholesaler will be about 25-30 piasters per kg with a net margin of about 20-25 piasters/kg. The retailer's price will be about L.E. 1 per kg which means that the retailer's margin is about 30 piasters/kg.

The low margin that the farmer realizes compared to the margins of middlemen indicates that the farmer has a serious problem in marketing fresh milk. The producer's share of the consumer pound is about 10%.

CONSTRAINTS FACING MARKETING OF FRESH MILK

The low price for the producer relative to the retailer is primarily due to (1) the long distance between the producing region and the consuming region, mainly the big cities of Cairo and Alexandria, (2) the lack of appropriate transportation facilities equipped with cooling systems, and (3) poor road conditions.

The sample also revealed lack of bargaining power for milk producers. The average number of cattle per farmer is 3 in one region and about 4.8 in the other region. The productivity of milk per head ranges between 5.7 to 6.4 kg of milk daily or an average of about 6 kg per head daily. The quantity sold as fresh milk is around 52% of total production which implies that only about 3-3.3 kg of milk per head is marketed daily. Multiplying by the average number of milk cows owned by the farmer in each region, it appears that about 10-16 kg of fresh milk will be marketed per family in the two regions. Such a small quantity can only be marketed either in the farmer's house or in the village market, particularly since there are no improved processing facilities.

The supply of milk in each region is greater than the demand because most of the people living in the area have some milking cows which provide them with their own milk requirements.

RECOMMENDATIONS

The small quantities marketed by each farmer, either in horticultural crops or milk, represent a constraint toward realizing marketing efficiency. Lower margins are obtained as a result of higher marketing costs and prices are relatively lower because of the lack of collective systems or power. Marketing systems in the new land should be established on the basis of marketing large quantities. The following suggestions could improve the existing marketing system:

1. One strong cooperative society for horticultural producers and another for milk producers should be established to market farmers' products. Each society should have a collecting center in a suitable location in order to minimize transportation costs and have access to highways. The two centers would be run by a board of interested producers of horticultural crops and of milk. One collection center would be equipped with a station for sorting, grading and packing horticultural crops. The other center should be equipped with refrigerators and provided with transportation that has cooling facilities.

2. Soft loans with low interest rates should be available for the two cooperative societies to purchase the assets needed.

3. A system should be set for each center to receive farmers' products and grade them.

4. Different marketing channels for the cooperatives should be investigated. They must have some distribution channels in big cities

which serve as direct retailing outlets and eliminate retailers' margins. In this respect, there must be some precontracts with exporting companies and processing companies.

5. Communication systems should be established in the area to get price information from different markets on a regular basis and to provide information about procedures for dealing with different companies.

6. In the short run, the government should stimulate the Kaha Company to obtain their needs from the existing cooperative society.

COMPARATIVE ECONOMIC ANALYSIS OF DESERT FARMING SYSTEMS IN EGYPT*

K. EL-KHESHEN and M. NASR
Desert Development Center,
American University in Cairo
Cairo, Egypt

Abstract Several farming systems (classification is based on economic considerations such as input/output relations, efficiency of input use) exist in the deserts of Egypt. An economic assessment of their performance is conducted with the aim of arriving at a comparative analysis of performance indicators with particular reference to the production process, input use, and returns to factors of production. The private and national costs and benefits associated with each farming system are calculated. The function of the systems is evaluated through the degree of goal attainment (public policy objectives: food security - input substitution - export promotion - income generation - maximum production). No one farming system was found to satisfy all objectives.

INTRODUCTION

Egypt is facing economic problems expressed in the inadequacy of available resources to produce food for her growing population. Deserts represent about 96% of the total area in Egypt. The food production potential of the deserts could consequently lead to growth of the national economy through food production, marginal resource utilization, export promotion and import substitution.

* DDC Projects Financed by IDRC, Ford Foundation and AID/Supreme Council of Universities.

Several farming systems (FS) exist in Egypt. In addition, the American University's Desert Development Center (DDC) has also developed some FS on its South Tahrir site. Furthermore, two FS typically used by smallholders and graduates were modified by the DDC.

It is worth mentioning that the current desert agricultural systems adopted in Egypt are either plant production systems, animal husbandry systems or a combination of both. The different plant production systems differ with respect to type of crops cultivated such as fodder crops, cash crops, legumes, fruit trees or a combination of several systems. Livestock products may be included in these.

This paper aims to compare different desert FS in Egypt in an attempt to define basic elements underlying the selection of the most feasible application in the desert from an economic point of view. In other words, the discussion is based on making an optimum choice with respect to efficiency in the utilization of the available resources in the desert such as irrigation water, labor, capital, management and technical know-how; in addition to considering the available natural resources and the preservation of the ecological balance in the desert agricultural climate. (The DDC's approach in formulating FS is based on consideration of these factors).

ANALYSIS AND RESULTS

This paper does not present a discussion of definitions of FS and farming systems research (FSR) since the literature on the subject is continuously growing. However, for analytical purposes, the definitions adopted are herewith presented so as to put the discussion into perspective. By now it has become universally accepted that farming is a system composed of interactions between natural resources, and agronomic, cultural, human, economic and technical variables. Thus, the emphasis in FSR is on interdisciplinary relations (multi-dimensional) rather than on single dimension aspects such as technology, agronomy or economics alone. Any classification of an interdisciplinary system can never sharply delineate where one example ends and the other starts, because of the multitude of variables and subvariables. The only practical approach is one which classifies the systems at hand according to the purpose in mind. The FS studied in this paper are of the sequential mixed output type subjected to dynamic socio-economic environments due to: (i) changes in demand for their products and/or (ii) shifts in resource impacts. They all have high crop/stock ratios, desert irrigation systems and a dry climate setting.

FSR categories in general as defined by Simmonds are:

a- FSR sensu stricto: FS as they exist.

b- FSR through on-farm research from FS perspective (FSR/ OFR/FSP): which is an attempt at altering existing FS to become more efficient performers.

c- New Farming Systems Development (NFSD): which attempts to create new designs of FS.

The examples which will be discussed according to the above classification are:

a- FSR sensu stricto:
- smallholders' farms (3-5 feddans)
- graduates' farms (20-30 feddans)

b- FSR.OFR/FSP:
- smallholders farms participating in the DDC's on-farm demonstration (OFD) activities.

- graduates' farms participating in DDC's OFD activities.

c- NFSD:

- DDC's FS (maxi models 1, 2 and 3)

Table 1 presents a comparison of 10 farming systems. The two NFSD perform best in terms of net returns per feddan, for both animal production alone and for combined animal/plant production. For the two established FS groups (i.e. smallholders 1, 2 and 3 and Graduates 1 and 2) the systems with a cash crop component are higher in net returns than systems based on cash crops alone or on a combination of fodder, livestock or fodder/field crops without cash crops. However, when viewed in terms of percentage of net returns/total costs, we find that ranking highest are graduate 1, graduate 2, smallholder 3, smallholder 1, and then smallholder 2. In general, a combination of livestock and cash crop components in the FS always improves the financial rewards. The same view is confirmed by the figures of the maxi farms where the least return and NR/TC ratio are in the fodder/cattle system, without cash crops.

In terms of efficiency of input use, only labor and water are considered as they are probably the two factor inputs to which the financial and economic standing of FS are most sensitive. Thus, the most efficient in terms of labor use and water use is the farming system OFD/GR1.

CONTRIBUTION OF DESERT FARMING SYSTEMS (DFS) TO THE NATIONAL ECONOMY

Table 2 presents a comparison of the desert farming systems in terms of system components and contribution to national economic goals. Examination of the table reveals:

TABLE 1: Comparison of budgets for different desert farming systems in Egypt

Farming System	Rotation	TVC	TFC	TC	TR	Ret. above TVC	1 NR Plant Prod.	2 NR Animal Prod.	1 + 2 NR Total	Ret/ man day	Ret/ m^3 Water	NR/ TC
Maxi 1 Fodder/cattle		203	175	378	352	150	-25	150	80	4.8	0.020	21%
Maxi 2 Fodder/veg./an. crops/ sheep		292	170	462	505	213	43	75	118	8.6	0.031	26%
Maxi 3 Fodder/field crops/ legumes/cattle, sheep		273	174	447	550	277	103	27	130	10.8	0.053	29%
S H 1 Fodder/cashcrop/cattle, sheep	W1/SL	515	150	665	720	205	55	152	207	2.4	0.033	31%
2 Cash crop	WC/SC	434	150	584	694	260	110		110	5.2	0.044	19%
3 Fodder/field crop	WL/SC	330	150	480	681	350	200		200	9.5	0.052	42%
G R 1 Fodder/cash crop/cattle, sheep	WL/SL	428	150	578	906	478	328	32	360	9.3	0.078	62%
2 Fodder/cattle, sheep	WL/S fallow	178	75	253	360	182	107	32	139	6.9	0.072	55%
OFD SH SH 1	WL/SL	830	150	980	1,303	473	323		323	6.8	0.065	33%
GR GR 1	WL/SL	647	150	797	1,290	643	493		493	10.5	0.099	62%

Source: (8), (9), (10)

<u>Note</u>: from TABLE 1

TVC = Total variable costs
TFC = Total fixed costs
TC = Total costs = TVC + TFC
TR = Total revenue
NR = Net revenue
NR plant prod. = TR = TC
Ret. ab VC = Returns above variable costs = TR - TVC
Ret/man day = Returns above variable costs/man day
Ret m^3 water = Returns above variable costs / cubic meter of irrigation water

Maxi = Research demonstration project sponsored by U.S. AID / Supreme Council of Universities and implemented on DDC sites

SH = Smallholders - Farmer settled in new land and allocated 3-5 acres
GR = Graduate - University or technical school graduate allocated about 20 acres in the new land.
OFD - On-Farm Demonstration conducted by DDC on farmers land

WL = Winter Legume
SL = Summer Legume
WC = Winter Cereal
SC = Summer Cereal

TABLE 2: Contribution of desert farming systems to the Egyptian National economy

Desert Farming System	Components of Farming Systems	Contribution to Macro-Economic Objectives of the Country: (1) Value of Agriculture Production — Value (L.E. million) (= plant + animal prod.)	% to the National Level
National level (Old Land)	Field crops (53%) + Fodder (30%) + Fruits (5%) + Vegetables (12%) + Animal Units 5.7 million	4822.5 (3616 + 1205.6)	100
Total area reclaimed	Field crops (49%) Fodder (35%) Fruits (85%) Vegetables (7.5%) + Animal Units 0.17 million	96.4 (72.3 + 24.1)	2
Reclaimed Western Delta	Field crops (39%) Fodder (24%) Fruits (29%) Vegetables (8%) + Animal Units 0.19 million	20.2 (15.2 + 5.0)	0.4
MAXI - Farming System No. 1	Fodder (100%) + Cattle (14 cows)	16.3*	0.3

TABLE 2: (Cont.)

Desert Farming System	Components of Farming Systems	Contribution to Macro-Economic Objectives of the Country: (1) Value of Agriculture Production: Value (L.E. million) (= plant + animal prod.)	% to the National Level
MAXI - Farming System No. 2	Fodder (50%) Field crops (30%) Vegetables (20%) Sheep (64%)	19.5*	0.4
MAXI - Farming System No. 3	Fodder (35%) (2 cows) Field Crops (35%) + 12 sheep Medicinal plants (30%)	21.5*	0.4
Smallholders - Farming Systems	Fodder (50%) Cash Crops (50%) Cattle (3%) Sheep (2%)	26.3*	0.5
Graduate - Farming Systems	Fodder (50%) Cash Crops (50%) Cattle (5%) Sheep (4%)	27.4*	0.6
On-Farm Demonstration Farming Systems	(for smallholders) Fodder (-) Field Crops (50%) Vegetables (50%)	28.8*	0.6
	(for Graduate) Field Crops (50%) Vegetables (50%)	29.6*	0.6

TABLE 2: (Continued)

Desert Farming Systems	Contribution to Macro-Economic Objectives of the Country						
	(2) Export Promotion	(3) Labor Utilization		(4) Food Security			
	% of Agri. Export of the total export	Labor Force		Degree of Self-Sufficiency**			
		% of total Agri.	1000	Wheat	Rice	Maize	Sugar
National Level (Old land)	16.3	100	4200	26	100	81	65
Total Area Reclaimed	0.51	32	1380	0.60	9.10	0.70	2.00
Reclaimed Western Delta	0.13	7	290	0.13	1.95	0.15	0.43
MAXI - Farming System No. 1	0.00*	4*	168*	0.00	0.00	0.08	0.00
MAXI - Farming System No. 2	0.02*	3.2*	134*	0.00	0.00	0.00	0.00
MAXI - Farming System No. 3	0.05*	3.1*	130*	0.00	0.00	0.00	0.00
Smallholders - Farming Systems	0.07*	6.6*	277*	0.01	0.00	0.01	0.00
Graduate - Farming Systems	0.08*	4.5*	189*	0.01	0.00	0.01	0.00
On-Farm Demonstration Farming Systems	0.08*	3.3*	139*	0.00	0.00	0.00	0.00

Source: (2), (3), (4), (5), (6), (7)
*Assuming the respective farming system is applied to the Entire Western Delta.
**Degree of Self-Sufficiency = $\frac{\text{Domestic Production}}{\text{Total Consumption}}$

1. None of the DFS contribute directly to food security since they do not include traditional food crops such as wheat, rice and maize, except in limited cases of small areas on some smallholders and graduate farms.

2. Most DFS include cash crops which result in higher private profits while not always contributing to national agricultural economic objectives.

3. The maxi DFS concentrate mainly on producing fodder crops required for dairy animals. This objective provides an indirect contribution to the macro objective of realising higher food security levels especially in animal products (Maxi 1 and 2). Maxi 3 contributes to exports of medicinal and aromatic plants.

4. The West Delta region is expected to contribute a higher proportion of agricultural exports as compared to other regions with different DFS.

5. The smallholder and graduate FS utilize labor intensive technologies which are more suited to private desert farms.

6. The OFD, graduate, smallholder and Maxi 3 contribute to raising agricultural production by 24, 27, 26 and 22% respectively. These percentages are higher than the current West Delta contribution.

CONCLUSIONS

From an economic viewpoint, FS may also be categorized as either labor or capital intensive, or land or water intensive (greenhouses are land saving). Thus, economists utilize the efficiency of input use in their classifications.

The results of this study show that labor intensive systems are more suited to Egyptian conditions due to the abundance of agricultural labor. The smallholder and graduate FS are two such systems. Some FS were found to contribute to national economic objectives (OFD, SH and Gr contribute to export promotion and the MAXI contribute indirectly to food security) while no single FS was found to contribute to all objectives simultaneously.

REFERENCES

1. Norman W. Simmonds. 1985. Farming systems research. World Bank Technical Paper No. 43, Washington D.C.

2. M. Nasr. 1976. "An analytical study of some agricultural economic development enterprises in the Arab Republic of Egypt" M.Sc. Thesis, Faculty of Agriculture, Ain Shams University, Egypt.

3. Ministry of Agriculture. 1986. Organization of development and land use of reclaimed land. Unpublished Data. Cairo.

4. Ministry of Agriculture and Reclaimed Land. 1986. Estimation of agriculture production for the new lands. Unpublished Data, Cairo.

5. CAPMAS, Central Authority for Public Mobilization and Statistics. 1982. Statistical Yearbook, Cairo. (in Arabic).

6. CAPMAS. 1981. Annual Report for Land Reclamation in Egypt, No. 41/502, Cairo.

7. H. Alderman, J. Von Braun, and S. Saker. 1982. Egypt's food subsidy and rationing system: a description. Research Report No. 34, International Food Policy Research Institute, Washington.

8. Desert Development Center, American University in Cairo. March 1985. Desert community aspects in AUC desert development system; phase one: initiation of the research unit, Cairo.

9. Desert Development Center, American University in Cairo. May 1986. Desert community aspects in AUC desert development systems, phase one, second progress report. Cairo.

10. Alexandria University, The American University in Cairo. December 1985. Applied research, demonstration and training in desert development.

MEANS AND OPPORTUNITIES FOR EFFECTIVE MULTI-NATIONAL AGRO-URBAN VENTURES IN THE EGYPTIAN DESERT

MAHMOUD AHMED ABDEL-LATIF
Department of Architecture,
Faculty of Engineering,
Assiut University, Assiut, Egypt

Abstract Traditional practices which rely totally on technologies, self-help, and financing solutions from indigenous sources, though useful under certain conditions, show little success when applied in desert areas. Alternatives, which depend heavily on imported technologies and foreign capital and personnel, appear to be more effective in completing the task, yet they entail a higher risk of negative impact due to underdevelopment of the socio-economic and political fiber of societies. Desert areas in developing countries, as in Egypt, can be developed more effectively using a proposed approach which gives fair consideration to domestic as well as foreign inputs to the development process. In the proposed approach, desert national development goals are better achieved through a larger multi-national venture development process which manifests itself in the establishment of permanent agro-urban desert settlements. Experts are invited to study the proposed approach and to accumulate knowledge toward a more elaborated formula of international cooperation to achieve local and national goals through multi-national development processes, in Egypt and elsewhere.

THE GREAT DEVELOPMENTAL CHALLENGE

During the next twenty years Egypt will have to face a great developmental challenge. Simply stated, the country must build the equivalent of another country in the desert, under complex, interrelated, and seemingly unfavorable, but possibly supportive, local, national and international societal contexts.

NEW DESERT SETTLEMENTS

New desert settlements will have to be built to accommodate an expected population increase of 25 million persons. In addition, in order to upgrade the deteriorating living conditions in the overcrowded urban and rural centers of the Nile Valley, 10 million persons must migrate to the new settlements. These 35 million citizens will be housed in 7,000,000 residential units which will be distributed over 80 new cities (200,000 to 1,000,000 residents each), some 2,200 villages, and 8,000 smaller settlements.[1]

Only 4% of Egypt's land is nonarid. The rest is desert with varying degrees of adversity, largely due to a hostile climate and the lack of accessible potable water. Almost all of the nonarid land is utilized as valuable agricultural land that can not be converted to other purposes or to urban activities. Therefore, desert sites must be selected for further settlements.

To maintain a reasonable level of permanency, the new settlements must be established in locations which are physically accessible and economically feasible.[2] Many sites in the Egyptian deserts, especially peripheral and coastal ones are easily accessible by land, air, and possibly sea. However, physical accessibility is still a major obstacle to desert development -- largely due to the heavy expenditure required in building new local, regional, and national transportation and communication networks.

Economic feasibility, which is a precondition for self-sufficiency and permanency of desert settlements, means that a dependable energy source must be found.[3] Energy will help find or bring potable water and support necessary agricultural and industrial production and other economic activities and services needed to sustain the life of the settlements. Traditional and new energies are available in Egyptian deserts, yet again, it is a question of enough capital and technology.

In order to become socially attractive, desert settlements must provide their residents with adequate protection against the hostile desert environment by maintaining a suitable livable physical structure. The settlement must provide adequate infrastructure including housing, roads, transportation facilities, public utilities, medical facilities, training centers, and recreational facilities. Planning and constructing such infrastructure requires sound financing to support procurement and management of capable manpower and equipment.

MAINTAINING FOOD SECURITY

While building new settlements in the desert, the country must also find ways to solve the problem of food shortage and to provide additional food for the population increase which is predicted. Even with rich countries, food self-sufficiency is a desirable goal. In Egypt, food security is achievable through agricultural expansion, horizontally and vertically. The country has a total of 5.9 million feddans of agricultural land producing only 40% of the required food supplies. The remaining portion is imported.[4] Reclamation of 6.5 million feddans during the next 20 years, or about 325,000 feddans a year, is required for self-sufficiency in food.

Programs will have to be initiated to increase the production of other food types such as meat and fish. As an example of what is needed, let us consider one item. Egypt now consumes about 450,000 tons of red meat per year with an average of 9 kg per person. The country produces 360,000 tons of the amount and 110,000 tons are imported -- a 35% deficit. If this level of consumption is to be maintained during the next 20 years, the country will need an additional 315,000 tons. Thus a total of 425,000 tons will be required to meet the demand for red meat.[5] Animal husbandry and poultry production are dependent on agricultural production to provide them with basic requirements of feed. Therefore, various food security activities must be incorporated in the future desert development process. Advanced food production technological systems such as the "controlled environment food-production systems" can provide the desert settlements with daily food requirements and possibly maintain enough surplus for outside marketing.[6] However, the success of these systems requires adequate planning and financing.

THE COST OF THE DEVELOPMENT PROCESS

It is not within the scope of this paper to provide an exact cost estimate of the described Egyptian development process; approximate figures will be enough to serve our purpose. Generally speaking, the cost of such development is the sum of three items: the cost of establishing future desert settlements, the cost of land reclamation, and the cost of supporting the existing population.

Table 1 summarizes population and cost estimates of five Egyptian new desert towns under development. The figures give a rough range of L.E. 3,000 to L.E. 4,000 (in 1985 price levels) as a development cost per resident.[7] Thus the new desert settlements to be established during the next 20 years are expected to cost L.E. 105 billion to L.E. 140 billion. One feddan of newly reclaimed land costs L.E. 4,000.[8] Thus, the total 6.5 million feddans to be reclaimed during the next 20 years will require L.E. 26 billion. To support the existing population during the development period, the country must provide an additional L.E. 160 billion.[9] Therefore, the overall expenditure required for the development of new settlements, including land reclamation, and maintaining the existing population during the next 20 years, is a sum ranging between L.E. 291 billion to L.E. 326 billion, roughly an expenditure of L.E. 15 billion to L.E. 20 billion a year.

TABLE 1. New community projected population and cost in the year 2000 (L.E. Million)

Community	Base Year	Target Population	Total Cost (Year 2000)
10th of Ramadan City	1983	500,000	1500.0
Sadat City	1985	500,000	1500.0
New Ameriyah City	1985	390,000	1170.0
6th of October City	1984	250,000	1050.0
15th of May City	1985	150,000	600.0
TOTAL		1,590,000	5820.0

Source: The National Urban Policy Study (NUPS), PADCO Inc. with Engineering Consultant Group, Cairo, Jan., 1982.

THE ECONOMIC SITUATION OF EGYPT

Since the 1970's, Egypt has been in difficult economic straits a situation typical of that of many developing countries. Over three decades, foreign assets have gradually decreased, leaving the country in a vulnerable economic situation and suppressing her development process. A country which once loaned L.E. 440 million to Great Britain during World War II, is required now to pay heavy foreign debts.[10] Egypt's foreign debts increased dramatically in the last 15 years from $1.4 billion in 1971 to $18.6 billion in 1982 and finally to $40 billion in 1986. Annual debts have reached an alarming rate equal to 70% of the country's income from exports.[11]

Discussion of some of the reasons beyond this situation may give insight as to how to deal with it more effectively. Many internal and external factors can be cited. A major internal deferrent is the centralized bureaucratic government and political structure of the country. Although Egypt changed from a monarchy into a republican government in modern times, the structure has been incapable of dealing effectively with problems associated with rapid economic, social, and physical transformation processes. The country has had ambitious goals for industrialization, land reclamation, desert settlements, social enhancement, and political strength, but inadequate planning and management of the country's development potential have resulted in using valuable resources to achieve low priority goals. Traditional practices have been inadequate in bringing about the transformation within the desired time frame.

A key external factor is the dominating world-system that has undermined Egypt's developmental potential and disoriented her resources while insuring the realization of the economic and political goals of the more advanced world, both Western and Eastern. This dominating world-system plays a major role, perhaps indirectly, in accelerating regional conflicts in the Middle East, through various forms of military aid and assistance, weaponry sales, and even through some direct intervention.[12] Major wars have resulted in heavy military and civil expenditures before and after each conflict. Eventually, the situation reached the condition described above.

MAJOR DEVELOPMENTAL QUESTIONS

It is obvious that at this time, and for many years, Egypt will not be able to self-finance projects and activities incorporated in the described development process. Yet, can the country afford not to do it, or even to delay the process any longer on the basis that enough and long-lasting capital is not available? Problems will only get worse and costs will increase. Because the country realizes that she is standing in the midst of the "must-do-it" stage and has already decided to proceed with the process, some way must be found to finance the unavoidable undertaking.

Since capital is only one variable in complex multi-variant process, a much wiser search should be directed toward finding a more comprehensive formula that addresses the wider issues of planning, financing, implementing, and even managing the whole process. More specifically, Egypt's developmental question is simply: how can the country plan a capital-intensive, large-development process with limited self-financing capabilities and under apparently adverse physical and social conditions on the local and national levels, but within an interactive, possibly more responsible and supportive, world-system.

THE SEARCH FOR A NEW NONTRADITIONAL APPROACH TO DEVELOPMENT

With limited resources, especially financial resources, people, and countries as well, tend to become more careful in allocating their money. Certainly, with limited resources, Egypt cannot afford a wrong start which will lead to the establishment of a few uninhabited desert towns or to the reclamation of a few thousand acres of low-productivity land while losing her precious supply of existing agricultural land. To continue spending the country's resources to expand overpopulated centers provides only a temporary and partial solution to housing problems but not to the entire developmental spectrum. Under all conditions, repeated acceptance of "easy-take, easy-spend" mismanaged big loans must not be allowed before a proper and comprehensive development approach is determined. This new approach to development must be different from what has been applied in Egypt for decades.

REVIEW OF TRADITIONAL DEVELOPMENT PRACTICES

During the last 30 years, with some exceptions, development has been carried out under what may be called a "Traditional Development" approach (TD). The TD approach uses simple methods, fairly basic equipment, limited energy, intensive labor, and low levels of capital investment. It is claimed that the TD approach is suitable for societies with large populations, poor economics, and low technical capabilities. The facts, however, reveal different conclusions.[13] The traditional development approach, after decades of application in Egypt, resulted in considerably low economic returns, brought marginal contributions to the overall developmental status of the country, and supported only a few small-scale desert settlements. In sum, the TD approach has not been able to effectively use the country's resources to better realize her goals, solve her problems, and satisfy her needs. This conclusion can be substantiated by reviewing the deficiencies of the TD approach in two areas: in land reclamation and in the establishment of desert settlements during the last thirty years.

LAND RECLAMATION UNDER THE TD APPROACH

In 1940, Egypt had a population of 15.7 million persons and agricultural land amounting to 5.7 million feddans. In 1986, although the population had reached 50 million persons, agricultural land had increased to only 5.9 million feddans. The outcome of land reclamation during that period was 1,000,000 feddans of low productivity land. During the same period, urban sprawl managed to swallow 800,000 feddans of Egypt's best arable lands. Thus, out of the reclaimed 1,000,000 feddans, there is a gain of only 200,000 feddans.[14] To further illustrate the point, traditional development practices applied in the New Valley Project in the Western Desert have resulted in the reclamation of 42,250 feddans from potential arable lands which equal 6.7 million feddans.[15] The New Valley Project started in the early 1960's.

DESERT SETTLEMENTS UNDER THE TD APPROACH

During the last thirty years, the establishment of new desert settlements falls into two distinct periods. Before the mid 1970's, desert settlements were built under the typical TD approach. The whole outcome of this period was a few small-scale, conventional settlements serving the limited land reclamation sites. Some of the settlements, as in the Al-Tahrir Province Project, were located on the fringe of existing urban areas. The project was a great undertaking at its time, yet the use of the traditional practices and techniques in agriculture and in other developmental sectors has produced much less than the potential of the project. Since 1961, the New Valley Project has attracted only 111,633 residents (1984 statistics). The potential for settlement in the New Valley is claimed to be about 50 million citizens.[16]

Since the mid 1970's, the country has employed new desert development policies aimed at the construction of new cities upon desert sites for the propose of conserving arable land and accelerating the process of decentralization of large urban population centers, mainly Cairo and Alexandria.[17] Plans have been prepared for several new desert towns. By the year 2000, six major cities are expected to accommodate 1,590,000 residents (see Table 1). A second generation of new desert cities is underway following a more dispersed regional approach.

The development approach used for the first generation of desert towns has been improved in terms of planning and design functions, but apparent problems in implementation strategies and management mechanisms remain. Worldwide planning teams have participated in producing comprehensive plans suitable for the given socio-economic and political inputs. However, the plans have not accounted for means and opportunities to secure large capital investments required over time. Only governmental financing and local private investment and industrial companies were invited. Very few joint ventures have been introduced with the exception of the privately supported community of Al Amal. Therefore, as local public and private investment capabilities have weakened due to the cited economic difficulties, development of the economic and social infrastructures of the new towns has slowed down.

Due to mistakes in selecting potential residents, lands of new towns were subdivided and then sold to citizens, many of whom did not have the motivation to be a resident in a desert town, especially during the initial construction stages. A recent survey indicates that hundreds of apartments have been locked in the new towns.[18] In Al Salihiah City, there are 2500 residential units but only 200 units are occupied. In the 10th of Ramadan City, out of 7396 tracts of land sold to citizens, only 1780 tracts are under construction in a slow conventional way. Locked apartments comprise 40% of the total apartment supply of the city.

In sum, even if the recent new towns reach their expected population total of two to three million residents, the pace of development will not be enough to provide shelter for the rest of the 35 million potential residents. A much more capable and efficient development approach is demanded.

A NONTRADITIONAL DEVELOPMENT APPROACH

Based on the above analysis, it is the contention of this paper that Egypt can effectively realize her developmental challenge only within a much larger multi-national process which is based on a new nontraditional enlightened approach to resource utilization, problem solving, need satisfaction, and goal realization.

CONCEPTUALIZATION OF EGYPT'S DEVELOPMENT PROCESS

The main logic of the approach is found in the conceptualization and methodology of a system which recognizes the interrelationships that tie the components of a system together with the components of other systems that are involved in the development process of the original.[19] Any system, let it be the whole universe, a country, a region, a city, or even a firm has two basic parts: the social and the physical environments. Every system evolves and changes over time in what may be called the "system's developmental process". This developmental process has four major components: the resources, the problems, the needs, and the goals prevailing in the system. As time passes, these four components interact together and with the components of other smaller and larger systems. At the end of any developmental period, sets of new components come into existence.

Figure 1 conceptualizes the meaning of the country's development process under this systems approach. As illustrated, the country's developmental process is a multi-dimensional continual process of societal change from an initial state of society -- a certain input of resources, problems, needs, and goals -- to a later state of society -- an output of new resources, new problems, new needs, new goals. The process contains several regional and sectoral development processes. Every component and step of the process is both internally and externally interacting with the components and steps of other processes taking place within or outside the country. The interaction extends outside the developmental period covering the past, present, and future conditions of the systems involved.[20]

The application of the above thinking to the development of Egypt requires that boundaries of the country's development process be extended outward as far as is required to determine which interrelationships with other countries and systems are significant and which are not. Comprehensive and systematized thinking will, then, most likely help create better grounds for a more effective and efficient use of internal and external, local and regional, and national and international resources to achieve: (1) better solution of problems of involved systems, (2) higher satisfaction of their needs, and (3) greater realization of their goals.

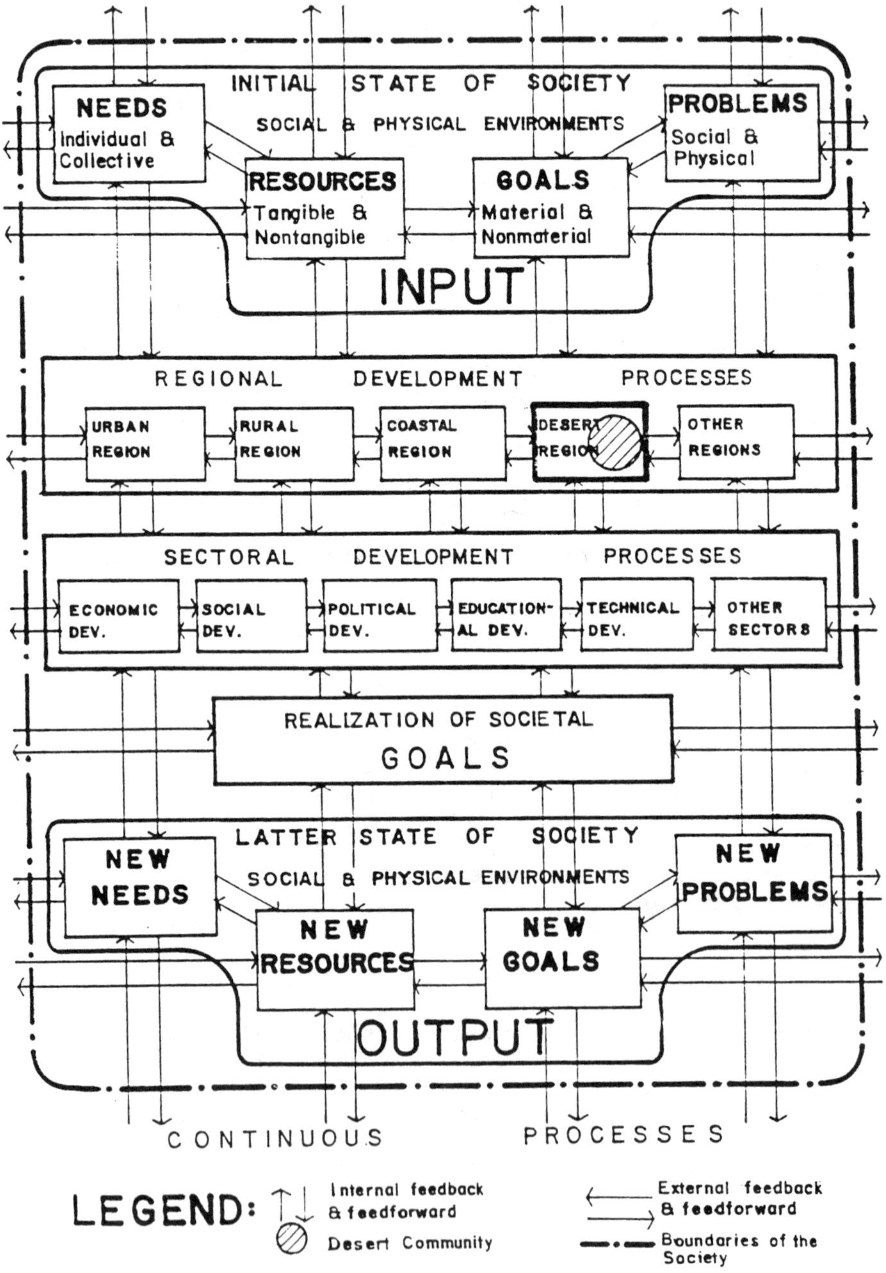

FIGURE 1: Conceptualization of development as a multi-dimensional, continual process.

A NONTRADITIONAL MODEL OF INTERNATIONAL DEVELOPMENT

The interactive function existing in the global system of today allows even conflicting countries to maintain a certain amount of understanding during certain periods of history. This phenomenon occurs out of necessity when countries see that such an understanding helps attain mutual benefits or prevent mutual losses. As conflict reduces between two countries, the area of understanding enlarges, and the chances become greater for fruitful cooperation and exchange of resources to achieve mutual goals. This model suggests that if an enlightened and sincere search is done to find areas of mutual interests, countries will, depending on the areas of understanding they can or try to offer, start serious joint and multi-national development ventures resulting in achievements which are even larger than those originally defined. It seems that this is the only approach to solving many global problems including Egypt's developmental challenge described above. Historically speaking, Europe and Japan, for instance, owe most of their recent advances and strengths to joining investment and assistance programs initiated after World War II with participation of the United States Government and large American industrial corporations.

A PROPOSED DEVELOPMENT PLAN

The nature of the development process of Egypt and the contexts described call for a fairly systematized comprehensive development plan. An attempt is made here to outline a plan which could serve as a base for further investigation. Experts are invited to study the proposal and to accumulate knowledge toward a more elaborated formula of international cooperation to achieve local and national goals through multi-national development processes, in Egypt or elsewhere. The proposed plan consists of four major stages: system definition, system integration, system implementation, and system evaluation and control. Figure 2 simplifies the structure, stages, and steps of the proposed plan.

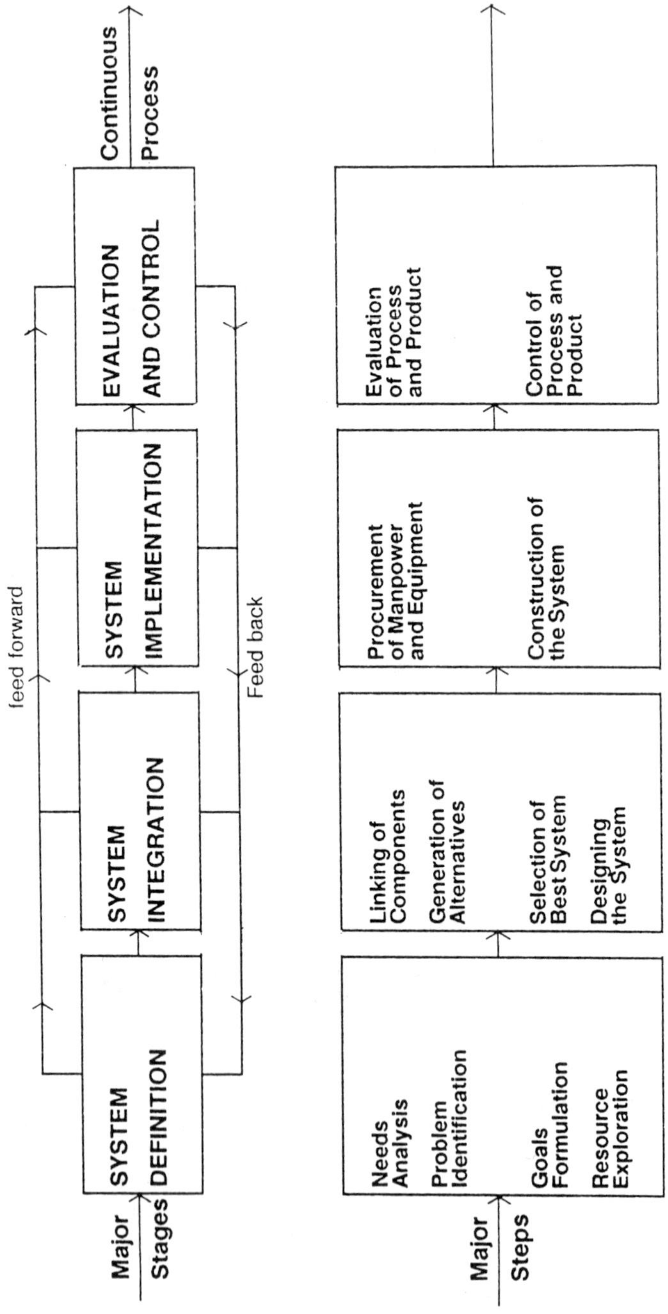

FIGURE 2: The structure of the proposed development plan

In the system definition stage, a worldwide search is conducted to identify potential countries and corporations which would be interested in participating in Egypt's multi-national development process. Egypt's principal trading partners: the United States, the Federal Republic of Germany, France, Japan, Saudi Arabia, and other resource-rich Arab countries are obvious participants. Large industrial corporations, financing institutions of the world, national and local private and public investment companies, and others are to be included.

Prevailing resources, problems, needs, and goals of the participating systems are defined thoroughly. Areas of mutual interests are identified. The Egyptian authority and other participants are informed of the analysis on a constant basis to provide better opportunities for more active participation and involvement. The stage must conclude with a preliminary view of what is there and how to use it. Specific statistics are then articulated to provide: (a) tentative numbers of agro-urban settlements and other development programs; (b) alternative locations and sites and priority of each location; (c) availability of natural resources; (d) types of industrial, agricultural and other economic and service investment potentials; (e) technological levels and methods for their transfer and assimilation; (f) cost of the overall undertaking and of individual programs and projects; (g) sources of finance and methods of investment per particular development project; and (h) a timetable for the overall process and for individual activities.

In the system integration stage, synthesis of the identified resources, problems, needs, and goals of the involved countries and institutions is made so as to reach an integrated formula which serves the interests of participants. During this stage, socio-economic and physical planing scenarios are generated including alternatives for location, size, type, cost, financing, and other characteristics of new settlements. The stage concludes with a selection of the best development scenario for implementation. It would be a good idea at this stage, based on the priority analysis, to designate a location for a first multi- national agro-urban desert settlement to be built as a prototypical model settlement. An elaborated plan would be prepared for the model settlement and the stage which should follow rapidly toward its implementation.

The comprehensive plan of the model settlement may be accomplished over a period of two years resulting in: (a) specific plans and programs based on particularized analysis of development objectives; (b) identification of major tasks to be performed together with capital, labor, energy, and other resource requirements; (c) schedules for each task; (d) organization of the required administrative and technical infrastructures for the entire program and for each task; and (e) evidence of available financial and other resources based on appropriate feasibility studies supported by tentative commitments of key development participants. Acceptance of the findings and recommendations of this stage would be negotiated with all countries and agencies concerned so that the first implementation element could be initiated at an early date.

Implementation of the overall national development process and of the model settlement will eventually begin. In order to override the shortcomings of traditional government infrastructure, it is advisable to create a special commission with an independent budget and autonomous administration powers. The commission will be coordinating the development of various projects and settlements.[21] Each new settlement should have a management agency comprised of teams of local and international experts to carry out various implementation activities.

Implementation of the overall process and of the model settlement and other projects entails several discrete steps, including: (a) establishing a management infrastructure; (b) negotiating binding financial commitments on an international basis; (c) arranging for final site selection and conveyance of available public domain property from the Egyptian government to the project authorities; (d) prioritized construction of major elements of the urban infrastructure for transportation and utilities; and (e) initiating phased development of publicly and privately financed agricultural areas, industrial districts and commercial centers, residential neighborhoods, community facilities, schools, hospitals, etc. Concurrently with development and construction, the Egyptian Government would establish the administrative and political infrastructure for local government.

During the various stages of the development process and at the end of the development period, evaluation and control activities are performed. The purpose of evaluation and control is to make sure that the plans and physical and social structures being implemented are at the highest level of adherence to the identified system components. The model settlement idea will serve the purpose of experimenting with new development techniques, thus allowing better evaluation and control measures to be developed and applied to the remaining projects.

CONCLUSION

The cost of implementing the ambitious development goals of Egypt will run into billions of dollars over two to three decades. The cost of not implementing the process will be a disastrous situation for a nation and, eventually, for other nations as well. A successful development program in the Egyptian desert could, on the other hand, demonstrate an enlightened international approach of understanding and cooperation toward solving the problems of overpopulation, agricultural deficiencies and food shortage, and poverty. Such a process should be in everyone's interest, but particularly in the interest of trading partners and traditional friends of Egypt.

REFERENCES

1. Salah Zaytoon, "Obstacles Affecting Egypt's Development Triangle", Al-Ahram Iktasadi, No. 928, January 1987, pp. 30-32.

2. Max Weber, The City, (Free Press, Glencoe, 1958).

3. Abdelmohsen M. Farahat, "Energy Environment, and New Communities in Hot-Arid Areas of the Middle East", (Ph.D. dissertation, Virginia Polytechnic Institute and State University, Blacksburg, 1980).

4. Salah Namiq, "The march into the desert to increase the agricultural production," Al-Ahram, September 26, 1986.

5. Isam Shabanah, "A seminar on ways to increase the country's meat production," Al-Ahram, October 10, 1986.

6. M. Jensen, Proceedings of the International Symposium on Controlled Environment Agriculture, University of Arizona Tucson, 1977.

7. PACDO Inc. with. Engineering Consultant Group, The national urban policy study (NUPS), Cairo, Jan., 1982, pp. III 45a III.45.b.

8. Yussef Wally, "New decisions to overcome obstacles against desert agriculture," Al-Ahram, January 15, 1987.

9. Based on figures of annual expenditure during the 1982-1987 and the proposed 1987-1992 five-year National Development Plans included in a presidential address given on May 1, 1986.

10. Osman Gheat, "The debts of rich countries," Al-Ahram, January 1, 1987.

11. Ahmed Baha El-Deen, Al-Ahram, January 14, 1987.

12. W. Lewis, <u>The Evolution of the International Economic Order</u> (Princeton University Press, Princeton, 1987).

13. For detailed review of literature and evidence see: Mahmoud Abdel-Latif, "Modern Desert Settlements: Requisites, Resources, and High Technology-- A Model Process of Strategic Planning," (Ph.D. Dissertation, Texas A & M University, College Station, Texas, 1985).

14. See reference 1.

15. Ministry of Information, The New Valley Province, October 3, 1985.

16. See reference 1, p. 31.

17. S. Sherrer, "New towns and satellite cities: a strategy for deconcentration & urban development," NUPS working paper, Annex 1, October, 1981.

18. Mamdouh Al-Walli, "In the New Towns: Apartments are Locked Without Residents, <u>Al-Ahram</u>, October 19, 1986.

19. M.B. Kline and L.M. Lifson, <u>Applying the Systems Approach to Development</u>, (Dowan, Hutchinson & Ross, Inc., Stroudsburg, Pennsylvania, 1974).

20. For more detailed analysis see reference 13 pp. 112-125.

21. The role played by the Royal Commission for Jubail and Yanbu in coordinating and implementing the Saudi Arabian two industrial desert cities has considerably contributed to the success of the Saudi projects, see: Mahmoud A. Abdel-Latif, "Settling the Desert with Advanced Industrial Technology: Two Recent Saudi Arabian Settlement Projects," <u>Ekistics</u> 311, March/April, 1985, pp. 162-175.

POPULATION, DEVELOPMENT, AND COMMUNITY OF SOLUTION

SAAD Z. NAGI
Sociology Department, Ohio State University, Columbus, Ohio

Abstract This paper examines the interrelations of population concerns, societal development, and forms of human communities. It identifies the nature of population problems from the viewpoint of national policies and the concerns and aspirations of populations in regard to their living conditions. It outlines varying perspectives on development and where population problems and aspirations fit. These two lines of analysis are brought together in terms of their significance to human settlements. The concept "community of solution" is introduced and some of its important implications are considered.

INTRODUCTION

The titles given this session and the conference at large suggest three broad concepts around which this paper is organized: population, development, and community. The attempt here is conceptual rather than empirical. The purpose is to further a common framework sensitized to the diversity of perspectives on these matters. The context of this conference, and conditions in Egypt, clearly suggest a sequence of relations among these concepts. Whether in response or in anticipation, population problems and concerns are pressing for accelerated rates of development. Addressing these problems requires the establishment of new settlements and communities. These in turn call for expanding development into the deserts which constitute large and very thinly inhabited areas. The following discussion is guided by this sequence of thoughts.

POPULATION PROBLEMS AND CONCERNS

From the viewpoint of national policies, the demographic structure and dynamics of a society present a number of complex and difficult problems. Unfortunately, most, if not all, of these problems are represented in Egypt to a varying extent. Let us examine some of the prominent features:

First, and most important, is the sheer size and rate of population growth in relation to current and potential resources. Warnings about population pressure were sounded hundreds of years ago by early analysts exemplified by Malthus' pessimistic projections.[1] These were to the effect that the rates of increase in population will be dramatically faster than those of expansion in resources, and that the ensuing imbalance would eventually lead to mass starvation. The theory of demographic transition provided explanations for three stages of population change: (1) high fertility and high mortality resulting in low rates of growth, (2) high fertility with lowered mortality leading to accelerated growth, and (3) low fertility and low mortality combining for low rates of growth.[2] While this transition promises the hope of an eventual stable low rate of expansion in the third stage, it raises serious concerns about unbounded growth in the second phase far out-pacing resources. Of equal concern is the potential for such imbalance to incur irreversible conditions before the emergence of the third phase.

Lest the realization of many aspects of these projections and theories in various societies instill a sense of inevitability, engendering hopelessness and chilling effects on action, let us consider some of the important qualifications and challenges. Malthus himself implored intervention to alter the course of his projections and to stave off the mass starvation that would otherwise ensue.[3] Furthermore, recent voices dismiss as myth that "hunger is caused by too many people pressing against finite resources."[4] Fully mindful of the need to limit population growth, these authors nevertheless estimate that "the world today produces enough grain alone to provide every human being on the planet with 3,600 calories a day."[5] These estimates do not include other common nutritional items. Of the many possible explanations of food shortages, the authors place most emphasis on the accountability of governments to their people, or rather the lack of it.[6] Adding to the weight of responsibility of governments are analysts who call the generalizability of demographic transition into question because of instances of deviation.[7] They urge careful documentation of historical events and trends within nations in the search for more refined explanations of population change. From this perspective, the role of governments and their policies, in terms of both what they choose to do or not to do, would loom highly significant. Before leaving this point, however, it should be emphasized that both the current population size and rate of growth in Egypt should cause serious concerns and motivate accelerated programs to reduce fertility. Table 1 illustrates the point.[8]

The second population problem is one of maldistribution. Imbalance in density can be the result of natural phenomena which render large portions of nations uninhabited. Such an imbalance can also be caused by high rates of migration from rural to urban areas in search for better employment opportunities, higher standards of living, and more developed services. When flight to urbanization is not matched by expansion in industrial or other suitable sectors of the economy, the consequences become detrimental both to the cities and to the new dwellers who remain in very marginal living conditions. The rural areas are not spared the negative effects of this migration pattern; it can seriously deplete the labor supply necessary to maintain agricultural productivity. A scant look at maps and census reports would quickly reveal the enormous influence of these two forces, natural phenomena and rural-urban migration, on the distribution of the Egyptian population. The first has led to a concentration of people onto a small fraction of

TABLE 1. Crude Birth Rates, Death Rates, and Net Production Rates, 1952-1978
(Per thousand)

Year		Crude Birth Rate	Crude Death Rate	Natural Reproduction Rate
1952		45.2	17.8	27.4
1960		43.1	16.9	26.3
1966	(June War)	41.2	15.9	25.3
1972		34.4	14.5	19.9
1973	(October War)	35.7	13.1	22.6
1974		35.7	12.7	23.0
1975		36.0	12.1	23.9
1976		36.4	11.7	24.7
1977		38.4	11.9	26.5
1978		38.7	10.6	27.1

Source: John Waterbury, *The Egypt of Nasser and Sadat*, p. 46.

the nation's territory and the second has resulted in excessive overcrowding of the cities and the overtaxing of their facilities and infrastructures. "Only 3.5 percent of Egypt's surface is inhabited although 15.5 percent has been designated as currently habitable."[9] Of the total population, urban dwellers have increased between 1947 and 1976 from 33.0 to 43.9 percent.[10]

Third, is the problem of fluctuation in the rates of population change. The succession of sizeable expansions and contractions in a nation's population can cause considerable dislocations in its institutions and severely test the stability and coherence of its policies. Consider the human wave referred to as the "baby boom" in the United States and its effects as it has been passing through the age structure. The pressure of the high points caused massive and hasty stretching of educational and other services before it made itself felt in the labor market. Concern now is over the low points and their influence on the adequacy of labor supply and support of income maintenance programs. Similar concerns now prevail in West Germany which is currently experiencing low points in a cycle of fluctuation. It should be noted that whether or not these fluctuations actually are cyclical remains an open question. In the foreseeable future, Egypt is not likely to experience a pendulous pattern in natural population change that is based on the balance between the rates of fertility and mortality. However, the size of in- and out-migration can generate similar although less pronounced effects. Large numbers of Egyptians are employed in the major oil producing states and in Southern Europe. Change in the economic fortunes of the host countries can cause an influx of returnees who would place demands on the local services and institutions. Age-specific analysis of potential fluctuations of this type would help identify the sectors of the economy and the society at large most likely to be affected. Also important in this respect is the occupational composition of in- and out-migrants as it may relate to the occupational structure in Egypt.

The effects of these and other population problems are manifested not only at the aggregate and structural levels of cities and communities. They are felt by individuals and families in most fundamental ways in terms of inadequacy in nutrition, housing, health care, educational opportunities, employment prospects, means of transportation and communication, and other aspects of daily life. Three approaches to alleviating these problems have been prominent in the agendas of nations: (1) reducing fertility through different types of family planning programs; (2) redistributing populations into new settlements of varying types, functions, and sizes as is being considered in this conference on desert development; and (3) raising the standards of living and literacy, and providing income security and health care necessary to improve the survival of infants. Indications are that the first two approaches, individually or together, yield less than satisfactory results unless combined with the third, that is, societal development.

DEVELOPMENT AND MODERNIZATION

The already massive and rapidly expanding literature on development conveys a variety of definitions and indicators. Development has been equated with modernization and we will use the two terms in a synonymous manner. Four traditions in the literature deserve special attention. These are centered around political, economic, other socio-cultural, and individual modernization and development. I shall briefly define these and then select a few important cross-cutting issues for elaboration, especially as they may relate to desert development.

Definitions of political modernization "focus on the differences between what are assumed to be the distinctive characteristics of a modern polity and of a traditional polity;" and is held "to be movement from the one to the other."[11] Huntington sees three broad trends in the process: (1) the rationalization of authority away from familial, ethnic, religious, and other traditional centers of power to a single national political authority; (2) the differentiation of legal, military, administrative, and scientific functions, and the development of specialized structures to administer them in ways subordinate to political authority but separate from the political realm; and (3) increased and meaningful participation in the various aspects and levels of politics by groups throughout the society.[12] Another feature includes an increased emphasis in the distribution of positions of power and authority on the basis of achievement and qualifications in contrast to ascription and interpersonal loyalties. Furthermore, means are developed to advance openness in conducting the affairs of the state and to strengthen the accountability of governments for their policies and of individual employees for their official actions.

Huntington identifies three emphases in political modernization which can be considered as stages in the process: centralization of power into national governments, expansion of participation in the political processes, and decentralization of rational secular forms of power.[13] Complex as is it, the centralization of power and liquidation of traditional authority seems to be far easier than expanding participation and redistributing power. Very few developing societies have succeeded in moving through the latter two stages. Neither extreme of power concentration or distribution would seem to be desirable. The question is where a balance can be struck that would allow room for local initiatives on the one hand and national integration on the other.

The meaning of economic development varies among analysts and over time. A host of features have been used to indicate this phenomenon: per capita income, per capita GNP, total GNP, industrialization, adaptation and adoption of technology, improved infrastructures especially in means of transportation and communication, efficiency, productivity, poverty, and inequality among others. None of these indicators individually or in narrow combinations conveys the complexity of the concept. In the most general way, economic development represents improvement in the total economic yield of a society and in the material well being of its people. It has become one of the most common objectives of policies in Third World countries.

The economic literature abounds with debates about conditions and approaches to economic development. Important among these are discussions about the significance of developed societies as demonstrations; the role and sources of capital formulation; policies oriented toward self-sufficiency versus those oriented to specialization in products in which a country has a comparative world advantage; protectionism versus free trade; policies related to the development of human resources; and the role of government in economic development. From the viewpoint of a social scientist who is not specifically trained in economics, I would like to bring up for discussion some of these issues as they relate to Egypt by making the following observations:

1. Apprehensions about economic hegemony, especially in the wake of newly gained political independence, caused Egyptian leaders in the fifties and sixties to shun foreign investments and concentrate on internal capital formation through the nationalization of large sectors of the economy, thus, creating state capitalism. This transformation has had considerable political as well as economic consequences. Coupled with an open policy of higher education and commitment to full employment, the result has been a bloated public sector of the economy seriously overstaffed with people whose qualifications are not necessarily matched with work requirements.

2. Later attempts to encourage the development of the free sector of the economy have been hobbled with formidable obstacles. Important among these is state ownership of primary industries which in effect become monopolies which determine the quality of products and their prices. For example, the garment manufacturers are bound by the quality and price of cloth produced in government owned factories. The same can be said for steel, glass and other products which have to utilize primary material from the non-competitive public sector industries.

   Free enterprise can and does lead to considerable economic inequality, and in itself is considered "evil" in a society saturated with egalitarian slogans. Concern about these inequalities, and their potential for politically destabilizing effects, engenders frequent shifts and changes in economic policies. The uncertainties created inhibit the investment of risk-taking entrepreneurs.

3. An important contribution of world system dependence analysts is a view of the states as open rather than closed systems. Economic interrelationships among countries raise the important question of whether national policies should be oriented to self-sufficiency or should concentrate on products and services in which Egypt has a comparative world advantage. The latter orientation calls for specialization and is predicted on open trade and symmetry in relations between nations. Neither premise prevails in today's world.[14] Furthermore, Egypt, like most other developing societies, is at a disadvantage in both respects.

4. Recently published data on agricultural productivity and on service and technology do not present a favorable picture of the situation in Egypt. When compared to a number of other developing countries, agricultural productivity in Egypt shows significant declines between the periods 1963-67 and 1978-82.[15] Service and technology capabilities were measured for several developing countries using studies based on: amounts and percentages of GNP allocated to research and development, capability for adopting technology from other countries, capacity to import scientists, capability for reproducing scientists, and major contributions to world science. Figure 1 shows where the different countries stand on this index in 1984.[16]

Modernization goes beyond political and economic institutions to refer to change in other aspects of the socio-cultural environment. These aspects contribute greatly to the quality of life -- a concept usually less emphasized in literature on developing. Important in this respect are advancements in education as reflected in the production and dissemination of knowledge, levels of literacy, and preparation of competent and productive citizenry; legal institutions and the degree

FIGURE 1. Developing Country Science and Technology Capabilities, 1984

I. Fully Institutionalized
India

II. Semi-Institutionalized

Brazil	Singapore
China	South Korea
Hong Kong	Taiwan

III. Partly Institutionalized

Argentina	Pakistan
Malaysia	South Africa
Mexico	

IV. Partly Institutionalized Petroleum Dependent

Algeria	Saudi Arabia
Iran	Trinidad and Tobago
Iraq	Venezuela
Kuwait	

V. Longshots

Barbados	Sri Lanka
Columbia	Thailand
Costa Rica	Turkey
Jamaica	

VI. Apparent Failures

Chile	Nigeria
Cuba	Philippines
Egypt	Vietnam
Indonesia	

VII. Others

From: McIntyre, The Political Economy of International Technology Transfer, p. 107.

to which they maintain justice, fairness, freedom from the arbitrariness of power, and equality of opportunity; health care and other human services as they contribute to the healthfulness of the population and to security in times of need and dependency; and the arts and their influence on aesthetics, taste, and civility.

So far, emphasis has been on the institutional aspects of development -- values, norms and organization. A corresponding social psychological dimension is that of _individual modernity_ -- the capabilities, motivations, and attitudes of individuals. On this, I will quote at length from Inkles and Smith:

> "A modern _nation_ needs participating citizens, men and women who take an active interest in public affairs and who exercise their rights and perform their duties as members of a community larger than that of the kinship network and the immediate geographical locality. Modern _institutions_ need individuals who can keep to fixed schedules, observe abstract rules, make judgments on the basis of objective evidence and follow authorities legitimated not by traditional or religious sanctions but by technical competence. The complex production tasks of the _industrial order_, which are the basis of modern social systems, also make their demands. Workers must be able to accept both elaborate division of labor and the need to coordinate their activities with a large number of others in the work force. Rewards are based on technical competence and objective standards of performance, strict hierarchies of authority responsive to the imperatives of machine production, and the separation of product from producer.
>
> In addition, modern institutions...require a greater acceptance of personal mobility, occupational and geographic; a greater readiness to adapt to changes in one's mode of working and living; indeed an innovator; more tolerant of impersonality, of impartiality, and of differences which may characterize the diverse backgrounds of fellow employees in complex organizations... There is little

tolerance for fatalism and passivity, but rather emphasis on persistent efforts and optimism."[17]

Clearly, modernization and development are highly complex phenomena which involve change in all areas of human thought, organization, and activity.[18] The processes involved are neither easy nor smooth. They frequently lead to conflicting results. For example, rapid economic change at certain stages of development may create political instability. Indications are that negative effects are most likely to occur in the least developed countries, no negative effects in countries at the middle of the continuum, and positive relations in advanced societies.[19] Huntington offers a list of consequences of rapid economic growth which may challenge political stability:

"1. disrupts traditional social groupings (family, class, caste), and thus increases "the number of individuals who are declasse...and who are thus in circumstances conducive to revolutionary protest;

2. produces nouveaux riches who are imperfectly adjusted to and assimilated by the existing order and who want political power and social status commensurate with their new economic position;

3. increases geographical mobility which again undermines social ties, and, in particular, encourages rapid migration from rural areas to cities, which produces alienation and political extremism;

4. increases the number of people whose standard of living is falling, and thus may widen the gap between rich and poor;

5. increases the incomes of some people absolutely but not relatively and hence increases their dissatisfaction with the existing order;

6. requires a general restriction of consumption in order to promote investment and thus produces popular discontent;

7. increases literacy, education, and exposure to mass media, which increase aspirations beyond levels where they can be satisfied;

8. aggravates regional and ethnic conflicts over the distribution of investment and consumption;

9. increases capacities for group organization and consequently the strength of group demands on government, which the government is unable to satisfy."[20]

From the viewpoint of public policies, the lesson to be learned is not to halt or even retard economic development but to prepare for its political consequences. Another important lesson concerns the great need for a clearer understanding of consistencies and inconsistencies in the process of development keeping in mind the strong interdependence of the various institutions of society.

This general discussion of the various dimensions of development and modernization suggest a number of cross-cutting issues with more direct implications to desert development in Egypt. First are questions about the direction of development. I suspect these "questions" will be the subject of extensive discussion, if not debate, in sessions on economic issues. Several options present themselves: spill-over satellite communities close to urban centers, resort communities in the coastal areas along the Mediterranean and the Red Sea to serve Egyptians in summers and Europeans in winters, agricultural settlements in areas reclaimed for cultivation, and industrial and mining communities depending on the availability of requisite resources. Heated debates are likely to revolve around two important concerns. One is over priority in the use of a vital and a scarce resource -- water. The question is whether to continue emphasis on extending cultivation into the desert, which requires a heavy utilization of water, or to resist what might be considered as a rural bias and orient desert development to less water-intensive forms of settlements. Implicated in this argument is a second concern over a question mentioned earlier about specialization in products and services versus emphasis on self-sufficiency in food and other essentials. It would seem that a high priority on population redistribution would call for urban satellite communities, specialization in economic activities would emphasize tourism and resort communities, and self-sufficiency would favor expansion in cultivated land. Assessing these and other priorities, and their short and long term effects, is as difficult as it is essential if a balanced direction for desert development is to be charted.

Second, water is not the only constraint; monetary costs constitute another limiting factor. It has been said that desert development is socially positive but economically negative. This very general statement should not be allowed to influence such a major policy without further specification and documentation. The costs of desert development should be weighed against the absence of such development. The cost-benefits should be assessed for the long term as well as the short term. Furthermore, the pattern of developments in Third World countries takes the form of demonstrations sponsored by high officials and favored with generous resources to evolve under near ideal conditions. They become showplaces for influentials in the country and dignitaries from aboard. However, they also become vastly too expensive to generalize to the large sectors of the population in need. Developments in the desert should be planned in realistic ways and scales amenable to generalization. A final argument is that it is not uncommon that funds are expended to attain valued social conditions.

The third issue relates to incentives for people to move to the new settlements. Important in this respect are both forces of pull and push. The forces of pull include economic and employment opportunities; land tenure and property rights; basic services in terms of electricity, water supplies, sewers, etc.; availability of other services such as health care and schools; and accessibility through adequate means of transportation and communications. Anticipatory socialization by increasing familiarity with desert conditions can help overcome misconceptions and fears stemming from misinformation and uncertainties. Educational curricula and the mass media can make valuable contributions in this regard by informing young and adult alike about potentials for desert communities, and inspiring a spirit of change. Many of the forces of push are already in place -- overcrowded cities and overtaxed facilities. Sound zoning policies and effective implementation in inhabited areas of the country would constitute an added incentive to relocate. The timing of such policies must be well articulated with the availability of opportunities in new desert communities.

The fourth and final example of cross-cutting issues that we shall discuss deals with the transition of new settlements from mere collections of residents into integrated communities. A community has been defined in terms of shared and institutionalized trust.[21] Important is the structure of authority and the distribution of power. As mentioned earlier, the initial shape in political modernization is one of centralizing power in governments away from familial and other traditional centers of authority, and the penetration of the bureaucracy into the daily affairs of citizens at the local level. This stage has much greater effect on villages and small settlements, which are usually governed through traditional authority, then on urban centers and their populations who are accustomed to dealing with the bureaucracy. One of the pronounced effects of this change is that which stems from the lag between undermining or eliminating the traditional authority and replacing it with an effective bureaucracy. The transfer creates a vacuum in power which often leads to local instability and the accumulation of unresolved problems. Front line government employees usually lack the necessary training to properly handle this transitional vacuum. Settlers in newly reclaimed areas face a similar situation. In the past, settlers of the newly reclaimed desert areas in Egypt were moved individually or in nuclear families. They left behind whatever authority structures and patterns of association that existed in their villages, and found themselves in immediate interaction with government functionaries. Adjustments have been fraught with problems.[22] Here again, it is essential that government employees in newly developed settlements be well qualified and trained in the tasks of building institutions and communities.

NOTES ON PLANNING

Discussions of development and modernization inevitably raise questions about the role of planning and that of government. Different positions on the issues involved are espoused by strong voices. In this city (Cairo), slightly more than thirty years ago, Gunnar Myrdal (who was recently awarded a Nobel Prize) stated that "it is now commonly agreed that an underdeveloped country should have...an overall integrated national plan.... All underdeveloped countries are now, under the encouraging and congratulating applause of the advanced countries, attempting to furnish themselves with a plan...the national government is expected to assume by means of the plan... responsibility for the direction of the entire economic development of the country."[23] Possibly underling this position is the assumption that the underdeveloped societies cannot follow the same path travelled by the industrialized nations toward development because of qualitative differences in economic conditions.

National planning and the almost exclusive reliance on government prescribed by Myrdal and others evoked opposition from prominent analysis. P.T. Bauer states the objections as follows:

> "Economic development requires modernization of the mind. It requires revision of the attitudes, modes of conduct and institutions adverse to material progress.... Comprehensive planning does not promote favorable changes in these attitudes and mores. It reinforces the authoritarian tradition of many underdeveloped societies which inhibits the development of faculties and motivations congenial to material advance. By continuing and extending state control over the lives of the population, central planning reinforces the subjection of the individual to authority. Such a development discourages self-reliance, personal provision for the future, sustained curiosity and an experimental turn of mind."[24]

In support of his own argument, Bauer refers to a Myrdal reinterpretation of the concept of planning as a "wholesale transformation of people's attitudes, values, and institutions, by compulsion if necessary."[25] Emphasis here is on remolding individuals and societies rather than on the state's control of the economy.

Arguments surrounding the role of government and planning are generally more than tinged with ideology. Ideological commitments tend to produce consistent but oversimplified positions. Also, at times, extreme positions are taken in order to emphasize particular viewpoints. In the complex real world of today's Egypt, however, the role of government in desert development is essential. It controls the land and water supplies, it develops the roads and infrastructures, and establishes and enforces laws conceived with property rights and zoning regulations. The question is one of providing sufficient room and incentives for private owners and entrepreneurs to participate in the economic undertakings, and for the civic sector to take part in shaping the social structure and the quality of life in the new communities.

William Petersen identifies three types of planning: ideological, deductive, and inductive.[26] He maintains that as planning "connotes the injection of rationality into a particular area of human life," it must be amenable to empirical evaluation.

> "With respect to utopian or ideological plans, first of all, evaluation has little or no meaning.... A utopian goal keeps receding as we approach it, and it cannot be used to measure progress along the road toward it.... Frequently, perhaps even usually, the utopia is less an incentive than an alternative to social reform. For in the eyes of the true utopian, anything that can be achieved is not worth attempting...."[27]

A deductive approach produces prior comprehensive plans which are designed before implementation begins. Such plans usually fail to accommodate unanticipated consequences. The literature, including that on desert development in Egypt, provides repeated illustrations of defects in this model.[28]

Inductive planning is continuous and begins with flexible guides rather than blueprints. This type of plan responds to unanticipated consequences of various steps and phases, and is modified on the basis of serendipitous knowledge and experiences acquired in the course of implementation. While this approach is close to "muddling through," when carefully implemented it yields superior results.

Effective planning and implementation require three elements which have been in short supply in Egypt. Essential are accurate and meaningful data. Intelligible examinations of the implementations of policies and instructive assessment of their impacts call for information which meets the norms of reliability and validity. The lack in such information opens the door to the influence of ideological biases and personal opinions. Addressing this problem entails more than the transfer of advanced survey technology; it calls for socialization into the ethics of research. The second element is comprised of long term commitment to development policies and to stability in implementation. Broad participation and careful planning can mitigate against the personalization of policies and the frequent shifts associated with change in leadership. Third, and finally, in a responsive bureaucracy promising policies often stumble to failure because of bureaucratic impediments to implementation. Modernizing the bureaucracy and its routines to increase accountability must be an essential component of the quest for development.

COMMUNITY OF SOLUTION

A society is made up of varying types of communities. Thus, planning and development must find expression at the community levels where individuals, families, and other forms of organization reside. Community has been defined by geographic boundaries, political jurisdiction, market or service centers, or in terms of social interaction and identification. A definition useful for analysis and planning is offered by Warren, who conceives the community "as that combination of systems and units which perform the major functions having locality relevance...: production - distribution - consumption, socialization, social control, social participation, and mutual support."[29] In performing these functions, he sees vertical ties which relate the community to the larger society and culture, and horizontal ties which relate the local units to each other to provide a sense of community.[30]

> "...a community does not consist of a number of units, some of which can be classified as vertical and some as horizontal, but rather that each of a community's units typically has both vertical and horizontal aspects. It is true that some types of units have a stronger vertical component...and others

> have a stronger horizontal component.... It may be likewise true that some individuals are more strongly oriented vertically...in their community participation, while other individuals...are more strongly oriented horizontally. But the attempt to identify a community's vertical units and its horizontal units is misleading. It is not the units which constitute the vertical or horizontal pattern; it is the vertical or horizontal aspects of the units - their relation, respectively, to extracommunity systems and to each other."[31]

Related to these distinctions is the concept "community of solution" which was introduced in the mid 1960's by the National Commission on Community Health Services in the United States. According to this concept, the boundaries of a community are those within which "a problem can be defined, dealt with, and solved."[32] In some instances, the elements of a problem might be found and effectively addressed at the local level. Many of the family-centered functions and similar systems of social support constitute appropriate examples. The preponderance of problems and issues in agrarian communities follow a horizontal pattern. In other instances, the elements of a problem and/or its resolution follow a vertical pattern that involves regional or national levels as in the case of certain aspects of health services, government operations, and industrial and business concerns. Exhibiting a strong vertical pattern also are bedroom communities developed on the fringe of urban centers where individuals and families are more oriented to these centers than to local affairs. Still in others, the vertical connections might transcend national boundaries. The resolution of many problems connected with Lake Nasser would by necessity involve Sudan as well as Egypt. The vertical connections of coastal communities along the Mediterranean might reach the European countries to whose citizens these resorts should be largely oriented. Local developers and operators of facilities in such communities must be concerned with the relative attractiveness of their communities and with meeting the requirements and tastes of potential vacationers, the modes of transportation and communication available to them, and the bureaucratic difficulties and red tape they might encounter. Such concerns might even create a sense of competition at the local level. In this sense, development and modernization can be expected to increase the strength and pervasiveness of vertical orientation possibly at the risk of weakening those of a horizontal nature -- a classic trade-off between material advancement and the cohesiveness of primary relations.

In conclusion, it should be emphasized that cross-national comparison would clearly indicate a dangerous negative spiral in the conditions of countries where the rates of societal development fall behind those of population growth. For various reasons, internal and international, Egypt has fallen behind. Overcoming the widening gap requires keen awareness of the enormity of the problem and extraordinary commitment to curbing the seemingly unfettered increase in population, to the redistribution of inhabitants of the overcrowded areas, and to accelerating the pace of development. The current openness in discussions and the earnest search for effective approaches bode well for the future.

REFERENCES

1. William Petersen. 1979. Malthus Cambridge, Mass.: Harvard University Press: pp. 46-48.

2. William Petersen. 1965. The Politics of Population Garden City, N.Y.: Anchor Books, Doubleday & Company, Inc.: p. 166.

3. William Petersen. Malthus. p. 46.

4. Frances M. Lappe and Joseph Collins. 1986. World Hunger: Twelve Myths. San Francisco: Institute for Food and Development Policy: p. 23.

5. Ibid, p. 9.

6. Ibid, pp. 4-6.

7. William Petersen. The Politics of Population: pp. 166-167.

8. John Waterbury. 1983. The Egypt of Nasser and Sadat: The Political Economy of Two Regimes (Princeton, N.J.: The Princeton University Press: p. 46.

9. Ibid, p. 41.

10. Ibid, p. 42.

11. On definitions of political modernization, we rely heavily on Samuel P. Huntington. 1969. Political Order in Changing Societies. New Haven and London: Yale University Press: pp. 32-36.

12. Ibid, pp. 34, 35.

13. Ibid, p. 146.

14. See for example Brian M. Pollin. 1985. Breaking trade dependency: A global simulation of third world proposals for alternative trade Regimes, International Quarterly, Vol. 29: pp. 282-312; and Albert O. Hershman. 1981. Essays in Trespassing: Economics to Politics and Beyond Cambridge: Cambridge University Press: pp. 27-29.

15. Richard H. Adams, Jr. 1986. Development and Social Change in Rural Egypt. Syracuse, N.Y.: Syracuse University Press: pp. 20-25.

16. John R. McIntyre and David S. Rapp, (eds.). 1986. Political Economy of International Technology Transfer. New York: Quarum Books.

17. Alex Inkles and David Smith. 1974. Becoming Modern. Cambridge, Mass.: Harvard University Press: p. 4.

18. Samuel Huntington. Political Order in Changing Societies, p. 32.

19. Ibid: 50-53.

20. Ibid: 49-50.

21. Bertrand DeJouvenel. 1963. Sovereignty. Chicago: University of Chicago Press: 123; quoted in Samuel Huntington, Political Order in Changing Societies, p. 28.

22. See for example: Helmi R. Tadros. The human aspects of rural resettlement schemes in Egypt. In Bernardo Berdichewsky, (ed.) 1979. Anthropology and Social Change in Rural Areas (The Hague: Mouton Publishers: 142-144.

23. Gunnar Myrdal. 1976. Development and underdevelopment. Quoted in P.T. Bauer, Dissent and Development. Cambridge, Mass.: Harvard University Press: 69-70.

24. P.T. Bauer, Dissent on Development: 84.

25. Gunnar Myrdal, Asian Drama. Quoted in P.T. Bauer, Dissent on Development: 73, 74.

26. William Petersen, The Politics of Population: p. 325.

27. Ibid: 326.

28. See for example, Helmi R. Tadros, op. cit.; Laila S. El-Hamamsy, and Hind Abou Seoud. 1962. The initial reactions of the first group of settlers to their new environment. Cairo: Social Research Center, American University in Cairo, and Hussein M. Fahim. 1983. Egyptian Nubians: Resettlement and Years of Coping. Salt Lake City: University of Utah Press.

29. Roland L. Warren. 1963. The Community in America: Chicago: Rand McNally: 167.

30. Ibid: 237.

31. Ibid: 241.

32. National Commission on Community Health. 1966. Health is a Community Affair. Cambridge, Mass.: Harvard University Press: 2.

DEMOGRAPHIC DYNAMICS AMONG SEDENTARIZING PASTORAL NOMADS AND ITS PLANNING IMPLICATIONS

AVINOAM MEIR
Department of Geography and Jacob Blaustein Institute for Desert Research
Ben-Gurion University of the Negev Beer Sheva, Israel

Abstract Many nomadic societies in Third World countries are sedentarizing. The associated changes in modes of production result in social transformation that may carry with it significant demographic consequences. An adaptation of demographic transition theory to the nomadism-sedentarism continuum is proposed by treating demographic change as a transition between demographic regimes associated with specific modes of production and suggesting that fertility increases and later declines, and mortality decreases along the continuum. Rising fertility during sedentarization is assumed to be a consequence of both social modernization and economic growth, which complement conventional theoretical statements. Data from several countries suggest that birth rates do increase along the continuum, but their decline at post-sedentarization, although possible, will depend on trends in the general rural sector. In several cases, mortality rates of sedentarizing nomads have been found to be higher than for nomads, suggesting deficient public health measures. Nevertheless, natural increase rates of sedentarizing nomads are considerably higher that those of pastoral nomads. This adaptation may help to achieve the goals of filling a gap in demographic transition theory and providing a conceptual framework for future research of more detailed and specific case studies. If the adaptation proves correct, then one of the lessons for controlling demographic pressure in marginal arid lands is that strategies for pastoral nomads other than sedentary ones should be sought. Solutions for pastoral nomadic development should preferably remain within the

> pastoral framework so that the demographic pattern of a pastoral society is maintained, thus preventing a possible population explosion among them if sedentary strategies are adopted.

***

In a series of three papers (Meir, 1984; 1986a; 1987), the author has argued for a demographic transition approach to understanding demographic dynamics of sedentarizing pastoral nomads. Also, an attempt was made to extract some general policy implications related to sedentarization of pastoral nomads. This paper aims at summarizing the main arguments and findings regarding the demographic transition approach and its policy implications.

DEMOGRAPHIC DYNAMICS OF SEDENTARIZING NOMADS

In a recent article (Meir, 1986a), an attempt was made to put the changes that take place among sedentarizing pastoral nomads within the framework of changing modes of production. The main argument presented is that each mode of production results in specific social and economic relations which generate a typical demographic pattern. As modes of production change along the nomadism-sedentarism continuum, demographic patterns change as well.

Assuming the predominance of livestock ecology in a pastoral mode of production, it is useful to draw upon Spooner's (1971, 1973) propositions that there is an optimum-size range of herd and flock of which members of the group are conscious, that there is specific division of labor, and that, consequently, there is an optimum numerical composition and age-sex distribution of the group, which is maintained. This suggests rational demographic behavior of pastoral nomads despite some contrary evidence (e.g. Rigby, 1969; F.A.O., 1972).

Maintaining the size and demographic structure depends primarily upon the balance between fertility and mortality. Spencer (1972), Legesse (1973), Irons (1975), O.D.I. (1976) and Roth (1986) have provided detailed accounts of restraints and self-regulating mechanisms that may affect mortality and fertility among pastoral nomads. The net result points, in general, to mortality rates that are higher among pastoral nomads compared to their sedentary neighbors. As regards fertility, the question of rational behavior must be treated in more detail. Drawing upon Caldwell (1976), two types of societies may be distinguished. A society with a children-to-parents flow of wealth is characterized as having a rationality of higher fertility, while a society with parents-to-children flow of wealth may have a rationality of lower fertility. It is suggested here that typical of Third World nations, there are two levels within the first group. The pastoral nomadic society, in which the amount of food energy available is lower due to longer food chains, may have a rationality of lower fertility compared to sedentaries, although in both, the wealth flows from children to parents. Thus, pastoral nomads may achieve a stabilized demographic pattern through lower levels of fertility, higher level of mortality, and, therefore, lower levels of natural increase compared to sedentary farmers.

Upon sedentarization, a major change in modes of production of pastoral nomads may take place which may be followed by changing social relations and demographic patterns. Summarizing demographically-related social changes from various contributions (e.g. Abu-Zeid, 1959; Petit, 1962; Galaty, 1981) reveals the following: weakening of traditional lineage and tribal identity; emerging links to land, to the state and government, and to the market division of labor; emergency of class structure, inequality and social polarization; ideological shift from a collectivist ethos to individuation of rights over resources; growth of individualism in decision making, of consumerism, and of exposure to Western educational values; and growth in importance of the nuclear family. Consequently, fertility rationality that conforms to traditional agriculturalists (Meillasoux, 1972) may be adopted by the sedentarizing nomads. The increasingly self-dependent nuclear family entering a new precapitalist agricultural mode of production tends to maximize fertility as larger family labor resources are required (e.g., Swift, 1977; Frank, 1981). Also, the social change resulting from sedentarization is responsible for a change in the prestige structure, and consequently, family size now substitutes for previous prestige symbols leading to a rationality of higher fertility (Douglas, 1966). Further evidence and theoretical explanations for rising fertility are provided by Tabbarah (1971), Henin (1972), Bongaarts (1981), and Bongaarts _et al_. (1984).

As fertility changes are primarily related to changing social formations, mortality processes are a function of economic growth (Woods, 1982). Indeed, sedentarization itself may be regarded as an economic growth process with technological change, diversification and rise of income sources, and rise in standards of living with exposure to medical and public health services. Mortality may therefore decline. Thus, a unique demographic transition process of sedentarizing nomads may take place through rising fertility and declining mortality.

The social transformation process and the hypothesized resulting demographic pattern carry the sedentarizing society into the rural social formation of Third World societies. We may now hypothesize that fertility may decline, as may be hypothesized for rural LDCs, especially in Africa, although in many of the latter countries, fertility has not yet declined. Exposure to the effects of Westernization (Petit, 1962), may cause nomadic children to become consumers rather than producers of family resources, due to involvement in the educational process (Denga, 1983; Ibrahim and Cole, 1978; Varlet and Massoumian, 1975). Consequently, there may be a reversal of intergeneration flow of wealth. Coupled with other socio-economic processes that result from the changing mode of production, children's utility to the family may be reduced. Also, it is possible that a demographic response process (Davies, 1963) may take place in which population increase may act as a possible threat to the achieved standard of living. Consequently, a rationality of lower fertility may be adopted by the sedentarizing population, and fertility may likely decline (see also Tabbarah, 1971). Given more intensive exposure to health services, mortality may continue to fall, with the outcome of accelerated slowdown in natural increase rates.

Meir (1986a; 1987) has assembled comparative vital statistics of pastoral nomads in Africa and the Middle East along the nomadism-sedentarism continuum which tend to provide support to the hypothesized process: apparently, crude birth rates rise along the continuum, from a range of 25/1000 - 48/1000 for nomads, through 34/1000 - 64/1000 for semi-nomads, to 41/1000 - 62/1000 for sedentarized nomads. Crude death rates fall along the continuum (although with some rise during sedentarization) from a range of 18/1000 - 36/1000 for nomads, through a range of 5/1000 - 26/1000 for semi-nomads, to a range of 5/1000 - 34/1000 for sedentarized nomads. The calculated natural increase rates yield doubling times of 63-23, 63-16, and 31-12 years respectively. Moreover, in at least one specific case, that of the Negev Bedouin of Israel, fertility of sedentarized nomads has already begun to decline from its peak of 62/1000 in 1970 to 53/1000 in the early 1980's, and is still declining.

Thus, in conclusion, there are indications that fertility increases and later declines, and mortality declines along the nomadism-sedentarism continuum. Rising fertility during sedentarization is assumed to be a consequence of both social modernization and economic growth, which complements conventional statements of demographic transition theory. This adaptation of demographic dynamics of sedentarizing nomads fills a gap in demographic transition theory and provides a conceptual framework for future research.

PLANNING IMPLICATIONS

There are several areas in which policy implications of the above discussion can be made; these relate to the problem of assembling demographic data on pastoral nomads and the implications of the general demographic trends themselves. The very scarcity of demographic information regarding pastoral nomads points to the rather low development priority assigned by governments in general to these people. In many cases, pastoral nomads are categorized as simply a rural population. Although in many respects pastoral nomads develop intricate symbiotic relationships with farmers, their social system may be wholly different, and may result in a different demographic pattern. This is not recognized by governments as far as collection of appropriate demographic information is concerned. Consequently, quite often the demographic regime of pastoral nomads along the nomadism-sedentarism continuum is assumed to have little variation. Given the possible demographic trends as suggested above, governments should resort to data gathering on nomads that explicitly categorizes them as a separate social entity until they become socially and economically subsumed into the rural and sedentarized sector. It is only then that such separate categorization may become obsolete.

Such a socio-economic process may not always take place, because sedentarization and involvement in sedentarized activities (such as farming) may not be the only development strategy available to nomadic societies. In some cases, as in East Africa for example (Rigby, 1969), development of pastoral nomads has meant development of livestock, pasture, and water resources without a necessary change in mode of life or social structure. In this case, Knowledge and understanding of population dynamics also becomes important, and must result in a separate categorization of pastoral nomads in demographic statistics. Such importance becomes more pronounced because a <u>pastoral</u> development policy for pastoral nomads without a necessary shift toward sedentary agriculture may help governments to avoid population increase in ecologically fragile dry areas where nomads usually live. Knowledge of the reasons why pastoral nomads avoid a rapid population increase and the mechanisms by which this is achieved thus become highly important; these have been reviewed by Meir (1986a).

The second area in which the discussion in this paper may yield policy implications is related to the trends in population dynamics themselves along the nomadism-sedentarism continuum. As discussed above, there may be a general trend toward rapid population growth as pastoral nomads sedentarize. This is especially evident from those nomadic societies which were shown above to have populations spanning the nomadism-sedentarism continuum. If these trends prove real in specific cases, the rapid population growth of sedentarizing nomads requires an appropriate development policy that is capable of accommodating this growth. This is especially relevant with regard to social services. For example, the case may be that sedentarizing nomads become exposed to endemic diseases from which they could escape in their pastoral nomadic life, leading to rising morbidity and perhaps even mortality rates. Providers of public health services should take into account, as part of a development policy, not only the requirement that these services should be delivered in a manner that is culturally and socially relevant (Meir, 1985), but also the fact that the population is expanding and is thus applying more pressure on service facilities. The same observation is relevant for educational services. Increasing fertility of pastoral nomads under sedentarization has already caught several governments (e.g. Israel) unprepared in terms of providing educational facilities in the appropriate quantities and quality (Meir, 1986b) for this population.

The possibly changing demographic pattern of sedentarizing nomads has important policy implications in other areas as well. Many sedentarizing pastoral nomadic societies do not necessarily give up livestock grazing totally. If sedentarization schemes are to succeed, veterinary services should constitute an important policy element; and this of course requires demographic understanding and information because the interrelationships between the demographics of population and livestock as maintained in the nomadic phase are carried over to a certain degree into sedentary life.

Also important from a policy perspective is the issue of land tenure and water allocation and rights. Sedentarization of pastoral nomads (either voluntary or directed from above) required allocation of agricultural land in suitable quantities and quality. Conflict with sedentary farmers may arise over land tenure issues which may put nomads at a disadvantage because of smaller land parcels compared to the farmer. This problem may be further aggravated given the likelihood of rapid population increase of sedentarizing nomads and desertification processes which reduce the overall quantity of arable land and irrigation water at the regional and national levels. This in itself may entail a further policy implication which concerns appropriate alternative employment schemes for those ex-nomads for whom land parcels may be too small to survive even on subsistence agriculture. Here, as well as in the other policy issues raised above, familiarity with demographic dynamics and magnitudes of sedentarizing pastoral nomads becomes vital. Finally, general population policy and family planning, which are important issues in many Third World nations, should also be considered based upon these demographic processes.

Clearly then, the lack of knowledge and understanding of demographic dynamics of sedentarizing pastoral nomads is disadvantageous to both nomads and governments. This situation further deteriorates under the environmental desertification crisis which has been affecting pastoral nomads in many countries in recent decades. The changing demographic pattern of these populations under normal or tragic circumstances should therefore become an important input for policy considerations.

REFERENCES

1. A.M. Abu Zeid. 1959. The sedentarization of nomads in the Western Desert of Egypt. Internat. Social Sci. J., 11: 550-8

2. J. Bongaarts. 1981. The impact on fertility of traditional and changing child-spacing practices, in H. Page and R. Lesthaege (eds.) Child-Spacing in Tropical Africa: Traditions and Change (London), 111-29.

3. J. Bongaarts, O. Frank, and R. Lesthaege. 1984. The proximate determinants of fertility in sub-saharan africa. Population and Development Review. 10: 511-37.

4. J.C. Caldwell. 1976. Toward a restatement of demographic transition theory. Population and Development Review. 2: 321-66.

5. K. Davis. 1963. The theory of change and response in demographic history. Population Index. 29: 345-66.

6. D.I. Denga. 1983. The effect of mobile group counseling on nomadic fulanis' attitudes toward education. Journal of Negro Education. 52: 170-75.

7. M. Douglas. 1966. Population control in primitive groups. British Journal of Sociology. 17: 263-73.

8. F.A.O. 1972. F.A.O. Expert Consultation on the Settlement of Nomads in Africa and the Near East. F.A.O. Regular Programme NO. 20 (Rome).

9. R.W. Frank. 1981. Mode of production and population patterns: Policy implications for west african development. International Journal of Health Services. 11: 361-87.

10. J.G. Galaty. 1981. Introduction, in J.G. Galaty and P.C. Salzman (eds.). Change and Development in Nomadic and Pastoral Societies. 4-26 (Leiden).

11. R.A. Henin. 1972. The level and trend of fertility in the Sudan. In S.H. Ominde and C.N. Ejioyu (eds.). Population Growth and Economic Development in Africa. 77-83 (London).

12. S.E. Ibrahim, and D.P. Cole. 1978. Saudi Arabian bedouins; an assessment of their needs. Cairo Papers in Social Science, Vol. 1, Monograph 5 (Cairo).

13. W. Irons. 1985. The yomut turkmen: a study of social organization among a central asian turkic speaking population. University of Michigan, Museum of Anthropology, Anthropological Papers No. 58.

14. A. Legesse. 1973. Gadda, three approaches to the study of african society. (New York).

15. C. Meillasoux. 1972. From reproduction to production: a marxist approach to economic anthropology. Economy and Society. 1: 93-105.

16. A. Meir. 1984. Demographic transition among the Negev bedouin and its planning implications. Socio-Economic Planning Sciences. 18: 399-409.

17. A. Meir. 1985. Delivering essential public services to arid zone nomads. In Y. Gradus (ed.). Desert Development: Man and Technology in Sparselands. (Boston), 132-49.

18. A. Meir. 1986a. Demographic transition theory: a neglected aspect of nomadism-sedentarism continuum. Trans. Ins. Brit. Geog. N.S.. 11: 199-211.

19. A. Meir. 1986b. Pastoral nomads and the dialectics of development and modernization: delivering public education services to the Negev bedouin. Society and Space. 6: 85-95.

20. A. Meir. 1987. Comparative vital statistics along the pastoral nomadism-sedentarism continuum: some policy implications. Human Ecology (forthcoming).

21. O.D.I. 1976. Human pastoral population. Pastoral Network Pap. Ic.

22. J. Petit. 1962. The attitude of the population and the problem of education in the sahara. In the Problems of Arid Zones. Arid Zones Research, XVIII, Proceedings of the Paris Symposium, UNESCO. 459-70 (Paris).

23. P. Rigby. 1969. Pastoralism and prejudice: ideology and rural development in east africa. In P. Rigby (ed.). Society and Social Change in Eastern Africa. Makerere University, Institute of Social Research, No. 4, 42-52 (Kampala).

24. E.A. Roth. 1986. The demographic study of pastoral peoples. Nomadic Peoples. 20: 63-76.

25. P. Spencer. 1972. Nomads in alliance: symbiosis and growth among the rendille and samburu of Kenya. (London).

26. B. Spooner. 1971. Toward a generative model of nomadism. Anthropological Quarterly. 44: 198-210.

27. B. Spooner. 1973. The cultural ecology of pastoral nomads. Module Anthropological No. 45 (Reading, Mass.)

28. J. Swift. 1977. Sahelian pastoralists: underdevelopment, desertification, and famine. Annual Review of Anthropology. 6: 457-78.

29. R.B. Tabbarah. 1971. Toward a theory of demographic development. Economic Development and Cultural Change. 19: 257-76.

30. H. Varlet, and J. Massoumiam. 1975. Education for tribal populations in Iran. Prospects. 5: 275-81.

31. R. Woods. 1982. Theoretical population geography. (London).

PROTECTING TRANSBOUNDARY DESERT RESOURCES

HELEN INGRAM
Department of Political Science,
University of Arizona

Abstract Preventing degradation of environmentally sensitive deserts is everywhere difficult and is especially so if jurisdiction is divided among nations. This paper examines a case in which the United States and Mexico jointly agreed to prevent smelter pollution in the Sonoran Desert area. Understanding the conditions for reaching the accord is important for future management of the Sonoran Desert and other environmentally-sensitive areas intersected by international borders.

THE SONORAN DESERT

As shown on the map in Figure 1, the Sonoran Desert straddles the U.S./Mexico border. Enormous development pressures have been brought to bear upon this region with naturally limited carrying capacity. A Sunbelt population explosion is occurring in cities within and closely adjacent to the Sonoran Desert. By the year 2000 the Phoenix metropolitan area is projected to grow to 2.9 million residents which represents an increase of 68% from its 1984 population level. Tucson's population is also projected to expand to 0.94 million during that time, an increase of 58%.[1] On the Mexican side, the urban growth is even more startling. Sonoran Desert cities such as Nogales and Hermosillo are expanding rapidly, partly due to the siting of "maquiladora" assembly plants that employ inexpensive Mexican labor. For instance, a Ford automobile plant which profits from very favorable export arrangements with the U.S., was recently located in Hermosillo.

Development pressures have inevitably resulted in environmental degradation. The Colorado River is overcommitted, and usually becomes a dry river

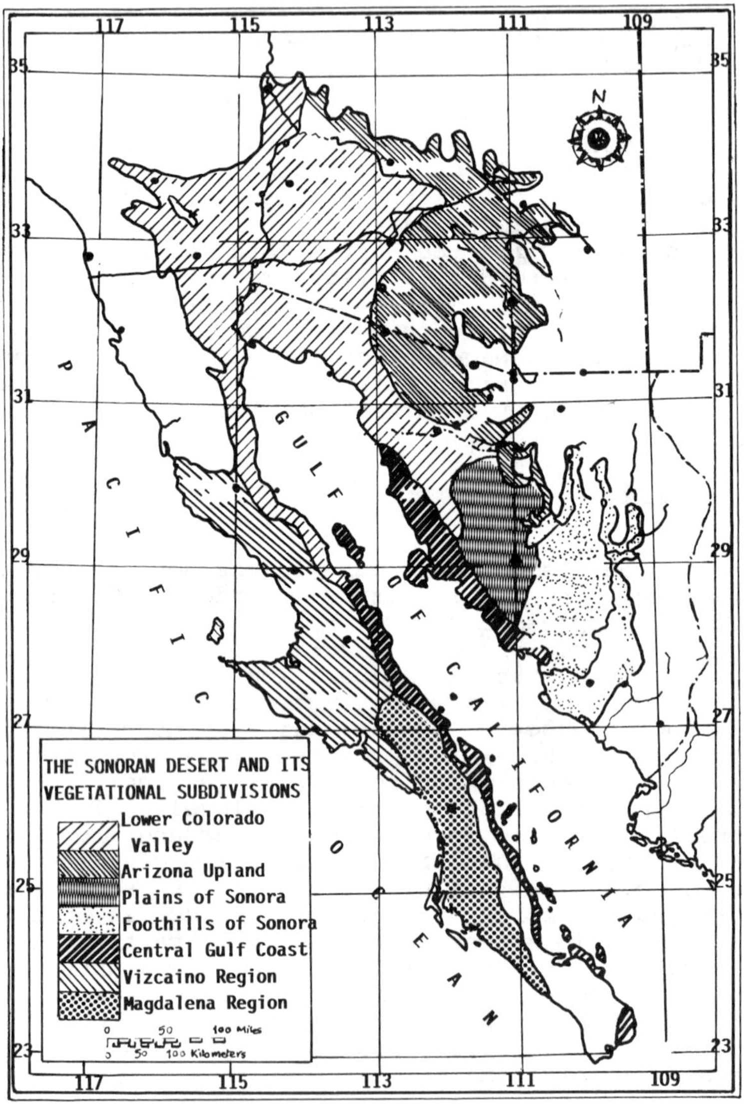

FIGURE 1

SOURCE: McGinnies, William G. Discovering the Desert. Tucson: The University of Arizona Press (1981).

bed or mere conduit of agricultural drainage water long before it reaches the Sea of Cortez. Shortages will increase when the completed Central Arizona Project moves water to Phoenix and Tucson and when the upper Colorado River Basin further develops. Diminishing quantities of water have affected quality, particularly salinity. Pesticide contamination also occurs. Groundwater makes up 60% of Arizona's water use, and currently the overdraft is in excess of 2.5 million acre-feet per year. Millions of acres of desert landscape have been converted to agriculture and to urban development. Farms have been purchased by cities for their water rights, and retired lands produce tumbleweed and dust storms. Urban and industrial expansion leads to air pollution. Neither Phoenix nor Tucson are meeting national air quality standards. Toxic substances are an increasing health hazard, and careless disposal contaminates groundwater in some areas.

Policies to protect desert environments are by nature difficult to legislate. The benefits of such policies are diffused to the general population and to future generations in such a way as to fail to generate strong support. The costs of environmental regulation are focused upon particular development interests which are politically motivated to resist regulation. Special and far more severe political problems arise when problems of natural resource depletion and environmental degradation extend across international boundaries. Those who may bear the costs of control may be located in a different national jurisdiction from those who benefit from action, making the mobilization of sufficient support for environmental quality more difficult. Furthermore, where shared resources are at stake, perverse incentives may operate against conservation. For instance, in the case of transnational groundwater, there may be a race to the bottom of the aquifer, each nation fearing that otherwise it will not get its fair share.

Agreeing to protect transnational resources is especially difficult for the U.S. and Mexico.

Rarely are two contiguous nations so different culturally, economically and politically. The asymmetry of power between the U.S. and Mexico means that the U.S. is always a great deal more important to Mexico than Mexico is to the U.S. Mexico fears any type of accord that may erode its autonomy. The U.S. ignores Mexico to the point of acting in a highly insensitive manner. Environmental protection laws vary. Fairly stringent environmental laws have existed and have been applied in the United States since the 1970s. Mexican environmental laws are newer, somewhat less strict, and have implementation problems. The ongoing economic crisis in Mexico makes it especially difficult to forego development or to spend money on environment. Governmental structure is vastly different in ways that impede mutual agreement. The federal system in the U.S. means that state governments influence a number of border decisions while centralization in Mexico concentrates control in Mexico City.

The impediments to agreement between the United States and Mexico are reflected in a history of natural resource disputes affecting the Sonoran Desert. As an upstream nation, the U.S. was slow in negotiating a treaty guaranteeing an allocation of Colorado River water to Mexico. Later, it was reluctant to respond to Mexican complaints that water deliveries were highly saline. Mexico has likewise been slow to clean up pollution that affects the U.S. The New River rises in Mexico near Mexicali and flows northward into California to the Salton Sea carrying highly contaminated water from inadequately treated sewerage and industrial discharge. Similarly, the San Pedro River flows north from Sonora into southeastern Arizona carrying highly discolored and contaminated water coming from the large copper works in the Cananea area. While it is clear that both nations will be increasing investment in groundwater facilities to meet growing population demands, only a very limited accord regulates pumping close to the border. There has been no apportionment of transboundary aquifers, and no mechanism for joint management.[2]

GRAY TRIANGLE SMELTER ISSUE

On January 14, 1986, the U.S. and Mexico signed an agreement to limit sulfur dioxide emissions from copper smelters in the Sonoran Desert region. This agreement represents an exception to more usual delay and frustration. It is important to know why this issue was resolved successfully and what it means for future bi-national action.

Copper production in the Southwestern U.S. has a long and extensive history, but a number of smelters have recently closed because of falling copper prices, increasing foreign competition and tighter U.S. pollution laws requiring heavy investment in controls. In contrast, activity has accelerated in Mexico. A new facility has been constructed at Nacozari and the smelter at Cananea is being expanded. In a worst-case scenario, these two facilities without pollution controls combined with the old smelter at Douglas would more than double existing sulfur dioxide emissions, perhaps to 3000 tons per day.[3] What has come to be called the grey triangle is shown in Figure 2.

The Copper Smelter Triangle

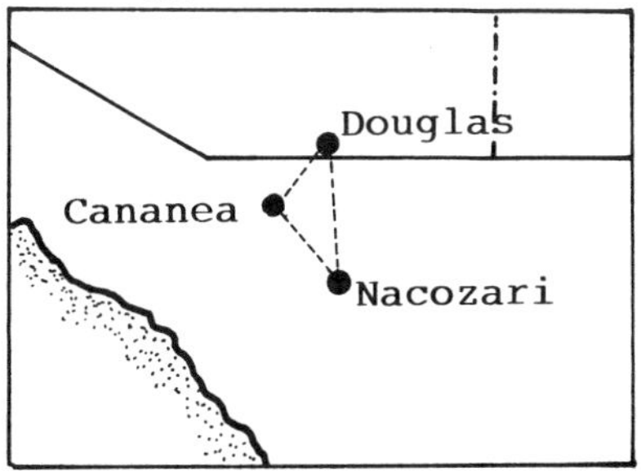

Figure 2

Joint action to prevent potential degradation of the air in the Sonoran Desert and adjacent areas was possible in this case because the issue had certain characteristics and developed in a certain way. The most important factors include:

Both Parties Gained Benefits From Agreement

Smelters on each side of the border were involved and each nation was the victim of pollution from the other. The Phelps Dodge smelter was one of the worst sulfur dioxide emitters in the United States. It was operating under an exception to the Clean Air Act and had a record of frequently failing to meet the conditions of its operating permit. A regular company strategy was to operate at full capacity only at night when winds carried its smoke plume over Mexico and away from U.S. air monitors. Mexican farmers were receiving direct payment from the Phelps Dodge Corporation to compensate for damage done to crops. Other health-related impacts were uncompensated. Damage to the United States was potential and severe. After having spent billions to clean up domestic air pollution, the Southwest faced worsening air quality and perhaps acid rain damage over the Rocky Mountains.

Overall Costs To Each Party Were Limited And Less Than Benefits

The Phelps Dodge plant was scheduled for closing in 1988 and negotiations with Mexico merely served to advance its closing date. Some 350 jobs in Douglas were terminated, but probably could not be saved over the long term anyway. While the Phelps Dodge Company had in the past shown great skill in keeping the plant going despite air pollution laws, domestic concern with adverse impact on health was growing. The economics of installing pollution controls on the new or expanded Mexican plants were much more favorable than retrofitting an existing plant. While some experts argued that the overall impact on the domestic copper industry would be negative, some of the costs of installing an acid plant could potentially be offset by utilizing byproducts to produce wallboard which could be sold in the United States. Appropriate technology was available, and the Mexicans contracted with a Canadian company to build the acid plant which was to be paid for by an international loan.

Negative Consequences In The International Community Would Come To Either Party Refusing To Make Concessions

As a relatively weak country, Mexico depends upon international law to protect its interests, particularly in relation to its powerful northern neighbor. In the Trial Smelter case, the International Court of Justice clearly articulated the principle that the nation of origin was completely responsible for preventing damage to other nations. Mexico was anxious to live up to its good record in observing international law. Furthermore, Mexico is highly dependent upon international loans for economic development and to survive the current economic crisis. Unless pollution controls were installed on Mexican smelters, interest groups in the U.S. would try to block loans.

The U.S. was in a difficult situation in terms of its reputation in the international community. It was already being criticized for causing acid rain pollution in Canada. To put pressure upon Mexico to install pollution controls without itself making some concessions would have adversely affected its image.

Strong Pressure Existed To Keep The Issue On The Agenda

Relations with Mexico are not a high priority for the United States, and decades have sometimes passed without objectively important Mexico-related issues gaining sufficient attention to be dealt with successfully. In the case of the grey triangle issue, U.S. environmentalists were mobilized to gain action. An enormously energetic and talented activist formed the Groups Against Smelter Pollution (G.A.S.P.) and lobbied tirelessly in the U.S. and Mexico to bring attention to the issue.

State Representative Were Supportive Of The Agreement

While by law only the national governments have the authority to negotiate international agreements, in terms of political power, border state officials can veto action. The states in the Colorado River Basin were for years able to block any agreement that would guarantee reasonable allocation of Colorado River water to Mexico. In 1944 a treaty was signed to insure Texas its share of Rio Grande River water and to gain Mexico's support in World War II. Colorado River Basin states allowed negotiations on salinity to go forward with Mexico only when it became clear that the federal government would pay the cost of a desalinization plant. Initially the Congressmen from Southern Arizona reflected the interests of the U.S. copper industry, but as the issue developed and came to be defined in environmental terms, positions changed. The former Governor of Arizona was personally sympathetic to Mexico and a strong environmentalist. He and officials from the Arizona Department of Public Health were in favor of coming to an agreement even at the expense of Phelps Dodge.

Other Federal Agencies Besides The Department Of State Were Supportive Of The Agreement

Where domestic interests are at stake, federal agencies such as the Departments of Commerce, Treasury or Labor have opposed and successfully thwarted international agreements favored by the State Department. In the case of the smelter triangle issue, the U.S. Environmental Protection Agency worked with the Department of State.

A Legal Framework Existed Through Which An Accord Was Facilitated

In August 1983, Presidents Reagan and De La Madrid signed the La Paz Agreement establishing procedures for settling environmental disputes along the border. Part of the impetus was a shared desire by the two Presidents to come together about something in the face of their strong disagreements about U.S. policies in Central America. At the time critics pointed out that the accord lacked substance and neither nation agreed to take specific actions. Further, the comprehensive joint management agency that many feel is necessary to protect groundwater and other resources was not established. However, discrete and specific issues such as the grey triangle were assigned to working groups that consisted of federal and state officials from both countries.

CONCLUSION

From a physical science perspective, delicate desert environments probably should be managed comprehensively and with extreme sensitivity to overdevelopment. Otherwise the limited carrying capacity of deserts is likely to be exceeded, their productivity will decline,a nd desertification will occur. However, actual management decisions are more often made in piecemeal, fragmented fashion and with sensitivity to a number of economic, political and other values that often eclipse the physical and scientific perspective. Preventing degradation of deserts is always difficult to accomplish through public policy, and becomes especially so when the desert is intersected by an international boundary. The case of agreeing to limit smelter pollution in the Sonoran Desert suggests that successfully negotiating bi-national accords to protect shared resources will depend upon the characteristics of the issue and the decision-making process. Agreements are more likely when the nations involved have mutual stakes, when the costs of inaction are perceived as greater than those of action, when political interests are mobilized to keep the matter on the agenda and when subnational governments and governmental bureaucracies do not act as veto points.

REFERENCES

1. Hogan, Timothy. 1986. Arizona population trends. In Arizona Department of Commerce, Arizona's Changing Economy. Phoenix: Commerce Press, p. 11.

2. Utton, Albert E. 1982. Overview. Symposium on anticipating transboundary resource needs and issues in the U.S.-Mexico border region to the year 2000. Natural Resources Journal, Vol. 22 (4): 741.

3. Varady, Robert G. 1985. Air pollution in an arid environment: The effects of copper smelting on the U.S. Mexican border region. Presented at the International Arid Lands Research and Development Conference, Tucson, Arizona.

APPENDIX A

CHRONOLOGY OF BI-NATIONAL AGREEMENT TO LIMIT SMELTER EMISSIONS

July 1982	World Bank's International Finance Corporation (IFC) approved $450 million package to rehabilitate and expand copper facility, Compania Minera de Cananea, at Cananea. This project would benefit U.S. construction and supply firms.
February 1983	U.S. copper industry took notice of the proposed project and began efforts in Congress to block it based on the arguments that copper produced would not be just for Mexican domestic use. On the world market, the increase in copper would harm U.S. copper industry. U.S. State Department, Treasury Department and IFC had found the project would not harm U.S. companies-- (law requires such a finding). At the time U.S. copper industry facing severe slump. Organized labor joined copper industry in opposing loan: cutbacks in employment in the industry. Environmentalists join coalition in opposition to loan somewhat later as issue gets redefined as environmental one, not just economic protectionism. Copper industry had noted lack of pollution control requirements on Cananea smelter and protested that they had to meet costly regulations. Also concerned about new smelter being built at Nacozari without emissions controls.

	McNulty, Congressman from Arizona, led Congressional opposition to the loan and also worked to pass other legislation to benefit copper industry.
May 1983	Hearing in Tucson examined environmental as well as economic aspects of project. Provided considerable publicity. Opposition to loan got time and opportunity to mobilize.
July 1983	Compania Minera de Cananea asserted project would go forward with or without U.S. participation. Would obtain alternative sources of financing. Plans continued without environmental controls.
August 1983	U.S. and Mexico sign La Paz Agreement establishing mechanism for addressing environmental problems along border. Movement toward this agreement proceeding independently of efforts by copper industry and environmentalists to control smelter emissions but provided framework that could be used to address issue.
July 1984	McNulty arranges meeting between U.S. Environmental Protection Agency official (Green) and Mexican officials from SEDUE (Secretariat of Urban Development and Ecology) to examine Douglas and Cananea operations to learn more about problem.
July 1984 - July 1985	Meetings held between U.S. and Mexican officials pursuant to La Paz Agreement to discuss agreement on curtailing smelter emissions--linking of curbing pollution on

	both sides of border. Pressure builds on Phelps Dodge's Douglas smelter to agree to meet EPA requirements.
Spring 1985	Question of denying state permit to Phelps Dodge for next year if it did not agree to an enforceable compliance schedule with EPA. Arizona Governor Bruce Babbitt identifies Phelps Dodge noncompliance as threat to reaching an agreement with Mexico on curtailing emissions on their side.
	Western members of Congress push for stronger bi-national agreement than the one being proposed.
July 1985	U.S./Mexican commitment signed to place limits on emissions at both Nacozari and Douglas by January 1, 1988 (when Phelps Dodge's waiver expires). Not a formal annex. Negotiations continue on formal agreement. Phelps Dodge given permit to operate for one more year while EPA studied Phelps Dodge's request to renew waiver of Clear Air Act standards.
April 1986	EPA said it would deny Phelps Dodge's bid for a waiver unless it could get a consent decree in which the company would guarantee to meet clean air standards or close the smelter by January 1, 1988.
June 1986	Nacozari smelter opens in Mexico without emissions controls but is in process of constructing acid plant to be operational by January 1, 1988.

July 1986	Arizona Department of Health director (Novick) denied Phelps Dodge a 1986 operating permit and issued pollution abatement order. Plant closed on July 9. Phelps Dodge had failed to obtain waiver from EPA.
July 29, 1986	Phelps Dodge plant reopened under terms of consent decree filed in U.S. District Court. Agreed to close January 15, 1987 and to meet interim emissions levels or face $100,000 fine for each violation.
September 1986	U.S./Mexico meetings to discuss emissions issue.
October 1986	Consent decree signed by Phelps Dodge, State of Arizona, and EPA. Douglas smelter to close in January 1987.
November 1986	U.S./Mexico agreement on toxic wastes signed (Third Annex to La Paz Agreement). Agreement to meet further on annex concerning smelter emissions.
January 1987	Douglas smelter closes. Fourth Annex to La Paz Agreement signed on January 14, 1987 to go into effect January 29, 1987. Places emissions limits on Nacozari smelter for March 1988 when acid plant due for completion and places limits on emissions at Cananea for when its production expands.

TECHNOLOGY TRANSFER THROUGH TRAINING: EMERGING ROLES FOR THE UNIVERSITY

HAROLD M. BERGSMA
New Mexico State University,
Las Cruces, New Mexico

Abstract The past twenty years of experience in development work abroad has additionally taught donor and recipient nations that the introduction of technologies, of ideas emerging from the vast research base of western universities, were frequently disfunctional when introduced cold into infra-structures which could not support such technologies. Additionally it became apparent that developed nations were progressing in their own technologies faster than underdeveloped nations could absorb them. Less prosperous peoples, have now, because of advanced communication systems, been bombarded with picture ideas of the prosperous peoples and their high living styles. Poverty in isolation was tolerable; poverty with the means to compare styles became intolerable and led to more dissent and dissatisfaction.

Simplistic development solutions were not the answer. It was important to satisfy the hunger of the people, but general development involved an integrated and complex interaction between people and technologies. This interaction required training and education, concomitant with the technologies being transferred and complementary to the process in formal institutional settings. Few rural adults were involved in formal education and, in order to become functional recipients of technology, needed to develop vocational and technical skills to use and maintain the innovations related to development.

Training, the development of usable skills for functional work and living, became a vital aspect of the development planning process. Education, the development of the mind of the individual in general knowledge areas, ranging from economics to literacy, became increasingly necessary among burgeoning populations which required world class knowledge in order to compete globally. The mechanisms for curriculum development, for training and educating, that is, the development of appropriate pedagogical skills for trainers, became a huge challenge for institutions of higher education involved with international work. Training people for productive work in the third world will require the special attention of universities in the training of trainers and the expansion of educational technologies to facilitate the technology transfer process.

INTRODUCTION

In the early sixties, much was written about the problems of introducing a technology from one country into a second country. Everett M. Rogers, in his now well-known book, Diffusion of Innovations, defined technology as "a design for instrumental action which reduces uncertainty in cause-effect relationships involved in achieving a desired outcome". He states that technological innovation creates uncertainty in the minds of potential adopters, particularly in regard to the expected consequences of adoption. The process of making decisions about adoption of both the hardware and software of a technology is a delicate one. The transmitter of an innovation, introducing, for instance, an improved model of an irrigation ditch, must understand the problem for the adopter-user and must try to reduce the new user's uncertainty, and at the same time, enhance the user's knowledge of the advantage of this technology. This, according to Rogers, will require careful communication of two types of information in respect to a technology such as a cement-lined irrigation ditch. First, there is information about the software of the technology, embedded in the technology itself and inherent in its production and use, which relates to the cause and effect relationship that leads to desired outcomes. Second, there is information about the innovation/evolution information which is the "reaction of uncertainty about an innovation's expected consequences".[1] Transmitters of change, in the past twenty years, have become sensitized to these concepts relating to technology transfer and have adopted more effective means to introduce new artifacts, new ideas, or new technologies. Communications have been enhanced to the extent that conferences are held to highlight how progress in various development programs has been achieved. Meetings are also held within projects which bring the adopters of innovation face-to-face with those who are introducing new technologies in order to improve the ways adoption can occur and to avoid pitfalls and past mistakes.

More specifically, the level of sophistication which has been achieved in technology transfer has increased tremendously in the last ten years. There is, however, an area of concern that requires additional work in all technical projects, and this concern is training. First, training needs to be given higher priority in the development of project papers and in the design of technical projects.[2] Second, the technologies of training and curriculum development, which have become more sophisticated, need to be adopted in training programs of the technical projects of universities, the seats of education and research in the world today. Colleges of educational technology within these institutions can play an important role in future years to bridge a gap that exists in providing expertise related to training technology.

IMPLEMENTATION

Assume that a developing country, or an area of development within a country, has been approached with a new technology and has been sold on the need for both the ideas and the hardware of the innovation. The first stages of the communication process for adoption of a technology, to use Roger's ideas, have been met. Both parties desire the innovation and plans for change are established. Here, precisely, is where the gap occurs. Users of the technology, such as farmers, businessmen, bankers, and even auditors, are not able to utilize the technology used because they lack know-how and lack the basic education that training could provide. Training is the prime means to bring about user-knowledge and user-adoption of a technology.

The writer recently worked with the national Pakistan Audit Department to provide training to future trainers, the auditors who would work with their staffs to introduce "software" innovations such as microfiche recordkeeping. What was being attempted was to provide prospective auditor-trainers with knowledge about new auditing technologies. This was done within a framework in which new knowledge about the technologies of the training was explored and practiced. These auditors gained information about how adult students learn, how to motivate adults, and how to structure basic information into a curriculum in order to simplify the tasks of teachers and clarify the roles and mission of the Auditing Training Institute.

Another example which comes to mind relates to the development of an On-Farm Water Management Program in Pakistan. The writer worked for the Pakistan government as an advisor on training and curriculum development. Irrigation engineers, agronomists, and sociologists were concerned about improving irrigation structures so that less precious water would be wasted by farmers. Beautiful structures were carefully engineered and built. Agronomists conducted research on how to increase crop yields by optimal water application techniques. Sociologists studied the problems of how to best establish "Water User Associations" among disparate social and economic groups. But there was a gap. Farmers and other water-users continued to use water in much the same way as they had before. (Some water is good; more water is better.) The problem (the gap) related to the need to teach farmers to learn new ways of using the ditches, i.e. the hard ware. The softwares of the technology, i.e. scientific application of water, which are the ideas inherent in the technology itself should have been the concern of the trainers also.

In the case above, farmers needed training. However, to achieve this, those who train the farmers also needed to be trained in appropriate techniques for transmitting information and skills. In addition, they needed help in developing a systematic approach to determine what should be taught. Vocational/technical training differs greatly from formal academic training, primarily because there are a limited number of skills or areas of knowledge required of users of new technologies. Various training models are now being used for this purpose. The model in Figure I has been utilized in a variety of technical training situations internationally and is applicable for use in most areas where training is needed in the technology transfer process.

HUMAN RESOURCE DEVELOPMENT

A great challenge exists for educators to train technicians who introduce innovations. Appropriate methods for packaging information (curriculum development) and motivating the learner-users about the new technology should be used. The focus here is on human resource development.[3] Frequently, dysfunctional development programs have been rooted in human learning problems.

Ishtiaq Qureshi, once Minister of Education in Pakistan, tells the story of efforts to incorporate training programs into economic development schemes soon after independence, in the early fifties. He tried to persuade an official high in the government bureaucracy to agree to sponsor a plan whereby the Ministry of Education would apply for aid from an American foundation. The reply given

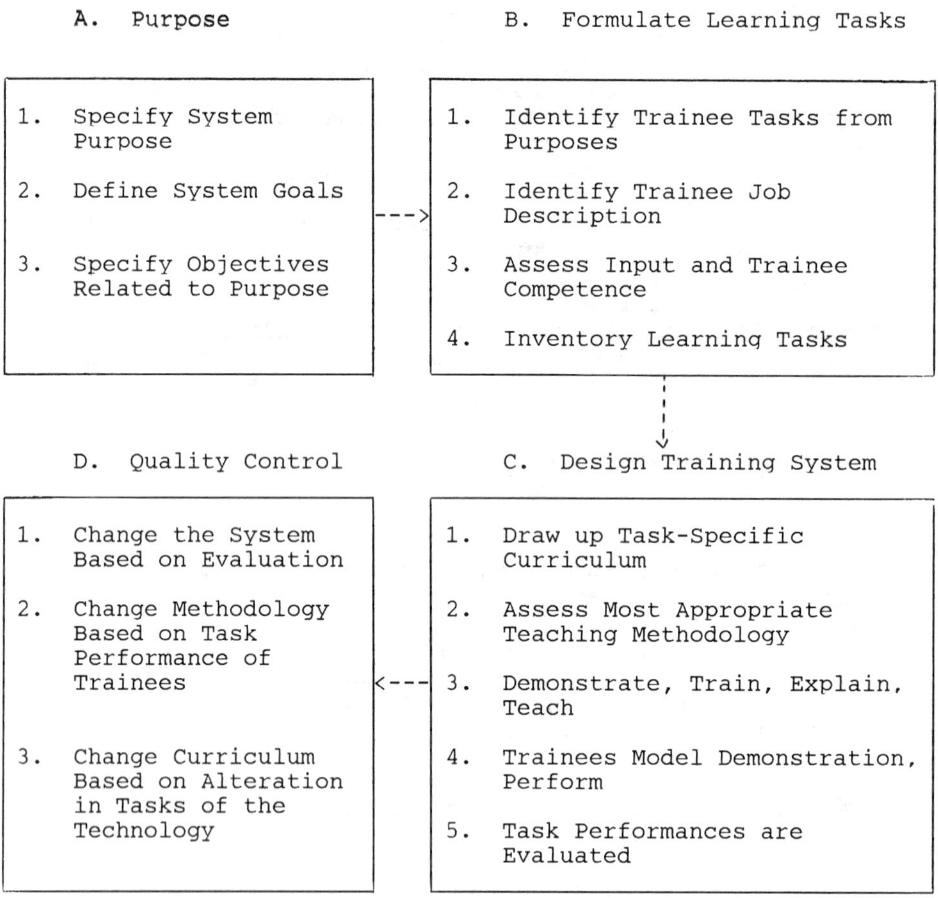

FIGURE 1. TECHNICAL TRAINING SYSTEMS MODEL

to him was, "We should first satisfy the hunger of the poor; luxuries like education will come later." What Qureshi reported for Pakistan was true for many developing countries in the fifties and sixties, and is, in many cases, still true.

Economic planners frequently looked at education and training as almost irrelevant in the entire process of economic development. During the past twenty years, particularly in sub-Saharan Africa, development plans did not bear fruit as many had hoped. Predictions for change did not, in many cases, work out as expected. People in the Third World frequently accepted the artifacts of technology, but were unable to use and maintain these optimally. Economic growth was tied to political realities and to those elites who were influential in maintaining their own status. Economic development was rooted in its "source of inspiration", to use Qureshi's words, ". . . the manpower to bring about and establish development".[4]

The last twenty years might be characterized as an era of technology transplant, as opposed to an era of human resource development. Communication systems in rich countries advanced rapidly with new technologies. These systems, in their advanced forms, including television, radio, computers, telephones, newspapers and movies, were introduced into undeveloped countries rather early in the development process. These became necessities for the elite. Soon the poor countries were bombarded with information about ways of life elsewhere and the living styles of the wealthy. With their own eyes, people saw how the rest of the world lived and made comparisons about their own life styles and obvious deprivation. Isolated poverty was tolerable when people had no comparisons and believed that their way of life was normative. Even the idea of others was rather foreign to the isolated village man. But when very rich and very poor, prosperity and poverty, are placed side by side, a politically dangerous situation emerges. Political change in the Third World has been rapid and frequently violent as evidenced in South

American and Africa.

The highly complex process of economic change and development was tied to the general standard of living and educational levels of the people. However, these standards and levels were not evenly spread. In particular, in every developing country the gap between the poor and rich grew wider. In Pakistan, for example, this gap was seldom bridged with communications or attempts at social integration between the rich and poor; rather the development process polarized the communities into distinct classes. Similarly, rich and poor nations, during the past twenty years, developed gaps. The richer nations, because of highly-developed economic structures, accelerated and developed their supporting technologies based on high levels of literacy and availability of general education for the majority of their people. Poor nations in comparison, lacking sophisticated infra-structures, feel farther behind. The elite who had received good educations in universities abroad, received not only power, but managed the economies within the political structures in which they worked. Conversely, rich countries experienced the development of an affluent middle class who became eager consumers of new inventions. In poor countries, the middle class was frequently very small. The masses remained in poverty, living on the land, subsisting, but not able to enjoy the benefits of a middle class which was consumption-oriented.

Economic development seen in this framework is certainly a complex process since humans and the resources available to them are a vital aspect of stability and change. Consequently, planners in the eighties are far more cautious and wary of simplistic development solutions, and of being too hasty with the introduction of technologies which can not be supported by poor countries. Experts working in foreign development settings come with their pockets bulging with technological know-how and equipment to do the job. Their biggest challenge relates to dealing with people, educating and training, so that the very technology they introduce becomes socially functional.

Farmers are now the target for international aid in most developing countries. It is well understood that any development of an industrial structure must have, as its support system, an agricultural structure which can feed the people and raise the standard of living of the population. Strategies toward this end are being suggested. A recent publication, Promoting Economic Development in Central America, states:

> The broad strategy is: First, to arrest the declines in incomes, employment, and economic activity by means of major balance-of-payments assistance; Second, to lay the foundation for long-term economic growth and the infrastructure necessary for export orientations; Third, to assure the widest possible distribution of the benefits of growth through assistance for improving health, education, and housing for the poorest groups; Fourth, . . . technical training, and the development of leadership skills.[5]

Concomitant with such planning comes development interruptions in the form of huge famines in Africa and frightening refugee problems in Afghanistan, Western Sahara, and South and Central Africa. Africa's growing dependence on foreign aid, along with many other Third World countries, creates additional interruptions in the attempts to bring development assistance in a way that recipients can support. John C. Whitehead,[6] Deputy Secretary of State, outlined a proposal which is a coordinated attempt to bring various international donor groups together to help solve the growing world crisis of hunger, particularly in Africa. Included in his plans are ways to help people help themselves, to educate, to train and to provide long-term education programs which are sector-wide in their scope.

Literacy education is a feature of most development plans. Along with development of the agricultural sector to support industrial growth, efforts are being made to increase literacy rates. Qureshi's story again comes to mind. It is important to satisfy the "hunger of the people" first, but general development involves an integrated and complex interaction between people and technologies, between satisfaction of the hunger for more information and more knowledge, and improved health.

Eric Ashbey, the author of the well-known Ashbey Report (1962),[7] was concerned with drawing up a manpower development plan for Nigeria. His writing was infused with the idea that education was the leaven for economic development. Education had to occur simultaneously with economic planning and the introduction of new technologies. Another economic expert at this time was Harbison who argued that "a plan for economic development must consider available resources of manpower trained to different needed levels and must make provision for such education and training."[8] Underdeveloped countries at that time faced a special problem. They had a limited educational base and a limited industrial support system. Training manpower for absorption by the economy was frequently an unpredictable and delicate affair. An over-emphasis in one area could result in imbalance and dysfunctional programs. Formidable expansion of secondary grammar schools and education facilities was a focus of the sixties.[9] Many of these schools today, in the eighties, are in shambles.

In highly developed countries, the sheer mass of the educational and industrial system insured that adjustment could be made fairly readily to supply manpower which would be absorbed by the developing technologies. In less-developed countries, there was danger in over-producing people with education which did not match national needs. Highly-trained people who could not be absorbed into a limited industrial/technical sector in the Third World became unemployed. In one sense, a person who was a technician but not employable was worse off than a common laborer looking for work.

Forecasting the type of person who would be needed to support developing industries became the hazardous job of planners, particularly planners who suggested changes in educational systems. In Pakistan, manpower planning for training occurred very late; their vocational/technical training programs were in their infancy. New technical schools require large capital investments for machinery and equipment. The machinery must be up-to-date and similar to what is used in industries. Funding needed to establish new institutions is simply not available. One need but travel from one technical training school to the next in Pakistan to see how poorly equipped these institutions are. Shops are meagerly equipped and many tools are in disrepair. The record of Pakistani industry in the maintenance of machinery has been very poor. In many instances, its main cause has been the lack of training and skill among technicians.[10]

Manpower training requires tools. Lacking tools, training tends to become purely theoretical and non-experimental. Two or three generations of graduates who are trained without learning adequate skills dilute the entire system. Standards go down rapidly. Frequently, one way out is to seek another injection of foreign aid to boost the system for a while.

In many instances, training of trainers is not a central aspect of the foreign aid package. Simplistic solutions have not worked. Short-term plans to "put out brush fires" leave the land scorched and arid. For instance, in the process of training technicians, it is important to remember that these people have a need to be broadly educated so that they have some understanding of their own society and its needs and a sense of personal responsibility to carry out tasks in functional ways. Narrow vocational training, especially in the early phases of development, is likely to be unproductive. Philip Foster discusses this action in much of his writing. His article, "The Vocational School Fallacy in Development Planning,"[11] reviews what occurred in India.

Educators developed technical training programs which put hundreds of children at spinning wheels and looms. Not only did such programs fail to remove the cause of poverty but a feeling of hopelessness was generated among those so trained.

It may be argued that the major concern of most developing countries is not to bring in sophisticated science and industry, but rather to raise the standard of living sufficiently so that poverty is reduced and the general standard of living of the masses is raised. Economic development in this frame of reference, requires continual attention to the development of training systems in all sectors related directly to the capability of countries to sustain this growth. Development in this sense is evolutionary, gradual and tied to on-going structures which exist in the societies in question. Development requires sustained time to occur.

W. David Hopper defines the development process as one in which "the pile of goods available to a population is increased at a pace much in excess of the traditional rate of growth or output,"[12] He goes on to state that development as he sees it is the "problem of engendering and supporting the transformation of the technological basis of production for . . . people, who were primarily dependent upon traditional agriculture and agrarian techniques, to methods based on modern science and technology. This transformation is the core of today's development concern."[13] Planning for training has its genesis here in the development of functional educational systems which allow for learning new ideas and developing knowledge to unleash new technologies. Another way to say this is that investment in human capital, building people, is the cornerstone of the development process.

EDUCATIONAL TECHNOLOGIES FOR DEVELOPMENT

An American university professor working as a

technical expert for a World Bank project in Pakistan a few years ago, was asked to help provide training to nationals in his area of expertise. His reply was, "Training is not in my terms of reference and is not on my job description. Sorry, I am here as a technical expert, not a trainer." Don Dwyer, the Executive Director of the Consortium for International Development (CID), said in 1985, that our technologies and expertise are assumed when we go abroad. Our major task will be to educate, to teach, and to train people. Universities are dedicated to this mission and, therefore, universities should have a central role in human resource development.

Institutions of higher education in this country are being requested to participate in the technology transfer process abroad at an ever-increasing rate. The expertise within these university systems is vast, impressive and flexible. Through US-AID, the World Bank and Asian Development Bank, among others, vast development proposals are being planned and funded. These proposals are frequently being staffed by university professionals. A tremendous challenge exists for these agents of change to consider the best possible means to introduce technical systems abroad. For technical educators, professionals in universities, a great professional challenge exists to become involved with foreign technologists and scientists in designing curriculum and training systems which will optimize the way people learn and adapt to new ways of living.

> Just as the primitive technologies of manufacturing and industry have been replaced by scientifically based procedures that provide a high degree of product control, so too is it anticipated by many that a scientifically based technology of education will emerge that will provide a high degree of control over the educational product. . .
>
> The concept of the development of a technology of education is somewhat different from that of providing a scientific basis for

> guiding the decisions the teacher has to make. A technology would, presumably, establish specific and well-defined procedures through which particular goals could be reached, together with packages of materials through which the goal could be achieved.[14]

Training people for productive work will require the attention of such teams working together to bring about change.

REFERENCES

1. Everett M. Rogers. 1983. Diffusion of Innovations. New York: The Free Press.

2. Dan L. Costly and Ralph Todd. 1978. Human Relations in Organizations. Los Angeles: West Publishing Company.

3. Sherry Sweetnan. May, 1986. Training proposals that sell themselves. Training and Development Journal. American Society for Training and Development.

4. I. Qureshi 1975. Education in Pakistan. Karachi, Pakistan: Ma Aref, Ltd.: 218.

5. R. McCormick. 1985. Prompting development in Central America Current Policy NO. 759. Washington, D.C.: Department of State, Bureau of Public Affairs.

6. J. Whitehead. 1985. The African Crisis (Current Policy No. 757). Washington, D.C.: Department of State, Bureau of Public Affairs.

7. E. Ashbey. December, 1962. Mission and Challenge of West African universities, Teachers' College Record: 227-230.

8. E. Ashbey. 1964. African Universities and Western Tradition. Cambridge, Mass: Harvard University Press: 227.

9. H. Bergsma. 1978. Education for International Development: Problems for Planned Change. Dialogue Series, New Mexico State University.

10. Op.cit., Quershi, p. 222.

11. P. Foster. October, 1975. Dilemmas of Educational Development: What We might learn from the Past. Comparative Educational Review: 378-402.

12. D. Hopper. 1983. New Horizons for Education in International Development. Proceedings of Presidential Symposium: New Horizons for International Development. Consortium for International Development. Arizona: 11.

13. Ibid: p. 12.

14. Robert M.W. Travers. 1972. Essentials of Learning: An Overview for Students of Education (3rd. ed.). New York: McMillan Company: 493.

THE ON-FARM DEMONSTRATION ACTIVITY: A SOCIO-ECONOMIC PERSPECTIVE*

SALAH EL-ZOGHBY AND HODA BALIGH
Desert Development Center
American University in Cairo

Abstract The on-farm demonstration activity is a means for achieving one of the main goals of the Desert Development Center (DDC). That goal is to improve the agricultural practices of the farmers in reclaimed desert lands in order to increase Egypt's self-sufficiency in food and feed production.

The objective of this paper is to report on the development of the DDC on-farm demonstration activity since its initiation in Summer 1985.

The paper covers initiation of the activity, selection of participants, objectives and operational basis of the activity; different programs functioning today within the activity; and some background information on the participants in the program. The paper emphasizes socio-economic variables and discusses some technical aspects that have had an effect on the success of the on-farm demonstration activity.

The on-farm demonstration activity is adversely affected by: participant's off-farm residence; intercropping, sharecropping, or long-term contracting (practices followed by the farmers as a means to increase their income); minimizing control areas; failure to keep control areas as control; discontinuation of financial assistance; non-availability of chemicals on the market and high prices of chemicals when available in the free market.

---

* Research is sponsored by the Ford Foundation.

INTRODUCTION

The on-farm demonstration activity of the Desert Development Center (DDC)[1] is carried out by the Center's agricultural specialists on the farmers' lands. The specialists work with the farmers throughout the year recommending appropriate practices for desert agriculture. The on-farm demonstration programs have been directed, until this time, toward increasing crop production and improving soil.

The major objective of the on-farm demonstration activity of the DDC is to improve practices of farmers in order to increase Egypt's food and feed production on the newly-reclaimed desert lands. Egypt's population is growing by 2.7% annually. Today, with a population of about fifty million, the country realizes a net increase of one million every nine months. Providing food for this ever-increasing population is one of the alarming problems facing the country. At this time, Egypt imports about 50% of its food, and with agricultural output increasing less rapidly than the population, the food gap continues to widen (Abdel Kader 1985: 4).

Thus, in addition to applied research and experimentation projects on its two sites in the South Tahrir area of Egypt, the Desert Development Center is demonstrating appropriate agricultural practices on the farms of settlers in the area.[2] The on-farm demonstration activity was initiated in Summer 1985. Although the activity has been in effect for only a little more than a year, much has been learned that will be considered in future implementation.

The purpose of this paper is to report on the development of the on-farm demonstration activity and discuss socio-economic variables (with some reference to technical aspects) affecting its success. The paper is divided into four parts: (1) statement of the objectives and the operational basis of the DDC on-farm demonstrations, (2) a discussion of the different programs operating today, (3) a presentation of background information about the programs' participants, and (4) a discussion of socio-economic and technical variables that have been found to effect the outcomes of the programs.

Data about the on-farm demonstration activity were obtained from (a) first-hand information gained as members of the planning team, (b) personal observations made on field-visits, (c) written reports of field-visits, and (d) a questionnaire administered to the on-farm demonstration participants.

OBJECTIVES AND OPERATIONAL BASIS OF THE ON-FARM DEMONSTRATION ACTIVITY

The on-farm demonstration activity is an attempt to transfer successful technologies from DDC sites to two major groups of farmers: the graduate farmers and the smallholders[3]. The graduate farmers are graduates of agricultural high schools or colleges. The smallholders have very little education. Both groups were allocated land from the Egyptian government and are on farms in the South Tahrir area.

The model for the on-farm demonstration activity has the following phases: (1) problem or research area identification, (2) experimentation, (3) demonstration and training, (4) evaluation and (5) diffusion. Having identified the problem and research areas, experimentation is carried out, if necessary on the DDC farms. The successful applications are then demonstrated on farmer's lands. At the same time, the farmer is being trained in the proper means of applying the technologies. When the program proves to be economically feasible and socially acceptable, diffusion of the program to the larger population is carried out. For example, agricultural bulletins for specific crops are prepared and distributed to the farmers in the area.

One of the first steps in an on-farm demonstration is the selection of participants. The variables considered when selecting the farmers to participate in the on-farm demonstrations are both social and technical. Social characteristics include the educational level of the farmer as well as his willingness to cooperate and follow recommendations. Technical variables included soil characteristics and the general conditions of the fields.

Selecting appropriate plots is the next step in implementing an on-farm demonstration. Areas that are satisfactory to both the farmer and the DDC specialists are selected as treatment and control areas on the farmer's land. In the treated area, the new technologies and the program recommended by the DDC are applied. In the control area, the farmer continues with the practices he applied prior to participation in the program.

The DDC on-farm demonstration interdisciplinary team of agronomists, economists, sociologists and anthropologists visit demonstration farms on a weekly basis. In addition, a liaison officer who resides in the area monitors the on-farm demonstration plots on a regular basis. Socio-economic monitoring is carried out periodically by the DDC to study the economic feasibility and social acceptability of the demonstrations.

As an incentive to participants, the DDC provides financial assistance. The expenses for chemicals, including costs of application, are paid in full by the DDC for the first year. Half of the cost of inputs is paid for the second year. In the third year of participation, the DDC does not pay the farmer but the DDC staff continues regular field visits and offers advice. Technical and socio-economic monitoring also continue. The farmer is assured that he will be financially compensated if he has followed DDC recommendations and if, for any reason, the yield of the treated area is less than the average yields that he reported, prior to participation.

ON-FARM DEMONSTRATION ACTIVITIES WITH GRADUATES AND SMALLHOLDERS:

Before beginning any on-farm demonstrations, the DDC conducted a survey of graduates in Summer 1984. The purpose was to collect background information about different farmer groups and to identify the major constraints that the farmers face, especially in crop production. The survey revealed that the problems of greatest concern to the graduate farmers were lack of water for irrigation, shortage of agricultural labor, and plant infestations (i.e., nematodes and weeds).

The first two constraints were on a macro level and the DDC did not have the means to solve them. In order to gain the confidence of the farmers in the initial stage of demonstrations on farmers lands -- the most skeptical and difficult stage -- problems were tackled for which quick successful results were assured. Based on previous success with weed control at the DDC's South Tahrir site, and the availability of qualified specialists in plant protection, priority was given to solving the problems concerning nematodes and weeds[4]. The fact that the South Tahrir area lacks any kind of extension service was an additional motivation for beginning the on-farm demonstration activity.

To combat nematodes in an area, at least two or three years of treatment are required. Therefore , a long-term Plant Protection Program encompassing the control of weeds, nematodes, fungus and bacterial diseases was designed and is being carried out by the on-farm demonstration team. The program focuses on application of nematocides, herbicides and fungicides.

The program was introduced in April 1985 with a slide presentation by DDC specialists for the graduate settlers of South Tahrir. At the presentation, graduates who were interested in participating in a Plant Protection Program were asked to volunteer for demonstrations on their fields.

Ensuing developments showed that things do not always go according to plan, however. Some of the farmers who expressed a desire to take part in the on-farm demonstration activity did not appear on the day when specialists had scheduled a visit to their farms. On the other hand, farmers who had not attended the presentations heard of the DDC's activity and asked to be included in the demonstrations.

The Plant Protection Program was initiated in Summer 1985 with six graduate farmers. The Program had positive results in the first season. Crops in the treated areas grew remarkably compared to those in the control areas on all demonstration plots[5].

Farmers on surrounding lands viewed the results with varying reactions. Some of the farmers who had asked to be included in the program originally, but who did not show up on the day of the specialists' visit to their farms, came, now convinced, asking to be included again. Some farmers came on field visits to see the specialists and to ask for different kinds of advice. Other farmers explained that during the season they observed the superior performance in the treated areas versus the non-treated ones and, in turn, applied the same practices to their farms themselves.

Based on achievements realized in the Plant Protection Program, the on-farm demonstration specialists started preparing comprehensive programs encompassing agricultural practices in addition to plant protection and fertilization. In February 1986, the second on-farm demonstration program, a Citrus Comprehensive Program was started with five graduate farmers (see Table 1).

Smallholders were brought into the programs in 1986. In Summer 1985, the DDC conducted a second survey. This survey queried smallholders and found the spread of plant diseases to be one of the major problems affecting their crop production. Fourty-two percent of the sample said that they lacked knowledge about preventive treatments for

TABLE 1: DDC on-farm demonstration programs (summer 1985-winter 1986/87)

PROGRAM	STARTING DATE OF THE PROGRAM	CROPS DEMONSTRATED ON		PARTICIPANT	PARTICIPANT GROUP OF FARMER	STARTING DATE OF THE PARTICIPANT	PARTICIPANT'S PRESENT STATUS IN THE PROGRAM		
		SUMMER	WINTER				PARTICIPATING	DISCONTINUED DATE	DISCONTINUED REASONS
Plant Protection Program	Summer 1985	Groundnuts	Peas	A	Gr.*	Summer 1985	X		
				B	Gr.	"	X		
				C	Gr.	"	X		
				D	Gr.	"	X		
				E	Gr.	"		End of Winter 85/86	Non-availability of capital
				F	Gr.	"		End of Winter 85/86	Absentee Land Owner
Comprehensive Program for Field Crops	Summer 1986	Groundnuts	Peas Wheat	A	Gr.	Summer 1986	X		
				B	Gr.	"	X		
				H	SH.**	"	X		
				K	SH.	"	X		
				L	SH.	"		End of Summer 1986	Not complying with the program
Comprehensive Citrus Program	February 1986	Limes		A	Gr.	Feb. 1986	X		
		Limes		B	Gr.	"	X		
		Oranges		M	Gr.	"	X		
		Oranges		N	Gr.	"	X		
		Oranges & Tangerines		P	Gr.	"	X		

* Gr. = Graduate Farmer **SH. = Smallholder

plant diseases. During the survey, smallholders were informed of the Plant Protection Program operating on the six graduate farms and the names of smallholders who wished to participate in this type of program were recorded. From those, three smallholders were selected. In Summer 1986, a third on-farm demonstration program, the Comprehensive Program for Field Crops started with three smallholders and two graduates.

The three programs, Plant Protection, Citrus Comprehensive and Comprehensive Program for Field Crops are operating concomitantly (see Table 1). Funding allows only a small-scale endeavor, however. The on-farm demonstration activity which started in Summer 1985 with six graduate farmers includes nine farmers at this time: seven graduates and two smallholders participating in three different programs. There are thirteen demonstration plots as two farmers participate in more than one program. Some farmers have left the program while others have joined. Cooperative participants continue with the program, while those who do not comply with regulations of the program are dropped.

PARTICIPANTS IN THE ON-FARM DEMONSTRATION PROGRAMS

Only a few generalizations can be drawn about participants in the on-farm demonstration programs because their number is small. Table 2 gives some background information about the participants. A total of twelve farmers have taken part in phases of the program; nine are presently participating. Of these nine, only two stated that they sought information about agriculture from other sources prior to taking part in a DDC program. Seven of the participants said that they had been following agricultural practices that they had learned from experience.

All nine participants reported that they changed some of their agricultural practices when they joined the program. The farmers had been

TABLE 2: Background information on the present participants in the on-farm demonstrations programs

PARTICIPANT	GROUP OF FARMER	AGE	EDUCATIONAL STANDARD	RESIDENCE ON-FARM WITH FAMILY	RESIDENCE ON-FARM WITHOUT FAMILY***
A	Gr.*	34	High School Diploma	X	
B	Gr.	48	B.Sc. in Agriculture	X	
C	Gr.	51	B.Sc. in Agriculture	X	
D	Gr.	31	B.Sc. in Agriculture		X
M	Gr.	48	B.Sc. in Agriculture		X
N	Gr.	43	Diploma		X
P	Gr.	42	Diploma	X	
H	SH.**	56	1 Yr. of School Education	X	
K	SH.	34	6 Yrs. of School Education	X	
TOTAL				6	3

* Gr. = Graduate Farmer

** SH. = Smallholder

*** Farmer commutes between the city where his family resides on his farm, on a more or less regular basis. The agricultural practices are usually carried out by a full-time hired farmer living on the graduates' farm.

using some fertilizers, herbicides, nematocides, and fungicides, more of less, before taking part in the OFD programs, but none of the nine were found to be using them correctly. All of the graduates (6) and one smallholder reported that they had applied fertilizers but, in all cases, the DDC specialists said that the farmers were applying them incorrectly. Herbicides had been used by four of the farmers, but again the specialists stated that they were not used efficiently. Nematodes and fungus received less attention from the farmer. Only one (a graduate) had applied a nematocide and another (a smallholder) spread manure with the belief that it would cause the nematode population to decrease. One graduate had used a fungicide, again incorrectly.

The following reasons were given for joining the program: to increase agricultural knowledge (3 graduates and 2 smallholders); to increase crop yield (2 graduates and 1 smallholder); and to have the inputs paid for (1 graduate).

VARIABLES AFFECTING THE SUCCESS OF THE ON-FARM DEMONSTRATION

Some socio-economic variables affecting the success of the programs can be identified. Place of residence has been found to be an important factor along with a variety of practices which the farmer chooses because of their financial implications.

The place of residence of the participant is an important factor. While all smallholders live with their families either on-farm or in villages adjacent to their farms, not all the graduates live on-farm with their families (see Table 2). For many graduates, the quality of life offered in South Tahrir is not satisfactory for their families. They chose to live where they have access to better schools or urban facilities. When the family is not residing with the farmer, he spends considerable time at their residence. Being absent much of the time, these participants often delay the scheduled program of the on-farm demonstration. They are often not available during field visits when specialists want to give them their recommendations and train them in how to apply the new technology. Because he was seldom available, one absentee landholder was eliminated from the Plant Protection Program at the end of the winter season of 1985/86.

Some practices which are followed by the farmers as a means to increase their income have a negative effect on the on-farm demonstrations. These are intercropping, sharecropping and contracting. In the case of intercropping, in order to compensate for the expenses spent on young citrus trees in the first years before fruit-bearing, some of the farmers intercrop different kinds of seasonal crops with the young citrus trees. The seasonal crop provides income from the land before the citrus trees mature. Because of intercropping, the program, which is planned for only one crop, has to altered. Quantities of different inputs are increased to serve both the citrus trees and the seasonal crop. This results in difficulties in technical monitoring. The economic monitoring is also affected, as input/output analyses are distorted.

Sharecropping, the partnership of two or more farmers in the cultivation of a crop, has also affected the technical and economic monitoring of the programs. For example, in Summer 1985, one graduate cultivated groundnuts on a sharecropping basis with a traditional farmer. The traditional farmer had weeded his groundnuts three to four times per season in the past, and was not at all convinced that applications of herbicides, the recommended treatment, would be sufficient for weed control. While the participating farmer believed in the new technology and was ready to apply it, his partner resisted adhering to the recommended program.

In another case, long-term contracting (for over a year) of a citrus orchard affected technical monitoring of the on-farm demonstration. The contractor is usually a merchant who takes over the application of all the agricultural practices including irrigation and pays the farm owner a yearly rent. Farmers consider contracting to be a means of keeping their orchards serviced when they lack time or are short of capital. An example is a graduate participant in the Citrus Program who works as a lawyer in the area in addition to farming his land. Due to lack of time he has contracted his citrus orchard. In two other cases, farmers had contracted their orchards prior to agreeing to participate in the DDC program. The DDC was not informed of this. In order to achieve maximum profit from the on-farm demonstration program, some farmers tend to minimize the control area. The smallholders especially tend to allocate a very small strip of land as the control area. This may make the control area invalid. It may be too small to measure output accurately or, if adjacent to the treated area, the control area may be sprayed unintentionally when the treated area is sprayed with chemicals. The team also encountered a problem in having the farmers retain the control area. It was often the case that, when the participant farmer observed the success of the treated area, he applied the chemicals used in the treated area to the control area as well.

The financial assistance offered as part of the program during the first two years of a farmer's participation was an incentive for some of the participants to adhere to the program. When full financial assistance was no longer given, two participants were reluctant to follow recommended practices.

The main obstacle for the on-farm demonstration activity, however, is one of supply on the national level. Chemicals are often not available or, when available in the free market, the prices are very high. For weed and nematode control, the DDC advocates chemical applications. Nonavailability of the chemicals in the market has often caused delay of the on-farm demonstration activity. At the beginning of the Plant Protection Program, some farmers who wished to be included in the activity could not obtain the needed chemicals and thus were excluded from participation. Other times, farmers were successful in finding the needed chemicals on the free market, but at very high prices. Prices were generally three to five times the official subsidized price set by the government. Due to lack of capital, one exemplary follower of the DDC program had to withdraw, regretfully, in the second year of the Plant Protection Program. He could not pay for 50% of the cost of the chemicals to be applied to his field.

CONCLUSIONS AND RECOMMENDATIONS

1- Prospective participants for on-farm demonstration programs should be considered carefully in order to guarantee the success of the program. Based on our experiences, we recommend that priority be given to:

   a) A full-time farmer, one whose only occupation is farming.

   b) A farmer who operates and manages his farm by himself, rather than through sharecropping or any other type of tenancy.

   c) A farmer who markets his own crops directly, not through a field sale contract.

   d) A farmer who resides on his farm with his family rather than one who resides on the farm alone while his family lives outside the area.

2- The government should give more attention to "old" new lands. To be able to achieve the desired returns from reclaimed desert lands of Egypt, the government should oversee providing these areas with the necessary needed inputs. The lists of agricultural chemicals prepared by the government do not coincide with the needs of the farmers in the South Tahrir area.

Cooperative societies, which are important institutions in the area, should express the demands of the farmers when preparing the list of chemicals requested from the government. The prepared list should be revised and approved, as needed, by a scientific agency which updates the list regularly with the latest scientific advances. The government would then be responsible for providing the cooperative societies in the area with the approved chemicals.

3) Farmers in the area lack accurate information about proper agricultural practices to follow and are desperate to improve their lot. We strongly recommend the establishment of an extension service office in the area of South Tahrir. Working in cooperation with such an office, the DDC could contribute successful programs from the on-farm demonstration activity to an on-going program which would serve a great majority of the farmers. In this way, diffusion of successful practices could be achieved on a large scale in the area.

FOOTNOTES

(1) The Desert Development Center (DDC) of the American University in Cairo operates on two sites: one in Sadat City and the other in the South Tahrir area. The DDC conducts applied research into alternative, economically viable, integrated approaches to arid land agricultural and community development.

(2) The site of the on-farm demonstration program is South Tahrir in Tahrir Province. Started in 1953, it is one of the oldest land reclamation projects in Egypt. Presently the population of South Tahrir is composed of different farmer groups such as agricultural college and high school graduates, smallholders (traditional farmers), and

personnel who previously worked in one of the following: a land reclamation company, a specialized or land reclamation cooperative society, or an Egyptian or joint-venture investor company.

The one-farm demonstration program is carried out mainly in El-Tahadi, the southwest area of South Tahrir. It consists of 37,078 feddans of sandy soil. A sprinkler irrigation system irrigates the area. Water is drawn from the Nubaria Canal, a branch canal from the Nile, as well as from ground-water. (Pacific Consultant 1980: 2.8)

(3) Beginning in 1977, graduate farmers were allocated twenty to thirty feddans of land. Today there are 498 families in the area. The smallholders were originally landless traditional farmers from overpopulated governorates of Egypt. They were allocated three to five feddans in the new lands. More than 2,000 smallholder families reside in South Tahrir.

(4) South Tahrir area, which is a major area in Egypt for growing groundnuts as a summer crop, is highly infested with various nematode species. Nematode may reduce productivity per feddan up to 60%.

For weed control, mechanized cultivation of weeds is not a popular practice among farmers in South Tahrir because the machinery is expensive and maintenance services and spare parts are difficult to obtain, especially in remote areas. On the other hand, the shortage and the expense of agricultural labor in remote and new land areas makes manual weed cultivation an expensive practice also.

(5) Brief results of plant protection program:

a) For Summer 1985, on groundnuts:

- Average yield of treated area: 14.75 ardab/fed.
- Average yield of control area: 7.25 ardab/fed. (excluding farmers #5 & #6)
- Average haulms value: 49.8 L.E./fed. (excluding farmer #6)
- Average net income increase: 429 L.E./fed.
- Treatment expenses: 43 L.E./fed. for weed control
  65 L.E./fed. for nematode control

b) For Winter 1985/86, on peas:

- Average yield of treated area: 1.92 ton/fed.
- Average yield of control area: 0.9 ton/fed.
- Average net income increase per feddan: 256.3 L.E.
- Treatment expenses for weed control range from L.E. 3-8
- Average treatment expenses for nematode and weed control: L.E. 49.0

REFERENCES

1. A. Abdel Kader. December 1985. Economic trends report: Egypt. The American Embassy in Cairo.

2. Desert Development Center Report 1985. American University in Cairo.

3. S. El-Zoghby, A. Bishay and B. De Roo. December 1984. "Planning phases for new desert communities in Egypt: An integrated approach". Paper presented at the VI International Rural Sociology Conference, Manila, Philippines.

4. Pacific Consultants. January 1984. Sociological Considerations. Tahadi: A case study of a land reclamation project in Egypt. Working Paper No. 2.

WOMEN'S PERSPECTIVE ON LIVING IN NEW DESERT SETTLEMENTS

BETTY M. WASS
Desert Development Center,
American University in Cairo, Cairo, Egypt

Abstract Over the past 25 years, reclaimed lands in the deserts of Egypt have been allocated to different types of settlers. These include smallholders, graduates, investment company employees, and private sector individuals. The problems that each group faces and the ability of each to adjust to desert farming should interest authorities who plan additional resettlement schemes.

This study analyzes part of a larger survey conducted in 1985 with smallholder farmers. It reports on the problems in desert living which were expressed by 132 smallholder wives and examines relationships between the number of problems expressed and other factors affecting their lives.

Most women expressed problems across a range of categories rather than concentrating on one type of problem. The problems expressed most frequently by the wives pertain to obtaining food, electricity and water, health care, and transportation. Significant relationships were found between the number of problems expressed and three independent variables: (1) amount of land farmed by the husband, (2) the husband's years of experience in desert farming, and (3) amount of education of members of their households. Fewer problems were expressed by those with more than three feddans of land, more than ten years tenure on their present farm, and relatively more education among members of the household.

INTRODUCTION

Egypt has accorded top priority to making desert lands productive. More agricultural land is a necessity if the growing dependency on food imports is to be lessened and the pressures on Egypt's burgeoning cities are to be relieved. With these goals in mind, reclamation and resettlement efforts began as early as the 1950's. Intermittently, different groups have been targetted for resettlement. Smallholders, graduates, investment company employees, public sector employees and members of the private sector are major groups who have been allocated land.

Reactions to resettlement efforts have been mixed. On the one hand, a stable population of settlers have inhabited the new lands for long periods of time -- some for 25 years. On the other hand, new cities on desert lands stand nearly empty because intended settlers are reluctant to leave the familiar surroundings of the overpopulated, deteriorating old cities. Uncertainties of life in the new lands and anticipated hardships deter prospective settlers.

To gain more accurate information about the realities of life in the new lands, settlers in South Tahrir, one of the earliest land reclamation sites in Egypt, were the subject of three studies by the Desert Development Center (DDC) of the American University in Cairo. From the studies, the DDC gained a better understanding of the characteristics and needs of the farmers, and identified the major problems perceived by them. This information guides the DDC in structuring outreach activities directed toward solving problems that are major impediments to a satisfying livelihood for the farmer. The surveys also provide data about the adjustment of different groups of settlers that should be of considerable interest to authorities who are planning additional resettlement schemes.

A survey conducted in Summer 1985 collected information from 144 smallholder farmers who represent a population of approximately 2100 smallholder households in South Tahrir. Wives were interviewed separately from their husbands and it is the qualitative information expressed by wives in this survey that is the focus of this paper. Interviews with 132 wives were completed by DDC staff members and trained interviewers from the International Center for Rural Development at Mariut.

SMALLHOLDER HOUSEHOLDS

In order to understand the perspective of wives of smallholders, it is necessary to first give a general view of characteristics of the smallholder households. The descriptive data which follow are specific to the South Tahrir sample.

The smallholder represents a type of farmer which has existed for centuries and is still the mainstay of Egyptian agriculture. Upon settlement on reclaimed lands in South Tahrir, smallholders receive three to five feddans of land. Some have increased their original allotments, yet the average size of all smallholder farms in the area in 1985 was 3.62 feddans.

Eighty-four percent of the smallholders indicate that they own their land. The rest are renters. In reality, the Government has promised ownership, but in most cases, has not yet given smallholders the official titles to their lands. Of the "owners", 31% rent additional land.

The smallholder households average 8.3 members. The selection criteria for smallholders who would be allocated land stipulated that they should be married and have a family of three or more. The smallholders studied tend to have four to six children and other relatives are often included in the household. The nuclear family predominates slightly, but extended family households are almost as common. In spite of the divergency of generations within a household, more than half of the population are below the age of 20.

Almost all of the household members who are able to work are employed on the farm. Only 5% of the households have a member working fulltime away from the farm. Agriculture is the lifeblood of the smallholders. The fact that almost all working members are involved in agricultural production indicates limited opportunity to do anything else. The new lands do not contain the range of economic and social activities found in the old lands. This lack of other employment is not necessarily disadvantageous to the farmer because most of the agricultural techniques used are labor intensive and large families are valued for their input in accomplishing the endless farm tasks that confront them. In such small scale farming, little mechanization is used.

The smallholder wives range in age from 20 to 70 years. The median age is 40. Eight-five percent of them have no formal education. The highest level of education attained by any of the 132 wives interviewed was six years and only one of the wives had reached this level. Their husbands had slightly more education on the average, but two-thirds of them also had no schooling.

Even though the educational level of women is lower than that of their husbands, women have an important voice in decision-making in the household. Ninety-five percent of the women interviewed said that they participate with their husbands in making decisions about farm as well as household operations. Women also have the major responsibility for small-scale selling of farm products. The settlers consider this activity to be an aspect of "farm" work. Nearly half of the households sell some dairy products, poultry or eggs.

There are additional facts which describe the lives of the smallholders. These include the following:

Smallholders usually live in villages and go to their fields either on foot or by donkey. Only about 50% have access to a post office and telephone in their village. Twenty-five percent live within five kilometers of a doctor or private hospital; 11% must go more than fifty kilometers to a doctor. About half have electricity and one-third have potable water in their houses.

PROBLEMS STATED BY WIVES OF SMALLHOLDERS

In an open-ended question, interviewers asked the smallholder wives, "What are the problems that you confront in the new community?" Two women expressed no problems and two stated more than twenty problems. The average number was 7.3. Table 1 shows the types of problems expressed by category, frequency, and percentage of the total number of problems stated.

TABLE 1. Problems of wives of smallholders on new lands.

Problem	Frequency	Percent
Animal production	5	0.5
Community facilities	10	1.0
Crop production	40	4.2
Cultural activities	29	3.0
Education	48	5.0
Electricity and water supply	190	19.8
Food supply	231	24.0
Health	143	14.9
Housing	16	1.7
Land tenure	7	0.7
Pests, vermin	40	4.2
Security (i.e., fire-fighting equipment, police)	13	1.4
Telephone, telegraph, post	24	2.5
Transportation	98	10.2
Veterinary service	1	0.1
General quality of life	65	6.8
Total	960	100.0

As can be seen, the problems expressed most frequently were those of (1) food supply, (2) electricity and water supply, (3) health needs, and (4) transportation. These are discussed below.

(1) Food supply

Under the category of "food supply", the most common complaints (from about half of the women) were of "no consumer cooperative in the village" and "no market in the village". "High prices for foods outside of the consumer cooperative" and "favoritism and bureaucracy in the consumer cooperative" ranked third and fourth.

It may be instructive at this point to make the distinction between consumer cooperatives and agricultural cooperatives. Most Egyptians take the existence of cooperatives for granted. One of the stipulations to smallholders who are allocated land is that they must agree to join an agricultural cooperative. An agricultural cooperative provides subsidized inputs to the farmers and, until 1987, the farmer was obliged to sell part of his production to the cooperative at the Government's set price.

To establish the agricultural cooperative system, geographical regions in the new lands were divided into "cooperatives" with each having a central village with main services such as schools and hospitals, and several small satellite villages. In South Tahrir, the smallholders were distributed over a total of 23 villages. Each of the main villages has a consumer cooperative for foodstuffs and, on the average, about 225 houses. The satellite villages have approximately 130 houses, but often do not have a consumer cooperative and have not attracted many private merchants. One woman in a satellite village complained of having to travel one hour to buy vegetables. Every settler is under the jurisdiction of an agricultural cooperative but may not have access to a consumer cooperative.

The concern for securing food reflects one of women's traditional responsibilities. In the corresponding survey conducted with the husbands in the smallholder households, obtaining food was not mentioned as a problem.

(2) Electricity and water supply

The problem ranking second for the wives is the electricity and water supply. Simply stated, if the electricity fails, the pumps that supply the water for irrigation cannot function, and the desert farmer cannot farm. Husbands as well as their wives said that obtaining water was a major problem.

A study by Mehanna, Huntington, and Antonius[4] of Egyptian irrigation stated that, "An irrigation system flows not only according to the dams and ditches of its physical structure, but flows also according to the social and political structures of villages and ministries." A farmer's relations with the operator of the pumping station and with his neighbors, as well as the proximity of his plots to the pumping station, are important factors which affect allocation of water to his farm. In the men's questionnaire, farmers were asked directly whether there were frequent cuts in electrical power that affected their supply of irrigation water. A majority answered affirmatively. Actually, no one said "No", but several declined to answer the question. Perhaps the latter felt that this was a sensitive area and their answers might be relayed to persons who would take offense from their responses.

The responses of wives indicate that they share their husbands' concerns about adequate water for farming. Electricity interruptions also affect household operations, but are less disruptive. The reason is that while fifty percent of the sample have electricity in their houses, a majority of these said they use it only for lighting, not for cooking, heating, or other uses. (They generally overlooked the use of electricity for radios and television sets which are owned respectively by 77% and 54% of the sample.)

Only 1/3 have potable water in their houses. This fact did not generate complaints, perhaps because water is available from a community supply and getting it affords them a social opportunity.

(3) Health needs

Health needs ranked third among problem areas, providing 15% of all complaints of the wives. The major concern under this topic was the quality of existing services. Hospitals or health units were said to be poorly equipped or to provide inferior care. Secondly, service from doctors was criticized as wives stated that they frequently were not present, not competent, or too expensive. Medicine was also an important concern in that either it was not available or, when available, it was not effective.

Other problems often expressed by women dealt with lack of facilities. They complained that there was no hospital, health unit, or pharmacy in their village. A few complained of no access to an ambulance, family planning unit, dentist, or first aid.

The incidence of pests and vermin such as mice, rats, and insects constituted 4% of the total number of all complaints. While these complaints were tabulated in a category separate from health needs, they are a health-related concern as well.

(4) Transportation

Ten percent of the problems expressed were about transportation. The quality of roads generated the most complaints. Many roads are unpaved or have had little improvement or maintenance since the basic infrastructure of the area was built in the 1950's and '60's.

A large number of wives stated that they did not have access to a bus or other public transport or that other available transport was too expensive. A few complained that services were infrequent or that they had to go a great distance to reach either public or private transportation.

* * *

Each of the problem areas should be investigated further to obtain more information about the specific problem and to suggest solutions for it. In this paper, however, the objective is to analyze the impact of the problems in total upon the smallholder wife and to find relationships between the problem level which she expresses and other variables that affect the well-being of her household. The problem level particular to each wife can be considered as one indicator of her satisfaction with life on the new lands.

METHODS OF ANALYSIS

The responses to the survey were classified in two different ways. The first was quantitative, i.e. according to the number of all problems stated. The second was by type of problem. In each the responses were ranked into high, medium, and low. A woman who stated a large number of problems which fell into only a few types would receive a high quantitative rank but a low type rank. Thus, the combination of number of responses and number of types resulted in nine combinations of high, medium and low designations: HH, HM, HL, MH, MM, ML, LH, LM, and LL. Examination of the responses revealed that 94% of the women were ranked HH, HM, MH, MM, LH, or LM indicating that most of the women expressed problems across a wide range of types rather than stating multiple problems of one type. Thus, to facilitate data analysis, the classifications were reduced further into three: high, medium, or low, (H, M, or L) according to the total number of problems stated.

The Chi-square test (X^2) of significance was used to determine whether systematic relationships exist between the number of problems expressed by the wives and a variety of independent variables. These were: (a) age of wife, (b) number of feddans farmed, (c) education of wife, (d) education of husband, (e) total education of all household members, (f) average distance from house to plots farmed, (g) husband's years on present farm, (h) tenure or ownership status, (i) number living together in household, (j) number in household who work on farm, (k) availability of potable water in house, (l) avail- ability of public transport, (m) accessibility of post office, (n) accessibility of phone, (o) building material used in residence, (p) size of residence in square meters, (q) distance to private doctor or hospital, and (r) the answer to the question, "Are you satisfied?".

Responses to the variables, 'age of wife' and 'Are you satisfied?' were obtained directly from the questionnaire administered to the smallholder wives. The structure of the survey made it necessary to deduce information about the rest of the variables from answers given by husbands in their interviews.

RELATIONSHIPS BETWEEN PROBLEM LEVEL AND OTHER VARIABLES

The X^2 test showed significant relationships between the number of problems expressed by the smallholder wives (H, M, or L) and (1) number of feddans farmed, (2) years of husband on present farm, and (3) total education of household members. Table 2 shows results for these variables.

DISCUSSION OF RESULTS

The effects of the significant variables found in

TABLE 2. Distribution of sample by problem level and significant variables (Chi-square)

Variable	Sample (No. of Subjects) High	Medium	Low	Total (N = 132)	df	Chi-square Value
Feddans farmed						
Up to 3.0	29	32	18	79	4	13.32**
3.1 to 4.5	2	12	14	28		
4.6 or more	5	9	11	25		
Years on farm						
Up to 10	26	28	15	69	4	11.57*
10 through 17	7	14	15	36		
18 or more	3	11	13	27		
Education of household members						
None	8	10	3	21	6	17.18**
Minimal	13	15	8	36		
Moderate	8	20	12	40		
Most	7	8	20	35		

* $p < 0.05$

** $p < 0.01$

the analysis as they may affect the perceptions of the wives are now discussed.

(1) Effects of Farm Size

A smallholder in Egypt is usually defined as a farmer with three to five feddans of land. The number of feddans farmed by the smallholders in the sample ranges from 1.4 to 13.75. While a few farmers have been able to increase their original allocations, 92% remain within the bounds of the definition in that they have five feddans or less of land.

The wives of the smallholders who have more than three feddans tend to express fewer problems in general about life on the new lands. Only 13% of the wives with farms of 3.1 feddans or more express a "high" complaint level. Of those expressing a "high" level of complaints, 81% have three feddans or less.

The landholding records of Egyptian agricultural cooperatives indicate that more than 80% of all landholders have access to less than three feddans of land (Adams, p.34). In no case did a smallholder wife say that not having enough land was a problem, but it appears that the psychological satisfaction which she derives from having more than the average extends so far as to affect the number of problems which she expresses regarding a wide range of factors.

Possibly the smallholder with more land is more prosperous. The information is not sufficient to confirm this, however.

(2) Effects of Desert Experience

The number of years that a smallholder has been on the present farm is an indicator of farming experience on desert lands. Most smallholders would have been members of farm families all of their lives, but the knowledge gained from farming on old lands is not easily transferable to the desert. The range for the sample is from five to 25 years on the present farm. The majority (55%) have been on their farms between five and ten years but 15% have farmed their plots for more than twenty years.

The most problems are expressed by the wives of the five to ten-year group. Seventy-two percent of those expressing a "high" level of complaints have been on their farms less than ten years. After eighteen years on the farm, only 11% are "high" complainers; 41% are in the "middle" level, and 48% express a "low" level of problems.

The fact that the number of problems expressed declines with length of time on the farm may be attributed to various reasons. Experience may enable the smallholder household to better cope with problems, thus achieving greater success, or expectations may be reduced over time.

New settlers go to the desert with determination to succeed. They may be expected to express fewer problems in their first years if they feel that acknowledging problems indicates that they personally are inadequate for the task at hand. Since none of the sample have been on their farms for less than five years, it is not known if the negative correlation between problem level and experience is constant or if the problem level gradually increases, peaks during the five to ten year period, and then declines. It is evident that after ten years on the farm, different explanations are possible: either satisfaction increases, aspirations, decline, or settlers find that expressing problems is futile.

(3) Effects of Education of Household Members

The majority of the smallholder wives and their husbands have no formal education. In most of the households, however, some other family member or members have some schooling. In only 17% (22) of the households, no one had gone to school. Nevertheless, the level of education for smallholder households remains low; only about one-fourth of the "educated" group have gone to school for more than five years.

For a comparison of households, years of education obtained by household members were added together and coded into four categories: none, minimal, moderate, and most. While the wives were mostly uneducated, it is nevertheless the level of education of the rest of their household that affected their perceptions of life on the farm. There is an inverse relationship between number of problems expressed by wives and total education of household members. Whether more education helps a smallholder household solve problems or primarily provides peace of mind is not known, but the majority of wives in households with the most education are "low" complainers. In contrast, thirty-seven percent of the wives in households with little or no education express a "high" level of problems. Only 20% of those in households with the most education are in the "high" problem group.

Adams (p. 142) provides insight into attitudes toward education in rural Egypt. In two villages in old lands which he studied, farmers often send one or two of their sons to school until they reach high school. It is a means of adding "income-generating capacity" to their family as they hope these sons will be able to secure jobs in the government bureaucracy. Government jobs may not pay well but they are a source of steady income when all else fails.

No doubt education is the most important factor separating social classes in areas of rural Egypt. Settlers on the new lands have role models of those who are educated in their communities. Bureaucrats employed in the cooperatives are appointed on the basis of educational achievement and represent a class who do not work in the fields. South Tahrir also has about 500 graduate farmers who have each been allocated twenty to thirty feddans of land -- considerably more than that available to the smallholder -- because they are graduates of agricultural technical schools or universities. Thus, the smallholder perceives the value of education.

On the old lands, the smallholder is not likely to send all of his children to school. Public education is free, but the costs of clothing, books, and foregone wages make it expensive for the poor peasant. Education is therefore deemed irrelevant for many sons and almost all daughters.

While girls were typically not sent to school on the old lands, smallholders in South Tahrir are digressing from this pattern. Ninety-three percent of the husbands who were interviewed said that they thought there was value in sending girls to school. Many of the smallholder household members in the sample who have some education are female.

The educational facilities and the quality of education available did not generate many complaints from smallholder wives. In contrast, in a study of women on new lands in North Tahrir, poor educational opportunities were the main cause of discontent.[2] The sample for that study was made up of women from a variety of occupational backgrounds including some wives of smallholders. As a group their educational level was higher than that of the smallholder wives of South Tahrir. With more education themselves, their expectations of education for their children was greater. The South Tahrir group who have very little formal education themselves are not prone to be critical of the educational system. The data suggest that the smallholder wives of South Tahrir gain satisfaction from knowing that members of their household have some education regardless of its quality.

CONCLUSIONS

When smallholder wives were asked, "Are you satisfied with life on the new lands?", nearly 70% answered, "Yes". The number of problems which they expressed, however, raises doubts about the degree of their satisfaction. The smallholders who settle in the desert gain a piece of land of their own in return for enduring a myriad of problems. Such difficult conditions demand settlers with a pioneering spirit.

Foremost problems mentioned by the wives are food supply and health care. They represent traditional women's spheres. The problems which the wives express seem to be appeals to the Government or some authority to alleviate conditions which the smallholder feels powerless to change. For example, while housing is also a woman's domain and, in the opinion of many

Egyptians, the smallholder's housing is substandard, wives seldom mentioned housing as a problem, apparently because they felt they themselves could do something about it! In this regard, when smallholder men were asked how they would spend money if there was any to spare, the third most frequent response (out of more than twenty) pertained to improving their family's housing.

The findings of this research show strong agreement among women about the major problems that they face on the new lands in South Tahrir. The research also suggests that there are other factors, not necessarily recognized by the smallholder's wife, that enable her to be more content with conditions in the new lands. They are not benefits that make her life personally more comfortable. These factors -- more land, more experience, more education for members of her family -- affect the well-being of her entire household.

REFERENCES

1. Richard H. Adams. 1986. Development and Social Change in Rural Egypt. Syracuse University Press, Syracuse, New York.

2. Gawaher Atef. 1987. The New Villages in the Reclaimed Lands: A Study on Women's Impressions Regarding the Level of Development of the Area. Unpublished Master's Thesis, The American University in Cairo, Cairo, Egypt: 73.

3. Desert Development Center. May 1986. Desert community aspects in AUC desert development systems, Report to the Ford Foundation.

4. S. Mehanna, R. Huntington, and R. Antonius. Irrigation and society in rural Egypt. In Cairo Papers in Social Science, Vol. 7, No. 4: 1.

SMALLHOLDER MANAGEMENT OF MIXED AGRICULTURAL RESOURCES IN DESERT REGIONS - PROBLEMS AND PROSPECTS

MARY MARTIN
Middle East Center, University of Pennsylvania

Abstract This paper is based on a study from 1974-1978 of smallholder operations in the region known as Khar-o-Tauran at the edge of the central desert in Semnan Province in northeastern Iran. The management decisions which affect, on the one hand, the viability of the household enterprise and, on the other, the prospects for sustained production are examined. The relationship of locally-made decisions to those made by government agencies removed from the area is a crucial variable to consider. Ideally, in desert and semi-desert regions, local decision-makers should have flexibility in options to overcome the problems of pastoralism and agriculture combined with potential for outside assistance in time of trouble that will not destroy the individual's control of his resources. The strategies for dealing with local and national conditions in a region of northeast Iran illustrate mixed farming problems and prospects in a particular setting. As an illustration of local level responses to particular social, political, and environmental pressures, the example has wider application in that it helps us to focus on key responses of local populations to environmental and socio-political problems.

* * *

Mixed agriculture increases options for survival in desert regions. It optimizes the use of unirrigated rangeland for animal production and, the irrigated and dry farms for fodder, family food and cash crops. When mixed agriculture is successfully practiced by families that own their land and animals, and have their individual or communal access to grazing land, there is potentially a situation where subsistence needs are met, excess produce is available for urban areas, and forced migration from rural areas may be stemmed. In these regions, however, outside support is a necessity, usually from the central government, to provide financial backing yet to allow for the flexibility of action necessary in an arid environment.

The case study presented here comes from an area in northeastern Iran where residents combine sheep and goat pastoralism with irrigated cropping of wheat and barley - a production emphasis of some antiquity in the region and in Iran. In addition, they also grow cotton, tobacco and fodder crops. Poorer families supplement their incomes by hiring out as seasonal shepherds for the transhumant Sangsari who winter on the rangeland in the area.

The region receives less than 250 mm precipitation. Agriculture is dependent on the qanat (in which an underground channel brings ground water out onto the surface by means of gravity flow), a system which, as of 1966, still provided water to 22% of Iran's villages.[2] In the specific region under consideration, qanats are only a few kilometers long and irrigate areas of less than 50 hectares, in contrast to other parts of Iran, such as the Yazd area, where they extend up to 43 kilometers. As a consequence, the amount of land under cultivation is limited. Water is owned in time units of 12-16 day cycles, and land in small units which can be irrigated with spade technology.

The nature of the agricultural enterprise itself is one that presents certain paradoxes for desert areas. Investment for irrigation has usually been provided by wealthy individuals. When outside investment is used to construct long qanats, control is rarely in the hands of smallholders. For deep wells, the situation is similar, for large investments by wealthy individuals or the government are needed.[3] When the government is the investor, mechanisms for repayment and maintenance, which are consistent with smallholder income and local social organization, are necessary.

The area discussed, known as Khar-o-Tauran, is on the border of modern Khorasan Province, an area where peasant proprietorship has always been more widespread than in other districts in Iran.[4] Khar-o-Tauran is distinctive in the region because of the small size of the land holdings per household and the importance of sheep and goat pastoralism. Residents own less than one hectare of qanat-irrigated land and, on the average, over 20 to 30 sheep and goats per herd. The role of smallholders in the production of sheep and goats is important nationally. For example, owners with less than five hectares control 27% of the national total of small ruminants.[5] In addition, wells for pastoralism may not need major capital investment and smallholders can begin small and work up to larger herds with the potential for access to larger grazing areas for these herds.

In the last century the Iranian government played a minimal role in agriculture and pastoralism, and local economic strategies were determined primarily by commercial opportunities and ecological factors. The twentieth century and the advent of the Pahlavi dynasty brought new changes in the relationship between agriculture and pastoralism and the government. Until 1962, when the government nationalized forests and rangelands, individually owned animals grazed in mixed village flocks. Hired shepherds oversaw the flocks on communal village land or on individually-owned grazing territories away from villages. From 1974-1978, the period of this study, the new government system had not yet completely replaced the pre-nationalization system so that the nature of local decision-making, although affected by the changes to come, did not yet reflect these changes.

Important strategies affecting pastoralism and agriculture in this period were related to local conditions and to policies imposed from the outside. Local strategies included changing cropping patterns to maximize water use, grazing away from the village as much as possible to increase per capita animal production, attempting dry farmed crops in years with excess precipitation, and supplementing incomes by shepherding.

The cropping pattern is determined by temperature and water limitations, pastoral needs for fodder and labor, government incentives, family consumption needs and preferences. Basically, the limited water supply from qanats is manipulated in order to grow a combination of staples such as winter wheat and barley, cash crops such as tobacco and cotton, and fodder crops such as sorghum, millet and alfalfa. Scarcity of water is a persistent problem. The problem varies according to both direct precipitation and the effect of the previous year's precipitation on the water table. When qanat flow is reduced, the villagers change their cropping emphasis in order to maximize productivity by, for example, growing less cotton and more tobacco; when the water flow is high, sesame or cumin may be planted. The major cropping pattern has changed little with the exception of the change in variety of cotton, and the cessation of opium cultivation after it was prohibited in 1955. The actual amount of land planted to a certain crop varies from year to year depending on water supply and the availability of family labor. There is some variation within the area based on differences in microclimate, soil, water salinity, and qanat flow. In all villages, however, the amount of land planted in the walled gardens remains constant from year to year and consists primarily of grapevines, fruit trees, tomatoes, and herbs.

Each farmer is responsible for his own small plots and individual farmers pursue various techniques to increase production. These techniques involve different combinations of the application of animal dung and chemical fertilizer, weeding, double cropping, and leaving land fallow. The dung available from livestock stables in the village is applied as fertilizer -- most intensively to those fields which will be double cropped. In this area of semi-desertic steppe, the amount of brush available for fuel is sufficient so that dung can be used as fertilizer rather than fuel.

"Fallowing" occurs in two ways. Irrigated fields are plowed and then left fallow, but dry farmed fields, in which the soil is much looser, are left fallow with the plant roots intact. In the first case, leaving fields fallow conserves moisture, and breaks up weed cultivation. The second situation prevents wind erosion by retaining the root structure and thus providing some stability in soils which would otherwise be vulnerable. Fields which are not planted and plowed do not "clump" as does soil which has had its structure modified by irrigation or application of dung. The latter are more vulnerable to the effects of trampling.

Dry farming is important in the region despite its undependability. Dry farmed wheat, for example, is preferred for bread baking. Barley, a crop which needs less moisture, is a good dry farmed crop, not for human food, as hypothesized by several writers on Middle Eastern agriculture,[6] but as winter fodder for sheep and goats. In Tauran, the need for barley as winter fodder has been linked to increased pressure on the range. Although this pressure appears to be a recent phenomenon in Tauran, it is not necessarily the case, for there may have been extensive pressure on the range in earlier periods. However, it seems that population growth in the twentieth century has been such as to begin to strain the resources available to man and animal.[7]

In mixed agricultural economies, it is usually necessary to keep animals close at hand so that family members can process milk products and tend to obligations in the fields at the same time. Keeping animal production high under these conditions may be a problem. One solution is to graze family animals in joint flocks which leave the village and return daily. Unfortunately, in

Tauran the common area around the village has less vegetation than areas further away because there have been multiple pressures on this land ranging from dry farming to fuel collection.[8] Despite current government regulation of these activities in the area, previous activities have depleted much of the shrub cover. A family choosing to graze their animals in the village flock will have reduced milk, meat, and wool production. One solution is to graze away from the village for several weeks in the spring. More drastic solutions involve renting or buying a grazing station away from the village so that animals can be kept there either over the summer when milk production is at its peak or, ideally, throughout the year. While this solves the problem of low animal production, the cost for labor is high since a permanent shepherd is necessary and the family must be brought out from the village for summer milk production. At the same time, various family members move back and forth to the village for the wheat harvest, irrigation demands, and weeding. Current labor costs and the non-availability of hired labor, in most cases, force the family to meet its own labor needs. For some families, cooperation with relatives provides a partial solution. The village-based relatives perform agricultural tasks while those at the grazing station oversee the animals and process their milk.[9]

Thus, in a mixed agricultural economy, the two forms of resource use, agriculture and pastoralism, complement each other but also compete with one another. The individual operator must make his own decisions about which course to take. As noted earlier, dry farming in good years can provide fodder for pastoralism. However, it is sometimes necessary to choose between dry farming and the increased milk production which would result from the spring graze. The decision will be related to whether or not a man has sons or other male relatives who can help him out (or in some cases, if he is wealthy, whether he can hire labor). In recent years, the availability of tractors for plowing in the spring has alleviated this problem to some extent. However, the tractors come from outside and are available only at certain times. A farmer wishing to plow at another time will have to use his own draft animals. Thus, in combining agriculture and pastoralism, there are important choices to make in the management of labor and the choice of crop, as well as the marketing strategy.[10]

In addition to strategies which are designed to increase production and maximize labor availability at the local level, there are means which aim to improve local conditions by manipulating outside forces, such as merchants and government officials. Local forces have attempted to improve conditions in several ways. They have attempted (a) to attract government investment in deep wells so that agricultural production would expand, (b) to obtain outside fodder for increasing animal production, (c) to modify government restrictions on fuel collection necessary for processing milk products, grape molasses, and wheat (bread baking), (d) to increase access to grazing away from communal village areas, (e) to encourage local construction and weaving operations which provide additional local labor opportunities, and (f) to obtain loans for houses for tobacco storage. The above mentioned attempts indicate an active interest in improving local conditions for pastoralism and agriculture by engaging government assistance.

Iranian rural organizations have been characterized as inadequate as "a vehicle for popular participation in local decision-making".[11] The attempt to maximize services to rural populations by combining small villages into larger ones[12] is but one example of a government strategy which undermines smallholder options. The major problems of mixed agriculture in the area under consideration are the lack of sufficient water and labor as well as the outside restriction of grazing flexibility without alternatives geared to arid environments. If those local strategies which promote sustained production can be encouraged while assistance to smallholders is provided where it is most needed (i.e., water resources, low cost fodder, veterinary services, flexibility in grazing regulations), local production can be increased and the strength of the rural economic base can be assured.

REFERENCES

1. This research was carried out during seventeen months of residence in the area during 1974-78 as a component of the Turan Programme for Integrated Ecological Research and Management under the aegis of the Department of Environment (1974-77) and the Department of Agriculture (1977-79) and coordinated by Dr. Brian Spooner of the University of Pennsylvania. For further description of the project see Brian Spooner and H.S. Mann, Desertification and Development Dryland Ecology in Social Perspective. (Academic Press, London, 1982).

2. J. Bharier. 1971. Economic Development of Iran 1900-1970. Oxford University Press, London.

3. In one study of smallholders in northeastern Iran with a sample of 100 farms in 300 villages near Nishapur, twenty-five percent had irrigated farms of less than one hectare. In this region, irrigation water was cited as one of the two factors limited production (the other was labor for weeding). 1978 estimates for the initial cost of a deep well were T 1,600,000 ($23,000). This size of well would provide water for 36 hectares -- more than the areas irrigated by a Tauran qanat. The financing of the initial investment, the technical problems of digging the well, installation of the equipment, scheduling water usage, maintenance and fuel preparation would be new issues needing training and experience. In addition, even with the 8% rate for government agricultural loans, all of these difficulties would need to be met in

Nishapur villages by groups of at least eighteen farmers for whom the initial investment would be several times their annual farm income, a factor that would be even greater in Tauran. See Mohammad Said Noori-Naini, An Economic Analysis of Resource Allocation Issues in Traditional Agriculture: A Case Study of the Nishaboor District of Iran, 1969-70. Cornell International Agriculture Mimeograph 61, Ithaca, New York: Cornell University, 1978.

4. See A.K.S. Lambton. 1969. Persian Land Reform 1962-66. Oxford: Clarendon Press, and Charles Issawi (ed.). 1971. The Economic History of Iran 1800-1914. Chicago: University of Chicago Press.

5. See Working Document #2 - Rural Geography. National Spatial Strategy Plan. First Stage Final Report. Tehran: Plan and Budget Organization (1976) p. 97.

6. See Huri Islamoglu and Suraiya Faroqhi. 1979. Crop patterns and agricultural production trends in sixteenth century Anatolia in Review (IIL3) Winter: 414.

7. J. Bharier in Economic Development of Iran, 1900-1970. (1971) London, Oxford, estimates that in the 1920's the rate of population growth rose from under 1 percent to 1.5 percent. By the 1950's it was up to 2.5 percent. Although currently the percentage of the overall population living in rural areas is less (79% in 1901 and 61% in 66), it is nonetheless greater than it has ever been (with the number of villages increasing from 15,200 in 1900 to 45,000 in 1966.)

8. See a discussion of why this is not a simple "tragedy of the commons" in M. Martin, "Conservation at the Local Level: Individual Perceptions and Group Mechanisms." in B. Spooner and H.S. Mann (eds.). 1982. Desertification and Development: Dry Land Ecology in Social Perspective. London: Academic Press: 145-170.

9. The milk processing process is described further in M. Martin, "Pastoral Production: Milk and Firewood in the Ecology of Turan," Expedition 22(4)" pp. 24-28.

10. M. Martin. January 1982. Case studies of traditional marketing systems: goats and goat products in Northeast Iran. In Proceedings of the Third International Conference on Goat Production and Disease. Tucson, Arizona, Scottsdale Arizona: Dairy Goat Journal.

11. Ahmad Ashraf. 1977. The role of rural organizations in rural development - the case of Iran. In The Role of Rural Organizations in Selected Countries of Asia. Kuala Lumpur: Asian and Pacific Development Administration Center: 74.

12. Nima Nattagh. 1986. Agricultural and Regional Development in Iran, 1962-1978. Wisbech (Cambridge) England: Mena Press Ltd.

VARIABLES AFFECTING POPULAR PARTICIPATION IN ORGANIZATIONS AND COMMUNITY DEVELOPMENT ACTIVITIES IN THE NEW DESERT COMMUNITIES IN SOUTH TAHRIR, EGYPT

SALAH EL ZOGHBY
Desert Development Center,
The American University in Cairo

ABDEL REHEM EL-HAYDARY
Faculty of Agriculture, Alexandria University

Abstract This study aims at identifying the degree of popular participation in community activities and in governmental and non-governmental organizations in some desert communities, as well as major socio-economic variables which affect this participation. Three different villages were selected in terms of population composition, farmers and non-farmers, presence of different institutions, and time and origin of each.

Data for this study were collected during the summer of 1986 through personal interviewing using structured questionnaires designed for the purpose of this study.

A sample of 92 persons was drawn randomly from the Housing Department List of each village. Different methods of statistical analysis were used in the study. The Guttman Technique was used to measure the degree of popular participation. Chi^2, simple correlation and multiple regression techniques were used to identify the relationship between socio-economic variables on one side and popular participation on the other. The step wise multiple regression technique was used to identify the most important variables which affect popular participation.

***

Many desert development schemes have not realized the importance of creating a community through settlers' participation. To get the settlers to organize themselves and participate in the development of their community, they must first have a sense of belonging to their new home and community. The sense of belonging is related to the settlers' degree of social adjustment after migration from the old to new land. Their adjustment, in turn, is influenced by the availability of secured land tenureship together with minimum requirements of services, accommodations, infrastructure, necessary institutions and organizations.

In, Egypt's South Tahrir desert community, citizens participation in community affairs was considered crucial from the start. In spite of the emphasis on popular participation in new communities, the vast majority of studies have focused on "old land" communities. To our knowledge, no sociological studies of this topic have been made for the new desert communities of Egypt. Accordingly, the authors undertook the present study to shed light on the important variables that affect popular participation in both organizational and community activities in new desert communities.

Definitions of popular participation are numerous and varied. In this paper, popular participation is defined in the field of rural development as the process through which rural residents voluntarily participate in making decisions about community development programs and provide resources (such as advice, finance, physical work, communication, organization, and supervision) that are necessary for the success of these programs.

OBJECTIVES OF THE STUDY

1. To identify variables affecting the level of popular participation in activities of organizations as well as in community -

oriented actions in the South Tahrir area in Egypt. These variables are pattern of settlement, gender, age, land tenure, status of residence, marital status, educational status and family size.

2. To identify opinions of change agents (formal leaders) and of change targets (farmers) about ways and means of promoting popular participation in community organizations and community development affairs.

GEOGRAPHICAL AREA OF STUDY

The South Tahrir area in general, and villages of Omar Makram, Baghdad, and El Maraka in particular were selected as sites in which to carry out this investigation because:

1. The South Tahrir area is one of the earliest pilot projects in the field of desert land reclamation, and settlement schemes.

2. The South Tahrir area comprises communities that vary in the length of their existence (old and new), and in the types of residents; beneficiaries (smallholders), graduates, and former employees in the South Tahrir Company).

SELECTION OF THE SAMPLE

The sample of the study consists of 92 heads of households who were selected randomly from the household records existing in the offices of the three villages. The distribution of the sample members is as follows:

The total sample represents about 2.5 percent of the total number of households in the three villages.

METHOD OF DATA COLLECTION

Personal interviews were conducted with sample members using a constructed questionnaire that was designed especially for the present study.

FINDINGS AND DISCUSSION

FIRST:

Socio-Economic Variables Related to Popular Participation: Correlation analysis indicates that there is no statistically significant relationship between level of popular participation in organizational and community developmental activities and other variables examined in the present study (Table 1)

TABLE 1: Relationships between Level of Popular Participation and Other Socio-Economic Variables.

Variables	Correlations Coefficient (R)
Pattern of Settlement	-.077
Gender	-.016
Age	.048
Land Tenure	.102
Length of Residency	.022
Marital Status	.083
Educational Status	.032
Family Size	.106

* Statistically Significant ($P<.05$)

But when these relationships were subjected to multiple regression analysis, age, educational status, and family size were found to have positive significant effects on level of participation, while size of agricultural holding was found to have a negatively significant effect on level of participation. Pattern of settlement, gender, marital status, and length of residency in the community were found to have no significant effect on level of participation. (Table 2)

TABLE 2: The Effects of Socio-Economic Variables on Level of Popular Participation.

Variables	Regression Coefficient	T
Pattern of Settlement	-.2771	-.873
Gender	-.1049	-.411
Age	.0204	2.178*
Marital Status	-.0733	-.420
Education	.1684	2.321*
Size of Family	.0508	2.442*
Size of Ag. Tenure	-.0014	-1.997*
Length of Residency	-.0040	-.394
R2 = 0.12		

* Statistically Significant ($P<.05$)

The positive significant effects of gender, family size, and educational status on level of participation that were found in the multiple regression analysis were anticipated and reinforce findings of many previous studies in the field of popular participation.

The negative effect of size of agricultural tenure on level of participation in the organizational and community developmental activities might be explained by the fact that the majority of individuals who have larger agricultural tenure are graduates or former employees in the South Tahrir Company. Their interests, needs and expectations are different from those of smallholders (beneficiaries). Therefore, they may not be interested enough to get involved with smallholders in common activities. Also, most of the graduates do not reside with their families on their farms, which might decrease their interest in participating in organizational and developmental activities.

The relative homogeneity of the sample with respect to the independent variables of pattern of settlement, gender and marital status might suppress the effects of these variables on level of popular participation. This may be reflected in the relatively small values of standard deviations of these three variables (Table 3)

TABLE 3: Means, Standard Deviations, and Standard Errors of the Study Variables

Variables	Means	Std. Dev.	Std. Errors
Pattern of Settlement	1.9239	.2666	.3173
Gender	1.8696	.3386	.2552
Age	47.3478	12.3087	.0094
Marital Status	2.8043	.5186	.1746
Education	2.5326	1.7632	.0726
Size of Family	6.4674	4.3182	.0208
Size of Ag. Tenure	125.3478	155.0712	7.2043
Length of Residency	21.2065	9.0269	.0102
Level of Popular Participation	1.5870	0.7860	.7726

SECOND:

CHANGE AGENTS (LEADERS) AND CHANGE TARGET (FOLLOWERS) SUGGESTIONS FOR INCREASING POPULAR PARTICIPATION IN ORGANIZATIONS

A. THE POINT OF VIEW OF THE CHANGE AGENTS

The official leaders of the different organizations in the study area were asked how popular participation in local government development activities planned by the organizations could be increased:

Their answers are summarized as follows:

1. Local organizations should plan developmental activities pertinent to and in accordance with the needs of the members of the society.

2. The impact of such activities sponsored by the different organizations should extend to all members of the society and not to a selected group.

3. The organizations working in the communities should consult and encourage members of the society by providing them with information about activities.

4. The organizations present should work on gaining the confidence of the members of the community by means of: a) Interacting more with members of the community, b) Encouraging members to freely express themselves, c) Facilitating means of communication between those working in the organizations and community members. and d) Demonstrating to the members of the community the existence of such activities and the results that such activities can generate.

5. Increasing supervision and evaluation of the activities of such local organizations by higher administrative levels.

6. Increasing the economic capabilities of the members of the society.

B. THE POINT OF VIEW OF THE CHANGE TARGET

The 92 members of the sample, most of whom were smallholders (beneficiaries), were also asked about the means by which the members of society can increase their participation in the developmental activities sponsored by the organizations in their community.

Their answers could be categorized as:

(1) what government organizations could do to increase participation in the organizations, (2) what members of the community could do to increase participation, and (3) other factors that affect the degree of participation.

Table 4, 5, and 6 indicate the incidence of responses in three categories. Respondents often expressed more than one suggestion.

WHAT GOVERNMENT ORGANIZATIONS COULD DO TO INCREASE PARTICIPATION AT THE LOCAL LEVEL

Table 4 indicates that community members who were interviewed emphasize that organizations should deal with important needs of the community. Both organizations leaders and citizens state that addressing local needs should be the first priority of organizations.

There is also strong agreement between the leaders and the community members in other responses. The second most frequent response of members is that the government organizations should allow members to express their ideas and discuss their problems. The same theme is expressed in responses of change agents when they advise that means of communication between the two groups should be improved.

The response which is third in frequency is also an aspect of communication. Members who were interviewed suggest that government organizations should hold meetings and group discussions to educate citizens about the importance of participation.

Members also place importance on allowing the whole community to benefit from activities. This response is listed as fifth. If these two items were combined into and item concerning impartiality toward recipients, the resulting item would be in fourth place among the suggestions from citizens. It was also seen as very important among

TABLE 4: Suggestions for ways government organizations could increase the degree of citizen participation.

Means	No. of Responses	%
Implement actions directly related to the specific needs of the community.	22	21
Allow members the chance to express their ideas and discuss their problems.	18	17
Inform and educate the community members about the significance of participation through meetings and group discussions.	17	16
Offer new ideas for informing the community of the available facilities and resources for solving their problems and meeting their needs.	11	10
Allow the whole community to benefit from activities, not just a selected few.	10	10
Change leaders in organizations and replace them with leaders who can more effectively deal with the community.	7	7
Increase supervision and inspection over organizations.	6	6
The government should set up various organizations in the communities and supply them with the needed resources to support them in serving their purpose	4	4
Organizations should not be biased toward certain individuals or groups in their efforts to provide services and activities.	4	4
Inform the community members about structures and functions of organizations.	3	3
Coordinate to avoid duplication among organizations.	2	2
Total	104	100

organization leaders as, it was their second most frequent response.

Two responses in Table 4 convey the idea that citizens want more help from the government in terms of supplying resources. Both the fourth and the eighth responses indicate that the interviewees want organizations to bring ideas, facilities, and resources to them.

WHAT MEMBERS OF THE COMMUNITIES COULD DO TO INCREASE PARTICIPATION IN ORGANIZATION

Table 5 summarize responses concerning what community members themselves should do to increase local participation in developmental activities sponsored by the different organizations in the community.

The response indicates that community members view organizations as a tool for getting things done when meetings of an organization are amicable. They say that participation in an organization increases when members have the opportunity to choose projects and set priorities for the projects rather than having projects imposed upon them by the government.

GENERAL WAYS AND MEANS FOR INCREASING PARTICIPATION

Table 6 lists other factors seen by members of the community as effective in increasing the degree of participation in developmental activities sponsored by organizations:

Money, education, and time are recognized as important factors which are associated with participation. Members of the sample recognize the influence of these factors on the degree to which one participates in organizations.

TABLE 5: Suggestions for What Community Members Should Do to Increase Local Participation.

Means of Procedures	No. of Responses	%
Amicable meetings of the community members.	41	42
Community members should designate priorities for projects.	19	20
The availability of local leaders who can effectively organize activities and distribute responsibilities and suggest projects and advise community members.	12	12.5
The awareness of community members of their needs and problems as well as means and ways of overcoming them through knowledge of similar circumstances.	12	12.5
The prevalence of community love and a sense of combatting errors.	5	5
Participation of community members in implementing organizational societies.	4	4
Participation and representation of community members in directing organizations.	2	2
Community members have to be conscientious and active.	1	1
The feeling of self appreciation and respect as one participates in beneficial activities.	1	1
Total	97	100

TABLE 6: Other Means of Increasing Participation

Means	No. of Responses	%
Availability of funds to the community members	36	30
Spread and increase of education	27	23
Reciprocity of trust between the community members and the government as represented by organizations	26	22
Availability of time	15	13
Settling in local communities and avoiding lots of travel for local dwellers	5	4
The lack of problems which are likely to occupy them and restrict their participation	5	4
An increase in the degree of religious expression can be substituted	2	2
Living in local communities (villages) not on farms as with graduates	1	1
Total	120	99

The remaining response among the top four responses pertains to mutual trust between the people and the government. When the trust is reciprocal, public participation in developmental activities increases.

We can conclude that there is strong agreement between the leaders of organizations and community members concerning how to increase participation of members. Two themes are prominent in the responses of both groups. One is the necessity for organizations to concentrate on local needs. The other is a desire and need for improved communication between leaders and their constituency. Finally, members rationalize that improvement of their own socio-economic status would make them more active participants in organizations. A certain amount of money, education, and time are seen as prerequisites for stimulating participation.

In summary, popular participation can be promoted by (1) increasing the degree of social organization, (2) encouraging the forming of positive attitudes toward participation, (3) improving the economic and social status of individuals, and (4) searching for and training informed local leaders who are capable of increasing social solidarity in developing communities. In turn, the ability of a community to meet its needs and solve its problems is increased.

REFERENCES

1. Samuel Bassilious. 1967. "al-igtimaa al rifi", Alexandria University Press.

2. M.I. El Ezaby. 1985 "Impact of Situational, and Orientational Factors on Residents Contributions to Community Field Structure: A Case Study". Ph.D. Dissertation. Iowa State University Library. Ames, Iowa.

3. M.I. El-Ezaby. 1986. "Al Musharaka Al-Igtimaiya fi Al-Mogtamaa Al-Mahali", in Gamee, N. and others: Kiraat fi tanziem wa tanmeyat al-mogtamaa al-rifi al-mahali. Library of Faculty of Agriculture - Alexandria University.

4. N. Gamee, et. al. 1986. "Kiraat fi tanziem wa tanmeyet al-mogtamaa al rifi al-mahali". Library of Faculty of Agriculture Alexandria University.

5. A.M. Shawky. 1986. "musharakat al-muatemien fi al tanmeyat al rifeya". Al yakazat al arabeya journal. Vol. 2, No. 10. Cairo.

6. Thayer Scudder, "A sociological framework for the analysis of new land settlements" in Michael M. Cernea, ed., Putting People First: Sociological Variables in Rural Development. New York, Oxford University Press for the World Bank, 1985, pp. 121-153.

7. United Nations Department of Economic and Social Affairs, Popular Participation in Decision Making for Development, New York, 1975, p.4

INDEX